CAMBRIDGE LIBRARY COLLECTION

Books of enduring scholarly value

Life Sciences

Until the nineteenth century, the various subjects now known as the life sciences were regarded either as arcane studies which had little impact on ordinary daily life, or as a genteel hobby for the leisured classes. The increasing academic rigour and systematisation brought to the study of botany, zoology and other disciplines, and their adoption in university curricula, are reflected in the books reissued in this series.

The Ferns (Filicales)

Frederick Orpen Bower (1855–1948) was a renowned botanist best known for his research on the origins and evolution of ferns. Appointed Regius Professor of Botany at the University of Glasgow in 1885, he became a leading figure in the development of modern botany and the emerging field of paleobotany, devising the interpolation theory of the life cycle in land plants. First published between 1923 and 1928 as part of the Cambridge Botanical Handbook series, *The Ferns* was the first systematic classification of ferns according to anatomical, morphological and developmental features. In this three-volume work Bower analyses the major areas of comparison between different species, describes primitive and fossil ferns and compares these species to present-day fern species, providing a comprehensive description of the order. Volume 3 describes, analyses and classifies extant species of ferns.

Cambridge University Press has long been a pioneer in the reissuing of out-of-print titles from its own backlist, producing digital reprints of books that are still sought after by scholars and students but could not be reprinted economically using traditional technology. The Cambridge Library Collection extends this activity to a wider range of books which are still of importance to researchers and professionals, either for the source material they contain, or as landmarks in the history of their academic discipline.

Drawing from the world-renowned collections in the Cambridge University Library, and guided by the advice of experts in each subject area, Cambridge University Press is using state-of-the-art scanning machines in its own Printing House to capture the content of each book selected for inclusion. The files are processed to give a consistently clear, crisp image, and the books finished to the high quality standard for which the Press is recognised around the world. The latest print-on-demand technology ensures that the books will remain available indefinitely, and that orders for single or multiple copies can quickly be supplied.

The Cambridge Library Collection will bring back to life books of enduring scholarly value (including out-of-copyright works originally issued by other publishers) across a wide range of disciplines in the humanities and social sciences and in science and technology.

The Ferns (Filicales)

Treated Comparatively with a
View to their Natural Classification

VOLUME 3:
THE LEPTOSPORANGIATE FERNS

F. O. BOWER

CAMBRIDGE
UNIVERSITY PRESS

CAMBRIDGE UNIVERSITY PRESS

Cambridge, New York, Melbourne, Madrid, Cape Town, Singapore,
São Paolo, Delhi, Dubai, Tokyo, Mexico City

Published in the United States of America by Cambridge University Press, New York

www.cambridge.org
Information on this title: www.cambridge.org/9781108013185

© in this compilation Cambridge University Press 2010

This edition first published 1928
This digitally printed version 2010

ISBN 978-1-108-01318-5 Paperback

Cambridge Botanical Handbooks

Edited by A. C. Seward

THE FERNS

VOLUME III

Cambridge University Press
Fetter Lane, London
New York
Bombay, Calcutta, Madras
Toronto
Macmillan
Tokyo
Maruzen-Kabushiki-Kaisha

CHRISTOPTERIS TRICUSPIS Christ.

Photograph from a specimen in the Herbarium, Royal Botanic Garden, Edinburgh. Reduced.
It shows a fourth lobe, indicating a leaf-architecture reminiscent of *Matonia* on the one hand,
and of *Phlebodium* on the other.

THE FERNS
(FILICALES)

TREATED COMPARATIVELY WITH A VIEW TO THEIR
NATURAL CLASSIFICATION

VOLUME III
THE LEPTOSPORANGIATE FERNS

BY

F. O. BOWER, Sc.D., LL.D., F.R.S.
EMERITUS PROFESSOR OF BOTANY
IN THE UNIVERSITY OF GLASGOW

CAMBRIDGE
AT THE UNIVERSITY PRESS
1928

PREFACE

THIS Third Volume completes a prolonged study of the great Class of the Filicales by the author: but the whole work only marks a stage in their detailed investigation. Historically there came first the period of discovery, diagnosis, and description, together with a provisional systematic grouping of the Class. The conclusion of this period may be held to have coincided with the publication of the *Origin of Species.* The state of knowledge of the Ferns at that time is fitly revealed in the First Edition of the *Synopsis Filicum* (1865), itself based upon the five volumes of Sir William Hooker's *Species Filicum*, of earlier date. Though the author of the *Synopsis* in his Preface makes no reference to Darwin's work, he does remark that "here as with other scientific systems those are the best characters which lead to a knowledge of the object sought for in the nearest and clearest way, keeping in view as much as possible its natural affinities." Nevertheless the goal of the *Synopsis* was primarily to form "a useful *vade mecum* for the travelling botanist and the cultivator of Ferns, and for ready consultation in the Herbarium." Thus was reflected in its pages the cataloguing aim of the earlier phases of Pteridology.

Already, however, the more penetrating spirit of Hofmeister was abroad. His *Vergleichende Untersuchungen*, published in 1851, revised and enlarged in the English Translation of 1862, was coeval with the appearance of the *Origin of Species*. These works of Hofmeister threw into coherent form the results of that laboratory enquiry which was needed to supplement the use of the hebarium, the garden, and the open country. In particular it brought into prominence the whole diplobiontic life-story, and provided a fresh stimulus for the observation of that wider field of characters the want of which in Cryptogams Sir William Hooker had deplored in his Preface to the *Synopsis* (p. xi). Not only was this ground traversed by the observers of detail in the living plant—particularly in Germany—but minute observation was soon to be directed more intensively than ever upon the correlative fossils, particularly in France and in Britain. The result has been to show with a high degree of certainty that those Ferns which have been designated by Von Goebel the Eusporangiatae represent a more archaic type than those which he had styled the Leptosporangiatae. Thus the way was being paved during the period succeeding the appearance of the *Origin of Species* for a grouping of Ferns that should reflect truly the broader lines of their evolution.

The results of this second period in the study of Ferns, together with that of Archegoniate plants generally, were summed up in *The Origin of a Land Flora*, published in 1908. Here a theory, based upon the facts of alternation, was advanced to account for their "diplobiontic" mode of life in relation to the passage of green organisms from an aquatic to a terrestrial habit. A critic of that volume once said, with some degree of justification, that the theory was out of date before it was published: and so it was as regards any general application to diplobiontic life, for this had already been demonstrated in Algae which had not invaded the land. Still this fact does not by any means rule out some close biological relation between alternation and that amphibial life which the Archegoniatae show (see Chapter XLIX). The whole subject is still open for further enquiry, and welcome light is being shed upon it by the writings of Svedelius, and others.

Incidentally the composition of the book on *The Origin of a Land Flora* demonstrated that the facts and methods current at the opening of the present century could not suffice for any full evolutionary treatment of the Class of the Filicales. Not only were further details necessary for an adequate comparison of Eusporangiate Ferns, but more particularly the tangled problem of the evolutionary lines of the Leptosporangiate Ferns was still too obscure in 1908 for any coherent treatment. Their phyletic relations were hardly more than hinted at in the *Land Flora*, pending further research. It was necessary in the first instance to widen the basis of comparison by the introduction of new criteria, each treated critically, so as to distinguish archaic features from those held to be of more recent origin. The conclusions thus attained were checked so far as possible according to the palaeontological evidence. The results of such study were first published in a series of Memoirs, I–VIII, in the *Annals of Botany* (1910–1923): and these supplied the material for Vol. I of this work. In Vol. II the phyletic method based on this wider comparison was applied to the relatively primitive Ferns: while the present Volume aims at a like phyletic treatment of the great mass of advanced Leptosporangiate Ferns.

In a sense then this work may be held as opening a fresh period in the Classification of Ferns. It is the sincere hope of the author that it may stimulate further enquiry, suggest the use of still other criteria of comparison, and perhaps lead finally to other conclusions than those here adopted. Readers who know the literature of the subject will be well aware that such general statements as are embodied here are really a summation of results from the widest possible sources, supplied by a host of investigators: and happily the spirit of enquiry is as active as ever. Advance rather than finality has been the aim of the author in summing up the results: for, repeating again the words of Stevenson used as a motto to the First Volume, "To travel hopefully is a better thing than to arrive."

F. O. BOWER

RIPON
1928

CONTENTS

The Author desires to acknowledge gratefully the assistance of the Carnegie Trustees, in the illustration of this concluding volume of his work. Readers will see how extensive the illustration of the three volumes has necessarily been, and will share the author's gratitude in proportion as they find the figures to aid the text.

F. O. B.

June, 1928.

INTRODUCTION TO VOLUME III

Noch immer gleicht die Systematik der Polypodiaceen einem nur teilweise gelicheteten Urwald, in welchem sich zurecht zu finden sehr schwer, das Verirren aber sehr leicht ist.
VON GOEBEL, *Ann. du Jardin Botanique de Buitenzorg*, Vol. XXXVI (1926), p. 107.

THE publication of a book that is liable to break off short as questions of special difficulty are approached requires some justification. This Third Volume on the Ferns is a case in point. Of the two preceding Volumes the first has laid the foundation upon which one may proceed in the phyletic study of the Class. In the second the method explained in the first has been applied to those Ferns which comparison has indicated as the more primitive in character, and are actually shown to have been of early existence by evidence drawn from the Earth's crust. But there remains for treatment that vast mass of genera and species which are collectively designated the Leptosporangiate Ferns, or in an older terminology the Polypodiaceae. These are essentially the Ferns of the Present Day. Not only are they more advanced in their organisation, but they are also much more numerous as living individuals, and as genera and species, than those previously treated; moreover they are more definitely standardised in their characters. Consequently they present a much more difficult problem to the morphologist who would attempt to group them along phyletic lines. A lifetime is too short a period in which to bring such a study to completion: and the first impulse may be to avoid any general statement that might be received as purporting to have done this. But the obligation that follows from intensive study provides some excuse for a volume, however incomplete, that aims at nothing more than the suggestion of lines for others to follow up by more exact enquiry: and so to complete, to amend, or even to negative.

At the close of the Second Volume the relations of the more primitive Ferns, including all those designated the Eusporangiatae, were suggested by a phyletic scheme which is here repeated. This scheme further includes an indication by dotted lines of the affinities of those more primitive types to the main body of the Leptosporangiate Ferns. The natural grouping of the latter will form the main subject of this Third Volume. It is indicated in the scheme that they fall into six large groups, which may probably be regarded from the phyletic point of view as distinct lines of descent. Each centres round some large and well-known genus, upon which its name has been based. The justification of this suggested grouping is to be found in the comparative examination of each phylum. All that is to be attempted here will be to range these large and comprehensive groups along the main lines of natural affinity. It must remain for those who come after, and possess the necessary time and specific knowledge, to work out the system towards its distal evolutionary twigs. It will be found difficult enough to compress within the limits of this volume what needs to be said of the main groups in general terms, together with some estimate of the probable relations of the genera one to another. But certain features of the scheme are already familiar. For instance, the affinity of the Davallioid and the Pteroid Ferns with the Dicksoniaceae is generally recognised. They may be held as marginal types advanced from a gradate to a mixed state of the

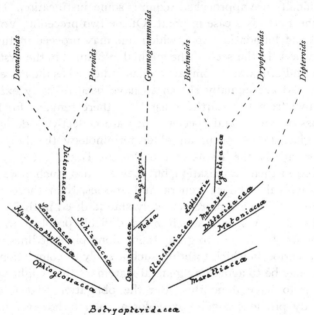

Phyletic scheme suggesting the inter-relations of the more Primitive Filicales: above are placed six main Phyla of Leptosporangiate Ferns, the dotted lines indicating their probable connection respectively with the Dicksoniaceae, *Plagiogyria*, and the Cyatheaceae.

sorus. Similarly the Blechnoid and Dryopteroid Ferns may be regarded as derivatives from a gradate Cyatheoid source, showing transition to a mixed type of the superficial sorus. The Gymnogrammoid and Dipteroid Ferns form phyla of less familiar relationship, as they stand in the current systematic works. The grounds for suggesting the affinities shown in the phyletic scheme will have to be set down in detail later in the text.

As the treatment proceeds it will become clear not only that there is evidence of polyphyletic progression to a mixed sorus, but also that in respect of the relations of the sori one to another, as well as in the manner of their protection, certain changes appear. One of these is the abortion of indusial growths previously present in one phylum or another: and so there may be brought into existence from several distinct sources examples of that unprotected state which was held as characteristic of the old comprehensive genus, *Polypodium*. Another change consists in the merging together of sori in linear series. This may happen either in the marginal sequence, giving the leading feature of the Pteroid Ferns: or in the superficial sequence, giving the leading feature of the Blechnoid Ferns. Again, in several distinct phyla the identity of the sori may be lost, by the spreading of the production of sporangia generally over the surface of the sporophyll. This gives the condition characteristic of the old genus *Acrostichum*. Such changes when proved to be polyphyletic can no longer be accepted as giving satisfactory generic characters, if the classification is to be one reflecting evolutionary results. In fact these old genera will have to be broken up in any phyletic classification. It will have to be realised, as indeed it already is in some quarters, that such old comprehensive "genera" as *Polypodium* or *Acrostichum* are not phyletic unities at all. The soral characters upon which they were based really represent *states* or *conditions* arrived at by evolution along a plurality of converging lines. Consequently an increased difficulty arises in the segregation of the distal evolutionary twigs. It is well thus to visualise some of these difficulties at the outset, so as to avoid misunderstanding as the treatment of the subject-matter proceeds. The fact is that a natural, or phyletic, Classification is becoming increasingly difficult, and is of necessity complex. For it is not always sufficient to observe the physical characters presented; we also require to have some reasonable view or knowledge of how the characters of the individual observed were arrived at in its Descent.

No attempt will be made to place all Ferns in their natural, that is their phyletic, relation to one another. There are not a few genera of so problematic a nature that they will have to be left under the designation "incertae sedis" (Chapter XLVIII). Nor will the treatment run into ultimate detail, except in some few instances where specific comparison appears to throw light upon the relations of genera, or even of families. What is undertaken is, on a basis of a wider comparison than has been generally practised hitherto, to lead from below upwards along reasonably safe and probable lines. But the work is offered as a "tentamen" only, not a finished task: and as such it is submitted to the judgment of those who shall come after.

Further, the treatment of the groupings will be as far as possible conservative. The author has much too high a respect for that intuition or special sense for affinity that is exemplified so often in the work of the great systematists, to allow him to neglect or needlessly to disturb the natural relations that they have already defined. These will be generally adopted, and changes will only be suggested where facts and comparisons, which had previously been unknown or undervalued, clearly dictate some other grouping. At the same time the relationships, as traced by previous writers, are by no means uniform. It is believed that the wider field of criteria here adopted may serve as a reliable guide in deciding some critical questions upon which authorities have disagreed. It is not in the spirit of the iconoclast that the systematic grouping of Leptosporangiate Ferns has been approached: but rather with a feeling of the deep respect that is due to the pioneers of classification, and with the highest appreciation of the work which they have carried through.

The author does not profess to a critical knowledge of the species and varieties of Ferns at large. Those who are happy in possessing it may find themselves at variance here and there with the conclusions as stated. He would ask them, wherever this is so, to try to reconsider the position along the lines of argument here advanced: and it may be that, with a slightly modified valuation of the relative importance of certain characters as compared with others, the suggested groupings may be found more worthy of consideration than they appeared to them at first to be.

It will be gathered from what is contained in these paragraphs that the work does not purport to be one of precise systematic treatment, but rather a morphological commentary upon the methods by which a natural grouping may be approached. From time to time statements will, however, be made of such tentative conclusions, general rather than specific, as seem to follow from the application of a method wider in its field than could have been available to the earlier writers. Perhaps these may aid the studies of Systematists of later date.

It is a matter for regret that the detailed knowledge of the gametophyte of Leptosporangiate Ferns is still very deficient. Its features are certainly more uniform, and offer less opportunity for comparative use than is the case in the more primitive Ferns. Nevertheless it may be that more extensive and detailed observation of them than has yet been carried out may in the future provide a fresh area of fact yielding material assistance towards phyletic conclusions. Here, however, the weight of comparison necessarily falls even more definitely upon the sporophyte than the available facts have allowed in regard to the Ferns of Mesozoic and Palaeozoic types.

In the naming of the species cited in these Volumes the *Index Filicum* of Dr Carl Christensen has been followed: and the author desires to acknowledge his profound indebtedness to the compiler of that most useful work. It is generally recognised as standardising the confused nomenclature of Ferns, while the tabular presentment of their classification in the earlier pages (I–LIX) gives in a most convenient form a condensed statement of the groupings of Diels. This should not be accepted as an expression of considered opinion by the author of the Index, but as an historical document of the position at the time when Engler and Prantl's Volume IV was published.

CHAPTER XXXVI

HYPOLEPIS, ETC.

IN Volume II, Chapter XXX, the Dennstaedtiinae have been described as a specialised Sub-Family of the Dicksoniaceae Their creeping habit, their long-stalked and usually finely divided leaves with open venation, their solenostelic vascular structure, sometimes polycyclic (Figs. 536–538, Vol. II), the origin of the leaf-trace often undivided (*Dennstaedtia*), but sometimes breaking up early into segments (Fig. 581, compare also Vol. II, Fig. 538), together with their dermal hairs and the absence of scales, except in *Saccoloma*, are all features of the vegetative system that compare especially with those of *Thyrsopteris* and *Cibotium*. The marginal position of the two-lipped sorus with its prevalent gradate sequence of the sporangia also points in the same direction. But certain species of *Microlepia* and *Dennstaedtia* have been found to show a departure from the strict basipetal sequence of the sporangia, and

Fig. 581. Transverse section of the petiole of *Saccoloma elegans* (×4). The vascular strands are black, the sclerenchyma dotted: the clear areas are ventilated parenchyma, connecting with the pneumathodes, *p, p*.

this is accompanied by some degree of flattening of the receptacle (Figs. 539, 540, Vol. II). The indusial flaps are unequal, the upper showing a tendency to merge into the general expanse of the lamina, while there are also steps towards a lateral linkage of the receptacles into a continuous chain (*Saccoloma*, Fig. 541, Vol. II). It was, however, noted that *Hypolepis* shows still more marked features of advance in the sorus; but the details of this interesting genus were held over to this volume: for it was thought that they would form a fitting introduction to the study of those of the Marginales which may be regarded as derivative phyletically from the Dicksonioid type. *Hypolepis* has in fact been selected as a favourable example for illustrating a source of those later transitions which a wide comparison of Ferns brings into view.

The genus *Hypolepis* Bernh. comprises 29 species, some of which have passed under various generic designations, such as *Lonchitis*, *Cheilanthes*, *Adiantum*, *Phegopteris*, and *Dicksonia*. This wide synonymy at once arouses interest, and suggests that it may be a synthetic or transitional type. It has been variously treated by systematists. Presl (*Tentamen*, p. 161) places it with *Lonchitis*, and close to *Cheilanthes*. Hooker (*Syn. Filic.* p. 128) assigns to it a similar place. Christ (*Farnkräuter*, p. 278) associates it with *Phego-pteris*, and especially with *P. punctata* (Thunbg.) Bedd. Diels (*Nat. Pflan-zenfam.* 1, 4, p. 277) places it in his Pterideae-Cheilanthinae, in near relation to *Cheilanthes* and *Llavea*. In all these decisions it appears that the chief weight of comparison has fallen upon the sorus, with its single apparently marginal indusium, and its usually subglobose form: little attention seems to have been paid to its probable origin in descent. But Sir William Hooker

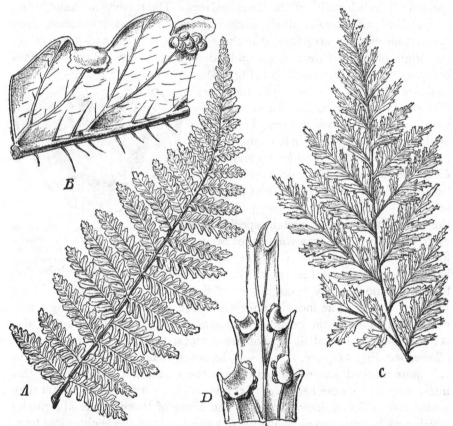

Fig. 582. *Hypolepis* Bernh. *A = H. tenuifolia* Bernh.: lowest pinna. *B = H. repens* Presl: segment with venation and sori. *C, D = H. Schimperi* (Kze.) Hook.: *C* = lowest pinna, *D* = part of a segment with venation and sori. (*B* after Baker, in *Fl. Bras.*: the rest after Diels, from Engler and Prantl.)

discusses also the habit of these plants (*Spec. Fil.* Vol. 11, p. 59). He points out how Presl in his *Tentamen* limits the genus to those species which correspond to the *Microlepia* group of *Dicksonia*. He also quotes John Smith as adopting this view, though at the same time comparing *Hypolepis* with some large-fronded species of *Polypodium*. He remarks that "their whole habit naturally indicates them to be a distinct group from the species which I retain as *Cheilanthes*" (Fig. 582, *A, C*). Mettenius (*Farngattungen*, V, p. 3) draws a clear distinction between *Hypolepis* on the one hand and *Cheilanthes*, according to the form of the spores, which he describes as spherical-tetrahedral ("kugelförmig-tetraedrisch") in the former, and spherical-quadratic in the latter ("kugel-quadrantisch"). He does not hesitate to bring forward this distinction, since it appears in all the species of the two genera, while other characters are very fluctuating. Moreover he extends it to *Plecosorus*, which he ranks with *Phegopteris* and the Aspidieae. If those who followed had given more attention to habit-characters and development, *Hypolepis* might have arrived sooner at some more fixed systematic position. Evidently Diels was dissatisfied with its place among the Cheilanthinae, for he remarks (*l.c.*, p. 277) that *Hypolepis* has in habit little in common with the other types of this series. To us who now realise that a very similar soral structure may arise along distinct phyletic lines, the habit-characters acquire additional value in these comparisons. It seems probable that the sorus of *Hypolepis* may have come from some bi-indusiate Dicksonioid source by abortion of the inner indusium, while that of *Cheilanthes* may have been uni-indusiate from the first (Fig. 582, *B*, *D*). Already Kuhn (1882) and Prantl (1892) had placed *Hypolepis* in the newly erected group of the Dennstaedtiinae. An examination of the genus, from the developmental point of view, shows that this is probably the natural place, though as its most detached member.

The habit-similarity of *Dennstaedtia* and *Hypolepis* appears in the creeping rhizome with relatively long internodes and ample upright-growing leaves, of high pinnation and finely cut: the vesture consists of hairs only: and each of the numerous sori is seated in a sinus of the margin. Both rhizomes are solenostelic. Gwynne-Vaughan has demonstrated the similarity of the vascular system of species of *Dennstaedtia* (Vol. II, Fig 536) to those of *Hypolepis* (Fig. 583). The latter, however, being smaller show no advance towards polycycly such as appears in the larger Dennstaedtias (Vol. II, Fig. 537), and notably in the erect *Saccoloma* (Vol. II, Fig. 538). But there are other indications of advance, for instance occasional interruptions of the continuity of the otherwise undivided leaf-trace are seen. These are of the nature of perforations, which appear also in the solenostele of some *Dennstaedtias*. They may be held as signs pointing towards that disintegration which is a marked feature in the vascular system of *Davallia*. Incidentally it may be noted that the supply to the lateral branches in *Hypolepis* is marginal in origin, not abaxial as in the Cyatheoid Ferns (Fig. 583).

The chief distinction between *Dennstaedtia* and *Hypolepis* lies in the sorus. That of *Dennstaedtia* is cup-like, and indistinctly two-lipped, while that of *Hypolepis*, though corresponding to it in position, has only a single lip, viz., the upper or adaxial, which curves more or less over the receptacle, and is rather membranous in texture (Fig. 582, *B*, *D*). It has been described as "formed out of the reflexed margin" (*Syn. Fil.* p. 128). The converse of this is probably the correct view, for we shall see that the indusium is by descent a superficial growth which has later become merged into the leaf-margin. Our comparison may start from *Dennstaedtia dissecta* (Sw.) Moore, a pinnule of which seen from below shows almost spherical sori, turned obliquely

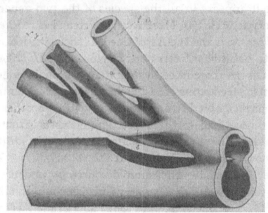

Fig. 583. Vascular system of *Hypolepis repens*, showing the departure of a leaf-trace (*L. T.*) from the solenostele, and the attachment to it of two lateral shoots, the one arising from the basiscopic margin of the leaf-trace (*l.sh.*), the other from the acroscopic margin (*l.sh'.*). (After Gwynne-Vaughan.)

downwards, and borne each on an anadromic veinlet (Fig. 584, *a*). The vascular strand terminates in the convex receptacle as an expanded mass of tracheides, and the sorus is enclosed by upper and lower indusial flaps (Fig. 584, *c*). The lips gape widely, pressed apart by the very numerous

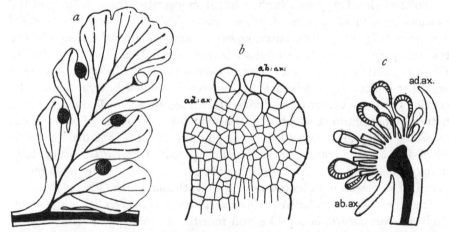

Fig. 584. *a*=a pinnule of *Dennstaedtia dissecta* (Sw.) Moore, seen from below, showing marginal sori on the apex of the anadromic branches of the veins (× 4). *b*=very young sorus, cut in vertical section, showing the marginal receptacle and superficial indusial flaps (× 150). *c*=mature sorus in vertical section (× 35).

adax=the upper indusial flap: *abax*=the lower. It will be noted that in *b* the upper surface is to the left, in *c* to the right.

sporangia. Various ages of them are intermixed, but the section shows that while stalks of old sporangia are grouped towards the centre, the young sporangia are chiefly near to the margin of the receptacle. A median section through a young sorus shows the oldest sporangium distal upon the convex receptacle, while others may be initiated below (Fig. 584, *b*). But though the basipetal sequence is thus seen at first, that order is not maintained. The section also shows that both the indusial flaps are superficial in origin, neither of them representing the margin of the pinnule, which is the receptacle itself. In fact the sorus of *D. dissecta* is of the Dicksonioid type, but with a more advanced state of the "mixed" condition than is seen in *D. rubiginosa* or in *Microlepia*

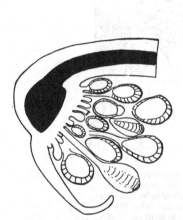 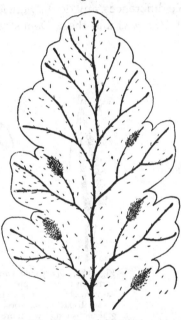

Fig. 585. Mature sorus of *Hypolepis nigrescens* Hk., cut vertically, showing its mixed character, and the absence of the inner indusium. (× 85.)

Fig. 586. Pinnule of *Hypolepis repens* (L.) Pr., seen in surface view. (× 10.)

(Vol. II, Figs. 539, *B*; 540). The basipetal sequence is, however, strictly maintained in *D. apiifolia* (Vol. II, Fig. 539, *A*).

The sorus of *Hypolepis nigrescens* Hk. presents in section a very similar outline to that of *Dennstaedtia dissecta*; but the curvature to the lower surface is stronger, and the sorus is more definitely of the mixed type, while the generic character appears in the absence of the inner (abaxial) indusium (Fig. 585). This may readily be interpreted as a more advanced state, derivative from that of the Dicksonieae, the absence of the inner indusium being related to the fact that its protection will be no longer needed where the curvature is strong. But in *H. repens* (L.) Pr., while the upper indusial flap may be reflexed in some examples, in others it is expanded in the plane of the pinnule (compare Fig. 582, *B*, *D*, with Fig. 586). The latter state appeared in my Jamaican specimens, in which the sorus is considerably spread along the vein, and this instead of terminating at the receptacle extends an appreciable distance onwards into the rounded marginal lobe that represents the upper or adaxial indusium (Fig. 586). It is this lobe which in typical forms of *H. repens* curves over and protects the sorus. A vertical section through such a sorus as that of Fig. 586, following the vein, shows how greatly the receptacle is extended and flattened, while the vein is seen continued far into the distal indusial flap (Fig. 587, *a*). If, in the presence of these palpable differences, any doubt were felt as to a real relationship between this type of *H. repens* and those Dennstaedtioid Ferns which have both indusia

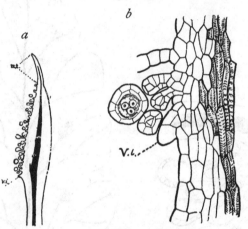

Fig. 587. *a*=young sorus of *Hypolepis repens* cut vertically: *u.i.*=upper indusium, traversed by a vascular strand: *v.i.*=vestigial lower indusium (×15). *b*=small part of the soral surface, including the vestigial indusium (*v.i.*), more highly magnified (×160).

present, the fact that a vestigial lower indusium may sometimes be found would remove it. Such a vestige is seen in Fig. 587, *v.i.* Often, however, the inner indusium is altogether absent, the sorus appearing superficial upon the vein, and distinctly intra-marginal. This is the characteristic of the Fern now designated, in Christensen's Index, *Dryopteris* (*P.*) *punctata* (Thunbg.) C. Chr. It has been variously ascribed to *Polypodium*, *Phegopteris*, *Hypolepis*, *Nephrodium*, etc. Many authors have noted how impossible it is to draw a definite line between certain Ferns ranked as *Polypodium* and *Hypolepis*. In particular Sir Joseph Hooker remarks on *Hypolepis* in the *Flora Tasmaniae* (Vol. II, p. 138) that "sometimes the reflexion of the pinnules' margin is so slight that the sorus is really naked, and then I cannot distinguish the genus from *Polypodium*, or the species *H. tenuifolia* from *P. rugulosum* Lab." Mr Carse, comparing such types on the spot in New Zealand, says (*Trans. N. Z. Inst.* Vol. XLVII, p. 85), "In some forms of *Hypolepis* the spurious involucre (=inner or lower indusium) is hardly or not at all developed, and sometimes it appears slightly in *Polypodium*, while the sori of the latter are frequently and distinctly marginal." The fact appears to be that these Ferns are variable in the features of the sorus as well as in its exact position. From a phyletic point of view there is no obligation to draw any sharp line between *Hypolepis* and the ferns styled *Polypodium rugulosum* or *punctatum*. That they are very closely related, notwithstanding soral differences, is shown by the habit-similarity, by the absence of scales and the presence of simple hairs, and by the practically identical vascular structure, as noted by Gwynne-Vaughan (*Ann. of Bot.* XVII, p. 694). We may take it that in this nearly related series those forms which are more conservative of the Dennstaedtioid characters have been designated *Hypolepis*, while the more advanced forms have been referred to *Phegopteris* or *Polypodium*; but still the series consists of Ferns naturally akin.

The sporangia of *Dicksonia* and *Thyrsopteris* are relatively thick-stalked, with complete oblique annulus consisting of numerous cells: in fact they have archaic features. Those of *Dennstaedtia* approach the usual Lepto-sporangiate type in having a long thinner stalk composed of three rows of cells, while the annulus is almost vertical, being partially or even completely interrupted by the insertion of the stalk, and it consists of relatively few cells (see Vol. II, Fig. 539, *C, D*). In *Hypolepis* also the sporangia have thin stalks and a vertical annulus. There is thus evidence of a progression in the detail of the sporangium itself from an archaic to a more modern type. These Ferns constitute in fact a series, which may be traced in a number of features, from such a source as *Cibotium*: but the vegetative characters remain more constant than the soral, and these features clearly indicate the unity of the whole series, as well as its primitive origin.

The systematic result of these comparisons, if classification is to be phyletic, can only be the removal of the Ferns designated *Polypodium* (*Pheg.*) *punctatum* Thunbg., and *Dryopteris* (*Phegopteris*) *punctata* (Thunbg.) C. Chr. from either *Polypodium* or *Dryopteris*, and their inclusion under *Hypolepis*. This follows from the argument given above: and a particularly distinctive fact is that here the dermal appendages are hairs and not scales, such as are a marked feature of the genera designated *Dryopteris* and *Polypodium*.

We may conclude then that the series *Cibotium—Dennstaedtia—Hypolepis* provides an example of phyletic advance within a circle of affinity which has been regarded by many systematists as a close one. The chief interest lies in the changes of soral character. In particular, (i) this series carries out in a very convincing way the transition from a typically gradate to a typically mixed state of the sorus. It will be seen later that a similar pro-gression may be traced in other sequences of Ferns, whether with marginal or with superficial sori. (ii) It illustrates a transition of the sorus from the marginal position to the surface of the·leaf, culminating in certain forms of *Hypolepis repens* that have been ranked as *Polypodium*. The slide of the sorus from the margin to the surface is again a feature that recurs elsewhere, for instance in the Pteroid Ferns. (iii) The series also demonstrates the partial or complete elimination of the inner or lower indusium: for the inequality of the two lips in *Dicksonia* (*Cibotium*), and more markedly in *Dennstaedtia*, leads to the complete elimination of the lower lip in *Hypolepis nigrescens*, while in *H. repens* various degrees of its abortion are found. Such abortion of the indusium is also illustrated in other Ferns. (iv) Together with these modifications of the sorus the type of the sporangium also changes. That of the Dicksonieae is massive, with a thick stalk, and complete oblique annulus of many cells: that of *Dennstaedtia* and *Hypolepis* is of a type usual for advanced Leptosporangiate Ferns with vertical interrupted annulus of few cells. But the spore-output remains relatively uniform. A broad com-parison of Ferns at large shows that all of these changes indicate advance. Here they all run substantially parallel to one another, and are probably correlative.

The Dennstaedtiinae prefigure in fact a number of those features which appear more deeply stamped upon the typically Leptosporangiate groups that are associated with the genera *Davallia* and *Pteris*. The creeping solenostelic rhizome of *Dennstaedtia* shows the first steps of perforation (Vol. I, Fig. 144, *B*), moreover the breaking up of the leaf-trace is already suggested in *Cibotium* and in *Saccoloma* (Fig. 581): both of these conditions are more fully developed in *Davallia* and in *Pteris*. The Dennstaedtiinae give no suggestion of the production of protective scales, except in *Saccoloma* which is relatively advanced in other ways: but scales are general in the Davallias, though they are not always present in the Pteroid Ferns. On the other hand, in *Saccoloma* there is a partial linkage of the marginal sori into a linear sequence, which is so outstanding a feature in the Pteroids. The flattening of the receptacle in the mixed sorus is foreshadowed in *Denn-staedtia* and *Hypolepis*, also the elimination of the lower indusium: this elimination does not become general in the Davallioid Ferns but it becomes a definite feature among the Pteroids. Further, there is the advance in *H. repens* to a flattened Polypodioid sorus: but there is in the Dennstaedtiinae no suggestion of any transition to an Acrostichoid state, unless the spread of the receptacle in *H. repens* can be so regarded. Lastly, the sporangia of *Dennstaedtia* and *Hypolepis* are of the advanced Leptosporangiate type, as are also those of the Davallioids and Pteroids. It will thus be seen how interesting is this intermediate Sub-Family in the morphology of those

Leptosporangiate Ferns which have actually marginal sori, or can be shown by comparison of early related forms to have been derived from such. This may be held as a justification for devoting a special chapter to *Hypolepis*: for though we rank it with the Dennstaedtiinae, it is actually in advance of them in certain features. It is a very suggestive type in a Sub-Family that takes its origin from a primitive source, where relatively massive sporangia were in the first instance of marginal position.

LEPTOLEPIA Mettenius, Kuhn

A brief note may here be added on this genus, which was included by Hooker in *Davallia* (*Syn. Fil.* p. 91), but was treated by Christ as a section of *Microlepia* (*l.c.* p. 308). It comprises two species, both Autralasian: the better known of these is *Leptolepia Novae-Zelandiae* (Col.) Kuhn. The creeping rhizome is covered with hairs, which are raised slightly from the surface each upon a multicellular cushion. It is traversed by a typical solenostele, with an undivided leaf-trace of wide horse-shoe form: these features are common for *Microlepia* (Gwynne-Vaughan, *Ann. of Bot.* XVII, p. 691, G.-V. slides 845–856). The sori are terminal on an abbreviated lateral vein: but the indusium, unlike that of *Microlepia* and *Davallia*, is attached only to the vein of the pinnule, and is free upwards, while its margin is eroded into tatters (see Fig. 588, *A*, *B*, p. 16). The sorus has not been examined developmentally. These characters collectively point to a place in the Dennstaedtiinae of Prantl, rather than with *Davallia*, which bears scales and has a highly disintegrated leaf-trace. *Leptolepia* appears to be even more primitive than *Microlepia*, which it most closely resembles (see Vol. II, p. 274).

MONACHOSORUM Kze.

Another example of elimination of the inner indusium is presented by the genus *Monachosorum*, which includes two Asiatic species: of these the better known is *M. digitatum* (B. 1) Kuhn. After wide vicissitudes of classification, and having received seven generic synonyms, it has been located by Diels as a substantive genus near to *Dennstaedtia*, which it resembles in habit, and particularly in the presence of bulbils in the axils of the primary pinnae. Detailed examination also supports this comparison (Studies VII, *Ann. of Bot.* XXXII, p. 56). The thin ascending axis bears laxly crowded leaves, with ferrugineous hairs, not scales. The rhizome contains a dictyostele not far removed from solenostely: the leaf-trace originates undivided as in *Dennstaedtia*, but it divides at once into two straps. The sorus is superficial without any indusium, and it is seated at or very near to the slightly enlarged end of a vein (Fig.587 *bis*). The sporangia are almost simultaneous in origin, while the annulus is interrupted at the insertion of the stalk. These characters, together with the anatomy and the presence of hairs not scales, justify the position assigned by Diels in proximity to *Dennstaedtia*, as a type that has become ex-indusiate, after the manner of *Hypolepis*.

Fig. 587 *bis*. Pinnule of *Monachosorum subdigitatum*, seen from below, with naked sori borne close to the ends of the veins. The sporangia have been removed from those on the right. (× 10.)

TRANSITION FROM DICKSONIOID FERNS TO THE DAVALLIOIDS AND PTEROIDS

The argument as to the importance of the marginal position of the sorus has been advanced in Vol. II, Chapter XXVI, and it has been seen how that position, so pronounced in the Schizaeaceae, is present also in the Hymenophyllaceae, Loxsomaceae, and Dicksoniaceae, while in the Dennstaedtiinae (Vol. II, Chapter XXX), and particularly in *Hypolepis* (Vol. III, Chapter XXXVI), it may be departed from by gradual steps, which run parallel with other signs of advance in organisation, anatomical as well as soral. These indications naturally prepare the way for the study of other Ferns either with the marginal position fully maintained, or gradually departed from, as in *Dennstaedtia* and *Hypolepis*. It will be seen that such changes are linked with others that lead to a condition characteristic of the more advanced Leptosporangiate Ferns. It is believed that such types are actually seen in the Davallioid and Pteroid Ferns, the discussion of which will be taken up from the point of view thus indicated.

These two groups have usually been treated apart. Diels places the Davallieae quite separate from the Pterideae (*Nat. Pflanzenfam.* I, 4, pp. 204, 254). The diagnoses are definitely systematic rather than comparative. The aim seems to be to distinguish between these two groups rather than to detect similarities, and so to arrive at diagnostic rather than at morphological ends. But if an evolutionary view be entertained, the fundamental idea being the origin of both of these groups of Ferns from some source such as the Dicksoniaceae taken in its widest sense, and if the possible progressions in respect of various criteria of comparison be examined, the probability will be seen to emerge that the two are really related, and that they mutually throw light upon one another. The leading diagnostic feature is that in the Davallioid Ferns the sori are usually—but not always—separate and marginal, with a double indusium: but in the Pteroid Ferns they are merged into linear sequences, and tend to spread on to, or even over, the lower surface, while the lower indusium becomes vestigial or abortive. It will be seen, through intermediate steps, how these states were attained. The Davallioid Ferns will be discussed first because they retain more persistently the individual marginal, two-lipped sorus, which is believed to have been primitive, as seen in the Dicksoniaceae.

BIBLIOGRAPHY FOR CHAPTER XXXVI

587. METTENIUS. Farngattungen, V, *Cheilanthes*. 1859.
588. HOOKER. Synopsis Filicum, p. 128. 1873.
589. DIELS. Natürl. Pflanzenfam. I, 4, p. 277, etc. 1902.
590. GWYNNE-VAUGHAN. Ann. of Bot. XVII. p. 694. 1903.
591. CARSE. Trans. New Zealand Inst. XLVII, p. 85.
592. BOWER. Studies VII, Ann. of Bot. XXXII. p. 50. 1918.

CHAPTER XXXVII

DAVALLIOID FERNS

THE Davallieae stand as Tribe IV in the systematic scheme of Christensen's Index, which restates in convenient form the arrangement of Diels (*Natürl. Pflanzenfam.* I, 4, p. 204). The Family is there held to include fifteen genera: but of those *Dennstaedtia, Microlepia, Leptolepia,* and *Saccoloma* were grouped with *Hypolepis* as the Dennstaedtiinae by Prantl, and this separation has been upheld here for reasons explained in Chapter XXXI, Vol. II. These genera have there been ranked as a Tribe of the Dicksoniaceae, but with *Saccoloma* and *Hypolepis* detached from the rest by somewhat more advanced features. The remaining genera of the Davallieae group themselves naturally round *Davallia*, which may be held as a central type of those more advanced Dicksonioid derivatives where the individuality of the sorus is as a rule maintained.

DAVALLIA Smith

The genus *Davallia* (excl. *Prosaptia*) includes about 60 species, as stated in Christensen's Index, which inhabit the warmer regions of the Old World. They are mostly of creeping habit, and are often epiphytic. The leaves are solitary, and usually highly pinnate, with open venation. The rhizome bears protective scales. The sori are seated separately on the ends of the veins, and are more or less intra-marginal, a condition which compares with what has been seen in some of the Dennstaedtiinae. But a distinctive feature is that the lower indusial margins are fused with the leaf-surface, giving a pocket-like form to the sorus (Fig. 588, *J, K*). Anatomically the genus *Davallia* itself is more advanced than any of the Ferns hitherto described. The rhizomes are frequently massive and fleshy, having little or no sclerenchyma. The vascular system is seen in transverse section of the rhizome to consist of a circle of meristeles, of which two are larger than the rest and flattened: these run parallel to the upper and lower faces of the rhizome. The circle is completed by a number of smaller strands which have been cut through in their course towards the alternately lateral leaf-bases. There are in fact two lateral rows of alternating leaf-gaps, and from the margin of each gap separate vascular strands arise, which represent a much-divided leaf-trace. Their nature as such is shown by their passage outwards, with varying anastomoses among themselves, into the bases of the petioles (Fig. 589).

There is some variety in the stelar structure of the Ferns which have been ascribed to *Davallia*: and the divergences from the highly segregated state

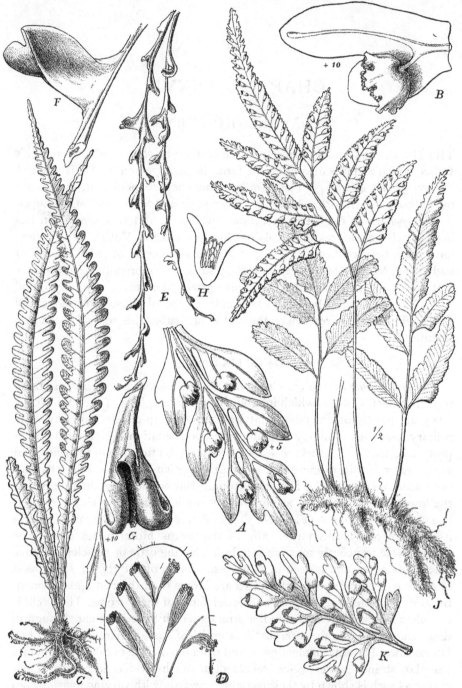

Fig. 588. A group of drawings of Ferns which have been ascribed to *Davallia,* but some now referred to other affinities.

A, B = *Leptolepia Novae-Zelandiae* (Col.) Kuhn (see Chapter XXXVI, p. 13): *A* = pinnule of the last order with venation and sori: *B* = fertile segment with sorus and indusium. *C, D* = *Prosaptia alata* Christ. *C* = habit: *D* = apical region of a fertile segment, on the right side the indusium removed to expose the sori (see Chapter XLVIII). *E–H* = *Prosaptia Reineckii* Christ. *E* = distal region of two leaves: *F* = lateral view of a pinna: *G* = frontal view: *H* = transverse section (see Chapter XLVIII). *J* = *Davallia pentaphylla* Bl., habit. *K* = *D. canariensis* (L.) Sm. a pinna of third order, with venation and sori. (*C,* after Hooker: the rest after Diels.)

are of comparative importance, suggesting for some of them a more primitive position than that of *Davallia* itself. For instance, in *Davallia contigua* (Forst.) Spr. (G.-V., slides 951–952) there is an almost perfect solenostele with occasional perforations, while each leaf-trace consists of two equal strands. This is, however, one of the species included in *Prosaptia*, and now referred to a position elsewhere (see Chapter XLVIII). Another exceptional type is *Davallia dubia* R. Br., which shows a large solenostele, and an undivided petiolar trace with incurved hooks (G.-V., slides 857–865). The synonymy of this Fern is a sufficient indication of its doubtful position in *Davallia*, and its probable Dicksonioid affinity. Putting such doubtful cases aside, it may reasonably be held that the vascular system of the typical *Davallia* is really derivative from a solenostele with undivided leaf-trace, as in *Dennstaedtia*, with leaf-gaps overlapping alternately, and the meristeles between them attenuated. Each leaf-trace meanwhile is divided to its base, coming off as a number of separate strands; but referable in origin to the type of trace seen in *Microlepia* (*Davallia*) *speluncae* (L.) Moore (see Fig. 159, Vol. I, p. 166).

Fig. 589. *Davallia dissecta* J. Sm. Rhizome, slightly magnified. $A=$ vascular system dissected, and flattened to a single plane: $o=$ upper meristele; $u=$ lower; $b=$ insertion of a leaf; $x=$ origin of a lateral shoot. $B=$ transverse section. (After Mettenius, from De Bary.)

According to the principles laid down in Vol. I, Chapters VII–X, so highly disintegrated a vascular system as that seen in *Davallia* indicates phyletic advance, while this accords with the presence of dermal scales, not hairs, covering the rhizome of true Davallias. .Such a type as *Davallia dubia*, with its solenostele and rusty hairs, will rightly take its place anatomically with the Dennstaedtioid Ferns, where Sir W. Hooker placed it (*Sp. Fil.* I, p. 71). The general conclusion will be that by its scales in place of hairs, and by its highly segregated vascular system, *Davallia* is a relatively advanced type.

The features of the sorus of *Davallia* also point towards phyletic advance. Their appearance is shown in Fig. 588, *J*, *K* representing species both of which fall under the section *Eu-Davallia* of Hooker (*Syn. Fil.* p. 94). The pocket-like sorus appears to be inserted upon the lower surface of the pinnule, but comparison with the Dennstaedtiinae and with other related Ferns suggests its derivation from a two-lipped sorus of the Dicksonioid type, by progressive inequality of the upper and lower indusia : the former becomes incorporated with the general leaf-surface, as in *Hypolepis*, the latter fusing at its margins with the expanse on which it is borne, and forming the closed and sometimes deeply protective pocket characteristic of the genus. The flat receptacle lying at the base of the pocket produces a mixed sequence of closely packed sporangia. When mature their stalks are greatly elongated, so as to lift the ripe capsules above the level of their fellows, and thus to secure the discharge of the spores.

The origin of the sorus has been examined by Von Goebel in *Davallia dissecta* (*Organographie*, p. 1143, Fig. 1134). My own observations were made on *D. pentaphylla* Bl., a species in which the mature sori are distinctly intra-marginal (Fig. 588, *J*). The question is whether or not the receptacle is truly marginal in the first instance in such species. First the leaf-margin becomes flattened, and the indusial flaps arise as superficial growths back from the flattened margin. The lower (*l*) (Fig. 590) takes precedence at first: but the more massive upper indusium (*u*), which is later assimilated to the general leaf-surface, extends by intercalary growth, and overtops it. Meanwhile sporangia appear upon the flattened receptacle between them. Thus the sorus in *Davallia* is of marginal origin even though it may appear superficial when mature. The more or less marginal position of the adult depends upon the varying activity of intercalary growth in the upper indusium. The further development of the sorus has been followed in *Davallia griffithiana* Hk.; here the receptacle remains flat, and the earliest

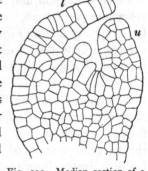

Fig. 590. Median section of a young sorus of *D. pentaphylla* Bl. The lower indusium (*l*) has run out into a single layer of cells, the upper (*u*), which ultimately forms the false margin of the leaf, is more bulky, and shows signs of intercalary activity. The first sporangium occupies a central position on the receptacle, and a later one is seen between it and the upper flap, thus indicating a gradate sequence. (× 250.)

sporangia arise median upon it, thus again suggesting a gradate sequence; but soon the sorus becomes mixed, successive sporangia being initiated promiscuously over the whole surface, pushing their way upwards between the long thin stalks of those already present. The sporangia themselves have the structure usual in advanced Leptosporangiate Ferns, with a vertical annulus of relatively few cells, and stalks composed, in part at least, of only a single row of cells (Fig. 591).

The facts thus detailed for *Davallia* indicate it as an advanced type in a sequence derived from a Dicksonioid source with marginal sori. The individuality of the sorus is maintained throughout the genus; the habit remains fairly stable in these creeping Ferns, with their much-divided leaves and open venation. While the typical Dicksonioid Ferns have prevalent solenostely usually with an undivided leaf-trace, hairs as dermal appendages, and marginal sori of the gradate type with conical receptacle, slightly unequal indusial lips, and massive often short-stalked sporangia; in *Davallia* we see an advanced state of disintegration of the vascular system, though clearly referable to a solenostelic origin, scales as dermal appendages, the marginal sori more or less markedly superficial in the adult state, of mixed type, with flat receptacle, and indusial lips often very different from one another, the upper being merged in the general leaf-surface, and with delicate, long-stalked sporangia. The sum of these features mark out *Davallia* as relatively advanced, though still essentially of the same type as *Dicksonia*. Comparison of the sorus of *Davallia* at an early stage of development with those of *Thyrsopteris, Cibotium, Dennstaedtia*, and *Microlepia* (Vol. II, Figs. 529, 534–5, 539, 540) demonstrates that in these related Ferns there has been a slide of the sorus, from the marginal position with definite indusial lips, to a superficial position, the upper and more massive lip being incorporated into the flattened expanse of the pinnule; and the change can even be followed in the individual development. It is, however, significant that the vascular supply terminates as a rule below the receptacle. The series illustrates the change in form of the receptacle. In the earlier terms it is conical with the

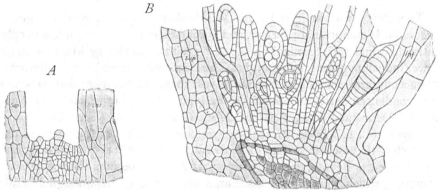

Fig. 591. *A* = young sorus of *D. griffithiana* Hk., in vertical section, showing the flat receptacle with the first sporangium lying centrally. *B* = an older sorus of the same, showing sporangia of different ages intermixed. *sup* = upper indusium: *inf* = lower. (× 100.)

vascular supply extending into it (Vol. II, Fig. 540): but in *Davallia* it is flat, the vascular tissue spreading out below the surface that bears the sporangia (Fig. 591, *B*). This has its close relation to the change from a gradate to a mixed sorus, and it has been shown that in this series there is a gradual change from a gradate to a mixed sequence of sporangia. Lastly, the sporangia of the series show a transition from a massive type with thick stalks and oblique continuous annulus to the more delicate structure of *Davallia*, with vertical interrupted annulus, and a stalk consisting of only a single row of cells. The parallelism that exists in respect of these various structural features confirms the reality of the progression from the Dicksonioid to the Davallioid Ferns, which the habit itself suggests.

Davallia is the centre of a plexus of Fern-types which have always presented difficulties to the systematist, as is shown by their varied synonymy. A special interest in any phyletic discussion attaches to such related forms, for not only will their similarities confirm the relation to some central type, but also their divergent features may illuminate the problem of their descent. The order in which the genera of the Davallieae are arranged by Diels appears to be quite arbitrary (*Nat. Pflanzenfam.* I, 4, p. 204). But if regard be paid to certain features of advance which they show, a natural grouping may follow. Such features are: (i) the relation of the sori to the margin or surface of the blade: (ii) the elimination of the lower indusium: (iii) the linkage of sori laterally to form coenosori. Parallel with comparisons according to these soral characters, those derived from the dermal appendages and the vascular anatomy may be held as adding grounds for criticism or confirmation.

(i) *Position of the sorus relative to the leaf-margin.*

Humata Cav.

This genus was merged by Sir W. Hooker as a section of the genus *Davallia.* It comprises 14 species, and is characterised by more or less deeply intra-marginal sori, terminal on the veins, with a circular or kidney-shaped lower indusium having free margins. The rhizomes bear scales, and alternate, coriaceous, and sometimes simple leaves: these may be regarded as small and condensed derivatives from a more highly divided type, such as is seen in the Dennstaedtiinae. Anatomically *H. heterophylla* (Sm.) Desv., and *H. repens* (L. fil.) Diels, present in transverse section of the rhizome the characteristic vascular system of *Davallia*, but with a leaf-trace of only two equal strands. Beyond the suggestion that *Humata* bears a condensed leaf-structure, which is frequently associated with a deeply intra-marginal adult position of the sori, the genus calls for no special remark (compare Diels, *Nat. Pflanzenfam.* I, 4, Fig. 112).

Nephrolepis Schott

This genus comprises 17 species of Ferns with upright, slightly scaly stock, giving rise to runners which are often tuberous. The leaves are simply pinnate, with open venation: they may often show continued apical growth. The stolons are leafless, and protostelic: but when enlarged into tubers they illustrate how disintegration into meristeles follows on increase of transverse section, even in the absence of foliar gaps (Vol. I, Fig. 182, p. 190). The normal stock is dictyostelic, with foliar gaps, and the leaf-traces consist usually of three strands. When a stolon turning its apex upwards enlarges into a leafy stock, the ontogenetic progression from protostely to dictyostely, as it is seen in the sporeling, is repeated in its vascular system: in fact, the stolon may be regarded as corresponding to a prolongation of the protostelic stage of the embryo (G.-V. MSS.).

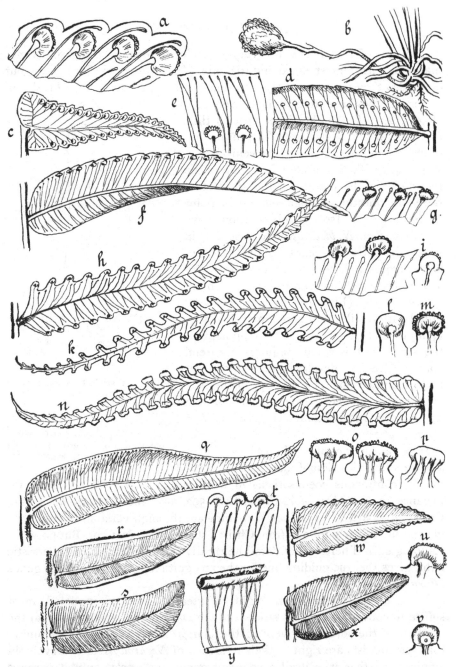

Fig. 592. Sori of *Nephrolepis*, illustrating their diversity of position and relation within the genus. (All after Christ.) *a, b, c = N. cordifolia* (L.) Presl, *a* = sori enlarged, *b* = rhizome, *c* = pinna, natural size. *d, e = N. acuta* (Schenkr.) Presl, *d* = base of pinna, natural size, *e* = sori, enlarged. *h, i = N. floccigera* Moore, *h* = pinna, natural size, *i* = sori from above and below, enlarged. *k, l, m = N. davallioides* (Swartz) Kunze, *k* = pinna, natural size, *l* = sorus from above, *m* = sorus from below, enlarged. *n, o, p, q = N. dicksonioides* Christ, *n* = fertile, *q* = sterile pinnae, natural size, *o* = sorus from below, *p* = sorus from above, enlarged. *t, u, v, w, x = N. abrupta* (Bory) Matt, *t* = part of a pinna, enlarged, *u* = sorus from below, *v* = sorus from above, enlarged, *w* = fertile, *x* = sterile pinna, natural size. *r, s, y = N. acutifolia* (Desv.) Christ, *r* = fertile, *s* = sterile pinna, natural size, *y* = part of the coenosorus, enlarged.

The sori of *Nephrolepis* are very variable in position, being sometimes marginal with almost equal indusial lips (*N. davallioides* (Sw.) Kze.), but often also they may be deeply intra-marginal (*N. cordifolia* (L.) Presl): in either case they are seated on the vein endings (Fig. 592, *c, k*). In the latter species they resemble in outline those of *Nephrodium*, but differ in being terminal on the vein (Vol. I, Fig. 224, *C, D*). So variable a relation of the receptacle to the margin in the adult state raises the question of that relation in point of development. Isolated observations have been made on *N. biserrata* (Sw.) Schott, in which the sori are distant from the margin: these suggest that the upper indusial lip, which in *Davallia* has been seen to grow by inter-calary activity (Fig. 590), is in *Nephrolepis* gifted with a still more active marginal seg-mentation (Fig. 593). The development of the sorus in this genus requires a careful comparative and developmental study: but provisionally it may be held as probable that all the species were derived from a source with marginal sori (Studies III, *Ann. of Bot.* 1913, p. 462).

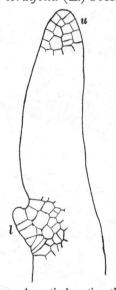

Fig. 593. A vertical section through the young sorus of *Nephrolepis biserrata* (Sw.) Schott, showing the great inequality of the indusial flaps. The lower (*l*) is markedly intra-marginal, the upper (*u*) appears very definitely as a continuation of the leaf-surface, and has an active marginal segmentation. (× 250.)

In certain species of *Nephrolepis* there may also be a lateral fusion of marginal sori, pro-ducing continuous coenosori, as in *Lindsaya* or *Pteris*: this is seen, for instance, in *N. dicksonioides* Christ (Fig. 592, *n–q*), and *N. acutifolia* (Desv.) Christ (Fig. 592, *r, s, y*). Such facts indicate the high variability within the genus which clearly follows from Christ's striking observations. But notwith-standing such differences of detail he rightly remarks that the characteristic habit gives the true guiding line for the recognition of this remarkable genus (*Farnkräuter*, p. 288).

These observations, which might be greatly extended in detail, must suffice to convey what every student of the Davallioid Ferns knows, that the position of the sorus relatively to the margin is very variable. Assuming that systematists are right in their reference of *Nephrolepis* to a Davallioid affinity, and that it is itself a coherent genus, its species exhibit perhaps more impressively than any other how diverse the relation of the sorus to the leaf-margin may be in nearly related Ferns. The facts may even suggest the facile conclusion that the relation of the sorus to the margin is a character systematically worthless. But the study of soral development in the Dicksonioid Ferns generally shows that in allied forms the marginal position is so tenaciously retained as to make it a reliable comparative feature, notwithstanding the exceptions. Instability in certain clearly derivative genera does not nullify the effect of constancy in many other related genera which are clearly primitive. Moreover, a comparison of the leaves where the sori are intra-marginal with those where they are marginal indicates that the former usually show relatively broad leaf-areas of condensation, while the latter for the most part retain their primitive cutting into narrow segments. In fact, widening of the leaf-area and shifting of the sorus from the margin inwards are features frequently related one to another.

OLEANDRA Cav.

This is a tropical genus of some ten species, with creeping and climbing shoots. It is usually placed in relation to *Nephrolepis*, and this may be accepted provisionally. It also has a kidney-shaped indusium covering sori superficially resembling those of *Nephrodium*. Until it has been fully investigated it must be ranked as a "genus incertae sedis." The difficulty in placing it illustrates once more how similar may be the results of homoplastic development among the later derivative types of the Filicales.

(ii) *Elimination of the lower Indusium*

It has been seen how in *Hypolepis* the lower indusium may be partially or even completely aborted in sori which have become superficial, while the upper indusial lip is merged into the expanse of the blade. A similar state, resulting in an apparently unprotected sorus, appears in certain other Ferns ascribed by systematists to a Davallioid affinity. An example is seen in:

ARTHROPTERIS J. Sm.

This is a widespread genus, but chiefly it inhabits the Old World, comprising four species. It has been placed sometimes with *Polypodium*, or with *Dryopteris*: but latterly with the Davallioid Ferns, in relation to *Nephrolepis*, with which in point of habit it has much in common. The rhizome bears scales, and has a vascular structure not far removed from solenostely. Two meristeles are seen in the transverse section, while the leaf-trace departs undivided, as is shown in a dry specimen in Kew. The punctiform sori are intra-marginal, and terminal on the tertiary veins. A kidney-shaped indusium is sometimes present, but it may be reduced, or even absent. These Ferns may be regarded as partially or completely ex-indusiate Davallioids (*Nat. Pflanzenfam.* I, 4, Fig. 110, p. 206). It thus appears that elimination of the lower indusium has happened also among the Davallioids, while the upper indusium is liable to be incorporated in the expanse of the leaf-blade, the change being essentially like that in *Hypolepis*, but starting from a different source.

(iii) *Linkage of sori to form Coenosori*

A special interest attaches to the progression leading from primitively separate sori to the fusion-sorus, partly because it is a change of this nature that accounts for the soral state of the whole series of the Pteroids; but also because there is ample evidence that the coenosorus has originated poly-phyletically among the derivatives of the marginal Dicksonioids. Already a partial linkage of sori has been described for *Saccoloma* (Vol. II, p. 272, Fig. 541, *A*): but numerous and more complete examples are presented by certain Davallioid derivations, such as *Nephrolepis* and *Diellia*, and these culminate in the condition seen in *Lindsaya* and *Dictyoxiphium*. The lateral fusion may involve not only the indusia, but the vascular supply is also connected up by commissures parallel to the margin of the blade, while sporangia may be seated upon these, so that the receptacle itself becomes continuous along the whole coenosorus. In the examples which follow, the linkage appears to be in close relation to condensation of the blade, marking a transition from a highly cut towards an entire form.

TAPEINIDIUM (Presl, 1849) C. Chr. 1906

This genus comprises four species from the Malayan region. It was founded to receive the Fern described in 1802 by Cavanilles as *Davallia pinnata*, but the species has passed under many synonyms, such as *Wibelia*, *Saccoloma*, *Microlepia*, *Lindsaya*, and even *Dicksonia*. This at once draws attention to it as a probable synthetic type among Ferns, with sori seated at or near to the margin. *T. pinnatum* (Cav.) C. Chr. is the species upon which detailed observation has been chiefly based. It is a Fern with a creeping rhizome which frequently dichotomises. The leaves are alternate, and simply pinnate, though the pinnae are sometimes forked, and may also be cut into separate pinnules at the base, thus indicating a relation with Ferns of more complex pinnation, from which it is probably a derivative. A type named *T. pinnatum* (Cav.), var. *tripinnatum* Rosenstock, from New Guinea, supports the idea of simplification from a Dennstaedtioid type of foliage. The rhizome is described by Diels as hairy (E. and P. I, 4, p. 216): but Prantl (*Syst. d. Farne*, p. 16) ranks it with his Davalliinae, with the words "Haare stets Zellflachen." The specimens at Kew show dark brown ramenta.

The anatomy as described by Gwynne-Vaughan is very similar to that of *Odontosoria aculeata*. Below each leaf-insertion there is a pocket that fades out downwards, giving a *Lindsaya*-structure till the next leaf-base is reached (Fig. 594). This imperfect state of solenostely compares closely with that of *Lindsaya* and *Odontosoria*, linking these forms with those that are fully solenostelic. As in these also the origin of the leaf-traces is marginal.

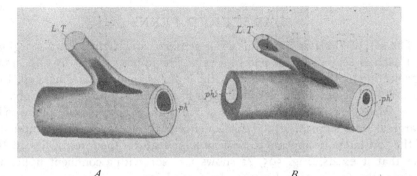

A   B

Fig. 594. A = diagram of the vascular system of the rhizome of *Odontosoria* (*Davallia*) *aculeata* (L.) J. Sm., including a node and the base of a leaf-trace, ph' = internal phloem. The external phloem is not indicated. The upper surface of the rhizome faces the observer. $B = Tapeinidium$ (*Davallia*) *pinnatum* (Cav.) C. Chr. Diagram as in A. The vascular system is curved so that the two cut ends face the observer obliquely, ph' = internal phloem. (After Gwynne-Vaughan.)

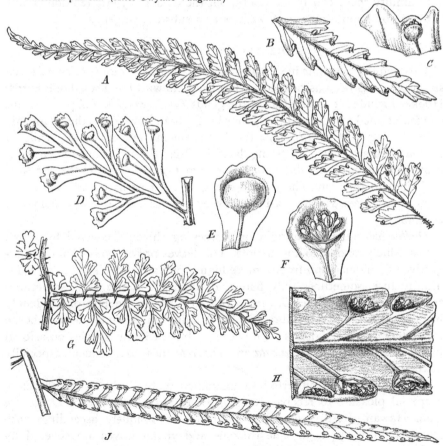

Fig. 595. $A—C = Microlepia$ *speluncae* (L.) Moore. A = secondary pinna, B = tertiary, C = part of a segment with sorus and indusium. $D—G = Odontosoria$ Presl. $D—F = O.$ *bifida* (Kaulf.) J. Sm., D = tertiary pinnule, with venation and sori, E = leaf-tooth with indusium, F = the same with indusium removed, to expose the sorus. $G = O.$ *aculeata* (L.) J. Sm., primary pinna. $H, J = Tapeinidium$ (*Wibelia*) *pinnatum* (Cav.) C. Chr., J = primary pinna. H = part of it with venation and sori. $A—C$, after Diels; $D—F$, after Baker in *Fl. Bras.*; H, J, after Fée.)

The sori of *Tapeinidium* are as a rule separate, and slightly intra-marginal, and each is seated on the end of a vein. The upper indusial lip appears as a projecting tooth of the leaf-margin, the lower as an intra-marginal cup (Fig. 595, *H,J*). The development of the sorus has not been followed, but probably, as in *Odontosoria* and *Lindsaya*, the origin of the receptacle will prove to be marginal, with signs of a gradate sequence of the sporangia. The confluence of the sori laterally appears to be a variable feature: the interest lies in the fact that it exists. Fig. 595, *H* shows two sori with a confluent upper in-dusium: the fusion may, however, be carried out in various ways. Separate sori are the rule, though fusion would be a natural sequel to leaf-condensa-tion. Such details of habit, anatomy, and sorus explain the wide synonymy of these Ferns, since they resemble a number of other types. Of the Daval-lioid affinity there can be no doubt, but the actual phyletic relation is best left open by maintaining *Tapeinidium* as a substantive genus.

DIELLIA Brackenridge

This genus is endemic in the Hawaiian Islands, and comprises some seven species. It has been associated by most writers with the Davallioid Ferns: Hooker included it as a section of the genus *Lindsaya* (*Syn. Fil.* p. 112). But it is maintained as a substantive genus by Diels and by Christ, the diagnostic points being that in *Lindsaya* the leaf-segments are unilateral, while in *Diellia* they are obliquely triangular: also that the indusium of *Diellia* is broadly adherent: this distinguishes it from *Leptolepia*, where the adhesion is limited to the vein. The three genera, together perhaps with *Saccoloma*, form a close nexus, the least specialised species having many points in common.

Diellia has either an upright or a creeping rhizome, covered by broad scales, which extend to the petiole. The leaves present various forms sug-gestive of condensation from a more highly pinnate state. The lateral veins usually fuse, sometimes only here and there (*D. falcata* Hk.), sometimes they are highly reticulate (*D. pumila* Brack.). The sori are more or less deeply intra-marginal, with a membranous lower indusium. They are usually borne separately on the widened vein-endings: but in many species occasional linkages occur, and in *D. falcata* and *D. erecta* these are frequent, especially at the bases of the pinnae.

The genus has not hitherto been examined anatomically: it was therefore a special pleasure to receive from Dr H. L. Lyon a supply of *D. falcata* from Hawaii. In this species the rhizome is obliquely ascending, with leaves spirally disposed. The rhizome and young leaves are covered by dark brown, closely imbricated scales, while others of more delicate form extend up the petiole. The stem contains a rather rudimentary dictyostele with sometimes two, sometimes three, leaf-gaps in the single section: in small stems the column of pith is small: but the *Lindsaya*-condition was not seen in any adult stem, though doubtless it occurs in the sporeling. The leaf-trace comes off as two equal straps: but these do not preserve their identity far, and sometimes they unite within the cortex of the stem. As they pursue their course upwards they become moulded in varying degree into a vascular tract of the type well known in the petioles of *Asplenium* and *Scolopendrium* (Fig. 596). But the outline is not uniform: sometimes the junction is effected by the abaxial margins, sometimes by the middle region of each tract. The latter is reminiscent of *Asplenium* or *Scolopendrium* (Vol. I, Fig. 157), the former of that of *Plagiogyria semicordata* (Studies I, Text-fig. 2). It may involve the peripheral vascular tissues only, while the xylem-tracts remain free: or the fusion may be complete. The absence of

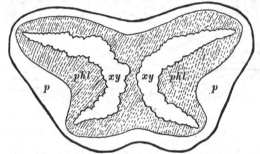

Fig. 596. Petiolar meristele of *Diellia falcata*, resulting from fusion of two separate vascular straps of the leaf-trace. The adaxial side is here directed upwards. The xylem-tracts (xy) are here separate, though they sometimes fuse. phl=phloem, p=pericycle enlarged.

precision in these fusions suggests that they follow upon a compacting of the blade from a more divided state. If that be so, then such fusions may be regarded as homoplastic in leaves subject to integration of the blade, rather than as giving any trustworthy basis for phyletic comparison. The pinna-traces are marginal in origin.

The fertile pinna of *D. falcata* shows a coarse reticulation of the veins (Fig. 597). The sori are distinctly intra-marginal, with the upper indusial lip assimilated to the surface of the blade, the lower membranous, with adherent margin. Sometimes the single sorus may be distal and solitary on a vein: but there are frequent signs of lateral fusion to form coenosori, with vascular commissures: the structure might accord equally with upgrade fusion or with downgrade disintegration of a coenosorus, but the former is the more probable interpretation (Fig. 597, *B*, *C*). The sporangium has a long stalk, composed in its basal part of only one row of cells. The annulus shows slight traces of obliquity, and has about 20 indurated cells, and a lateral stomium. The number of spores in each sporangium appears to be 48 to 64: they are oval with a prickly reticulate wall.

The genus as a whole is marked by great instability of leaf-outline. In *Diellia Mannii* Robinson, the leaf is diffusely bi-pinnate, with narrow segments: it resembles that of *Odontosoria*. From this state various steps of condensation of the pinnae are illustrated, such as are shown by Diels (E. and P. I, 4, Figs. 114). Sometimes various degrees of pinnation may be seen in the same leaf: these lead through a reduced secondary pinnation (*D. alexandri* Diels), and a simple pinnation with broad expanse (*D. erecta* Brack., or *D. falcata* Brack.), to an almost cordate pinna (*D. pumila* Brack.). With this condensation goes an increasing anastomosis of the veins, ending in a complicated network in the broadest. At the same time the sori recede from the margin to the lower surface, and they tend to be linked into

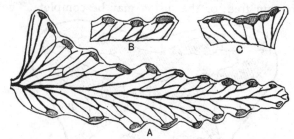

Fig. 597. *Diellia falcata* Hk. *A* = a single pinna with coarse reticulation, and sori intra-marginal, sometimes solitary on the vein-endings, sometimes showing fusion to coenosori. *B*, *C* show advanced examples of soral fusion. (× 2.)

coenosori, especially towards the base of the pinna. Such steps, illustrated within the genus *Diellia*, may be held as concomitants of condensation from a type of leaf such as that of *Odontosoria*. A like progression may be traced homoplastically in other Marginal Ferns.

The comparative conclusion from these facts is that *Diellia* originated from some Davallioid type having diffusely branched leaves. It appears to be related especially to *Lindsaya* and *Odontosoria*, and in its insular isolation to have carried out, independently of other related Ferns but homoplastically with them, a progressive condensation of leaf-form and of soral fusion. This genus, endemic in the Hawaiian Islands, where it is rapidly becoming rarer, deserves a full comparative examination from this point of view before its species become extinct. It would probably provide a convincing instance of endemic variation, homoplastic with similar variation elsewhere.

ODONTOSORIA (Presl) Fée

This is an old genus established by Presl in 1836, which comprises 19 species. Though included under *Lindsaya* in the *Synopsis Filicum* (p. 109), and by Christ (*l.c.* p. 295), it is maintained as a substantive genus by Diels (*Nat. Pflanzenfam.* I, 4, p. 215). A wide synonymy indicates the close relation of these Ferns to other Davallioid genera, but such difficulties are best resolved by maintaining the species of *Odontosoria* as constituting a substantive genus under Presl's old name. As representative types may be taken *O. retusa* (Cav.) J. Sm., frequent in cultivation, of robust habit, and with leaves of limited growth: *O. aculeata* (L.) J. Sm., the Bramble Fern (Vol. I, Fig. 44), and its ally or possibly variety, *O. fumarioides* (Sw.) J. Sm., both of which have unlimited apical growth of their straggling and climbing leaves. They were included in *Stenoloma* Fée, or *Lindsayopsis* Kuhn (Fig. 595, *G*).

The species of *Odontosoria* are mostly rhizomatous with highly divided leaves, the segments being triangular-wedge-shaped, with free veins diverging like a fan. The dermal appendages are paleae of varying width: in *O. fumarioides* they are continued upwards into a single hair-like row of cells (Gwynne-Vaughan, *Ann. Bot.* 1916, p. 502). Anatomically the genus shows fluctuations between full solenostely and the *Lindsaya*-state. The relatively robust rhizome of *O. retusa* is included in Gwynne-Vaughan's list of typical solenosteles (*Ann. Bot.* XVII, 1903, p. 691; G.-V., slides 977–986). Its undivided leaf-trace is like that of *Microlepia*, and the pinna-traces are marginal; but *O. aculeata* (G.-V., slides 866–880) has a structure like that of *Tapeinidium* (Fig. 594, *B*). There is in fact an incomplete solenostely with a sclerotic core filling a deep pocket (Gwynne-Vaughan, *Ann. Bot.* XVII, p. 712). In *O. fumarioides*, which is sometimes held as a variety of *O. aculeata*, Gwynne-Vaughan found a well-defined solenostele (*Ann. Bot.* XXX, p. 502). On the other hand, he describes a *Lindsaya*-structure for *O. chinensis* (L.) J. Sm. (G.-V., slides 914–929), and for *O. clavata* L. (G.-V., slides 931–932). Thus the genus fluctuates between solenostely and the *Lindsaya*-type of stele. Possibly size may be a determining factor, and it is worthy of note that Plumier describes the caudex of *O. aculeata* as no thicker than a writing pen (*Sp. Fil.* I, p. 191), while Gwynne-Vaughan notes those of *O. fumarioides* as fairly stout (*l.c.* p. 502). In accordance with their climbing habit the petiolar strand in the Bramble Ferns is condensed in a manner parallel with that of *Lygodium* or *Gleichenia*. This has already been discussed and illustrated in Vol. I, p. 171 (Fig. 165).

The highly branched leaves bear deltoid pinnules with forked venation, and the sori are marginal. They appear fused laterally along the width of the pinnule, and are protected by upper and lower indusial flaps. A vascular commissure traverses the marginal receptacle, connecting the vein-endings. The development of this coenosorus has been traced in *O. retusa* (Studies III, p. 459). First the margin of the pinnule becomes flattened (Fig. 598, *A*), while the segmentation and localised growth form projecting angles which are the upper and lower indusia (Fig. 598, *B*). The first sporangia appear about the middle of the flattened receptacle that lies between them: others appear laterally, but mostly on the adaxial side of those first formed. Thus

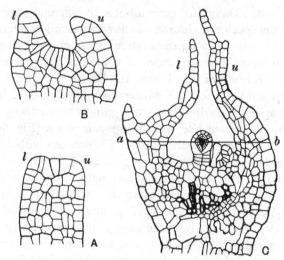

Fig. 598. Development of the sorus of *Odontosoria retusa*. *A* = a young stage where the margin is flattened, *B* = a later stage with the indusial flaps advanced, *C* = a still later stage with the oldest sporangium central on the receptacle, while other sporangia and hairs originate laterally, indicating a gradate succession, *u* = upper indusium, *l* = lower. A section along the line *a, b* shows that the first sporangia are constantly median. (*A, B*, × 250; *C*, × 100.)

the sorus is actually marginal, and is at first gradate: but a mixed condition appears later, though it is never far advanced (Fig. 598, *C*). The sporangia accord structurally with this condition of the sorus, for they have a slightly oblique annulus consisting of a continuous sequence of cells that extends uninterruptedly past the insertion of the stalk (Fig. 599). These facts show that the sorus of *O. retusa* is not far removed from the type of the marginal Dicksonioids, though it is liable to lateral fusion into coenosori and shows a transition to the mixed state.

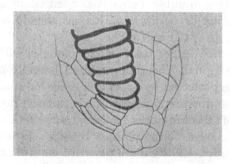

Fig. 599. Base of a sporangium of *Odontosoria retusa* showing the oblique annulus with its induration stopping short of the insertion of the stalk. (× 125.)

LINDSAYA Dryander, 1793

Lindsaya is a large genus, distributed over both hemispheres, with some 90 species as now reduced. Sir W. Hooker enlarged it by including as sub-genera various Ferns now held as constituting substantive genera. The rhizome bears superficial scales, and it contains a type of stele that is characteristic of the genus, and bears its name (Fig. 600). It has been fully described and discussed in Vol. I, p. 146, and it may be held as a general character of the genus (Studies VII, *Ann. Bot.* XXXII, 1918, p. 13). The leaf-trace is undivided: thus the whole vascular system is relatively primitive, and its retention by Ferns which in other respects show advance was probably promoted by the small size of the wiry rhizomes.

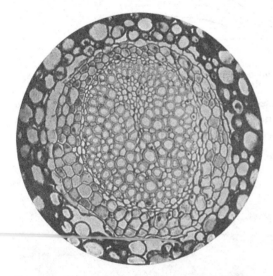

Fig. 600. Transverse section of a stele of *Lindsaya linearis* Swartz, showing the characteristic structure with the central phloem included in the otherwise solid tract of xylem. G.-V. collection, slide 992. (× 125.)

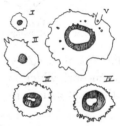

Fig. 600 *bis*. Outlines of transverse sections of the rhizome in *Lindsaya* and *Odontosoria*, all drawn to the same scale. The stele in each is shaded. The central pith is clear.

I = *Lindsaya clavata*, stelar diameter ·5 mm.
II = *L. trapeziformis*, stelar diameter ·9 mm.
III = *Odontosoria aculeata*, stelar diameter 1·8 mm.
IV = *O. fumarioides*, stelar diameter 2·2 mm.
V = *O. retusa*, stelar diameter 3·0 mm.

The incidence of the size-factor in relation to the structure of the rhizomes of the genera *Lindsaya*, *Odontosoria*, and *Diellia* is illustrated by actual measurements taken on the longer axis of the elliptical transverse section (Fig. 600 *bis*). In *L. clavata* with a measurement of ·5 mm., and in *L. trapeziformis* of ·9 mm., the *Lindsaya*-type of stele is seen. In *Odontosoria aculeata* with a stelar diameter of 1·8 mm., there is solenostelic structure with a small central column of pith: in *O. fumarioides* with a diameter of 2·2 mm., the pith-column is larger: and again it is of larger size still in *O. retusa* and *Diellia falcata*, each with a measurement of approximately 3 mm. Thus the larger steles are consistently solenostelic, while the smaller have the *Lindsaya*-structure. The conclusion appears to be that while *Lindsaya* retains permanently in its smaller rhizomes the state passed through rapidly in the ontogeny of these and many other Ferns, the larger types pass on to the full solenostely characteristic of so many rhizomes of larger size.

The leaves are variously pinnate, with fan-shaped or trapeziform pinnae and segments. For the most part the venation in *Lindsaya* is dichotomous and free: but certain species, grouped as §§§*Synaphlebium* by J. Smith, have anastomosing veins, though with the same habit and texture as the rest. In this respect they may be held as more advanced. Others have the blade condensed to a simple expanse, though with the veins free (*L.* [*Schizoloma*] *reniformis* Dry., and *L.* [*Schizoloma*] *sagittata* Dry.).

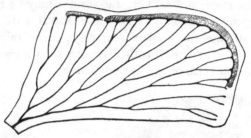

Fig. 601. A single pinna of *Lindsaya lancea* (L.) Bedd., showing the sori fused laterally to form an almost continuous intra-marginal series, or coenosorus. (× 4.)

The sori of *Lindsaya* are fused in varying degree in the several species into coenosori, which may extend along the margin of a large pinnule with or without interruptions (Fig. 601). The coenosorus appears intra-marginal,

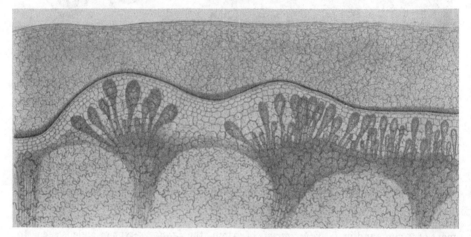

Fig. 602. A small part of the marginal region of a pinna of *Lindsaya lancea* (L.) Bedd., showing an incomplete lateral fusion of the sori. To the left is an almost isolated sorus of the *Saccoloma* type: to the right more complete fusion is seen, while vascular commissures connect the receptacles. (From a drawing by Dr J. McL. Thompson.) (× 70.)

with a continuous flap of a membranous lower indusium, while the upper appears as an extension of the blade. Below the receptacle the veins are connected by a vascular commissure, which is also liable to interruptions (Fig. 602). The development of the sorus has been traced in *L. linearis* Sw.

Here as in *Cibotium* the receptacle arises from the actual margin of the blade, into which the marginal segmentation is directly continuous (Fig. 603, i), the indusial flaps arising as superficial outgrowths. The upper (*adax*) is the stronger from the first, and elongates into the false margin of the blade, while the receptacle is tilted slightly towards the more delicate indusium (*abax*). The first sporangia appear at the apex of the receptacle (Fig. 603, ii),

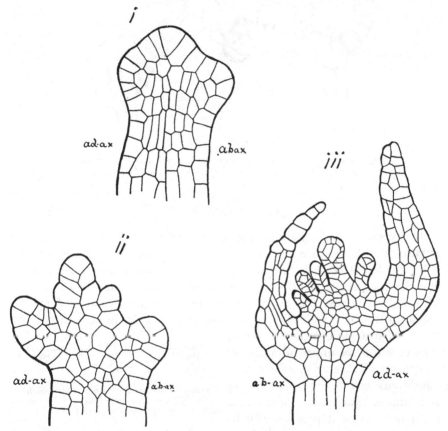

Fig. 603, i—iii. Vertical sections through the young sorus of *Lindsaya linearis* Sw., in successive stages of development. *adax* = the adaxial side, *abax* = the abaxial side of each. The first sporangia are marginal, and a gradate sequence is established at first, followed by a mixed condition. (× 150.)

and are thus marginal as they are in *Thyrsopteris* or *Cibotium*. There is a slight gradate sequence of the sporangia subsequently formed, especially towards the lower indusium (Fig. 603, iii), but it merges later into a mixed state, which becomes quite marked in later stages. The sorus being mixed, the oblique annulus loses its mechanical importance. It would therefore seem natural to find that the sporangia exhibit varying degrees of obliquity of the annulus. The induration of the cells stops opposite the insertion of the stalk, and frequently the series of its cells is interrupted there (Fig. 604). The number of spores in the sporangium is low, viz. 24 to 32, an indication of advance.

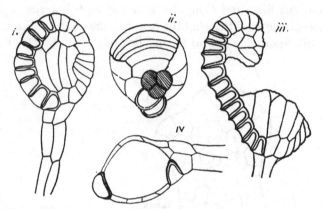

Fig. 604, i—iv. Sporangia, of *Lindsaya linearis*, seen from various points of view, showing the incomplete annulus. (× 125.)

DICTYOXIPHIUM Hooker

Such types as *Lindsaya sagittata* or *reniforme* give the nearest points of comparison with the rare and peculiar fern, *D. panamense* Hk., figured by Sir W. Hooker (*Gen. Fil.* Plate LXII), and compared by him with *Lindsaya*. Here the coenosori extend along both margins of the entire and reticulate leaf-blade, with a vascular commissure traversing each receptacle. A special peculiarity lies in the absence of the upper or adaxial indusium: but the abaxial is present, and the receptacle is tilted towards the upper surface, over which its mixed mass of sporangia may be slightly extended (Fig. 605). Pending more detailed examination of this rare fern, it may be held provisionally to

Fig. 605. Transverse section of a sporophyll of *Dictyoxiphium panamense* Hk. with the adaxial surface directed upwards. (× 18.)

be an end term of the coenosoric Lindsayas, but here while the receptacle is marginal, the upper indusium is abortive, and the numerous sporangia are of mixed origin. The three-rowed sporangial stalk corresponds to that seen in *Lindsaya*.

Unfortunately these comparisons of Davallioid Ferns are necessarily based upon the features of the sporophyte. There is as yet no sufficient knowledge of the details of the gametophyte in these ferns, so often rare and local. But there is no reason yet apparent for thinking that the facts when acquired will be subversive of the conclusions based upon the richer and more dependable data drawn from the sporophyte.

COMPARISON OF THE DAVALLIOID FERNS

The difficulties of the systematist often provide the comparative morphologist with his best opportunities. The Davallioid Ferns present as profuse
a synonymy as any other group: and this itself shows how divergent have
been the views of systematists in the disposal of a tribe richer in species
than in the materials for their ready diagnosis and classification. But from
the point of view of the practical student of evolution this means that,
whenever the necessary facts shall be to hand, there should be the opportunity
for tracing comparatively progressions which may finally illustrate not merely
detailed changes, but steps of very material advance. To guard against mere
comparative theorising it is necessary to keep a critical eye upon physiological
probability. The suggested progressions must be functionally probable as
well as comparatively possible. It will appear that the progressions which
comparative study of the Davallioid Ferns brings to light satisfy this demand
The changes that are traced comparatively in them are palpable amendments
upon the more primitive state of the Dicksonioid Ferns.

The leading Dicksonioid characters that serve in this comparison with
their derivatives are such features as the creeping habit, the highly branched
frond, the dermal hairs not scales, the sustained individuality of the persistently marginal and gradate sorus, the superficial origin of the two-lipped
indusium, the oblique annulus, and robust sporangial stalk: also internally,
solenostely, and an undivided leaf-trace. These features are all shared more
or less perfectly by those Dicksonioid derivatives, such as the Dennstaedtiinae,
which are held to be relatively primitive. But the comparative study of the
Davallioid Ferns shows how these features are liable to be gradually modified
and obliterated. Nevertheless their general correspondence with the main
type, and even the detailed similarity in some of them, point clearly to the
conclusion that the changes are not differences of kind, but modifications
often resulting from minute progressive shades of change. Thus the transition may be traced from the primitive Dicksonioid-Dennstaedtioid type to
the full Leptosporangiate character of the most advanced Davallioids.

The modifications from a Dennstaedtioid type seen in *Davallia* itself
involve the presence of scales in place of dermal hairs: a higher disintegration
of the vascular system, and especially of the leaf-trace: the sorus tends
towards a superficial position, with the margin of the lower indusium adherent
to the surface of the blade, with which the upper indusium is merged: the
flattened receptacle bears a mass of sporangia of mixed ages, though still
showing traces of a basipetal sequence: the sporangia themselves are attenuated, with vertical annulus and long stalks consisting only of one row of cells.
All these are recognised on the basis of the general principles enunciated in
Vol. I as features of advance. But the lines along which they have been

acquired from their presumable Dennstaedtioid ancestry are recognisable
by comparison. They mark the transition from a gradate to a fully Leptosporangiate state, but with the individuality of the sorus retained.

There are, however, certain further details of the sori that claim special
attention: the vegetative characters may be held as subsidiary. Of the three
modifications specially noted in the sori of Davallioid Ferns, the first is the
passage of the receptacle from the originally marginal position of the
Schizaeoid and Dicksonioid Ferns to the superficial position, which is more or
less marked in the Davallioids. This may properly be held as a change of
advance. The gradual steps are illustrated by such a series as *Thyrsopteris*
(Vol. II, Fig. 529), *Cibotium* (Vol. II, Figs. 534, 535), *Davallia* (Fig. 590),
Lindsaya (Fig. 603), and *Nephrolepis* (Fig. 593). Until the fuller developmental
details are to hand for the last of these, it must remain uncertain whether the
receptacle itself has ever in this group of Ferns slid from the margin to the
surface in point of initiation. In all the other examples the receptacle is
actually marginal in origin: but in all of them, in more or less degree, the
sorus shifts towards the lower surface in course of the individual development,
and may, when fully matured, appear as though seated far from the margin
(compare Figs. 588, 592). As in *Hypolepis* the upper indusium may be
assimilated to the general expanse of the blade, and this is a considerable
factor in producing the final result.

The developmental facts in these Ferns disclose a remarkable tenacity in
retaining the marginal position for the receptacle. It will be found that in
the Pteroids there is a like tenacity up to a point, but that in *Pteris* itself the
sorus has definitely passed from the margin to the surface of the blade: and
that this is so not only in the course of individual development, but also in
point of initiation (Chapter XXXVIII).

The second change, viz. the elimination of the lower indusium, is already
prefigured in *Hypolepis* (Figs. 585, 586, pp. 10–12). There can be little doubt,
in presence of the vestigial remains of it in certain specimens of *Arthropteris*,
that there also the inner indusium is subject to reduction or even to complete
abortion. The question seems not so clear in *Monachosorum*: there the
argument is comparative, without the aid of vestigial evidence. There is thus
reason to believe that the lower indusium may be eliminated in Davallioid
Ferns, as it is in *Hypolepis*, and as it will be seen later to be eliminated in
the Pteroids.

A third change is the linkage of sori laterally to form coenosori. This is
a very natural consequence of condensation from a more highly branched
leaf-structure. The highly pinnate leaves of Dicksonioid Ferns retain the
individuality of their sori: but in various Davallioid Ferns, and particularly
in *Lindsaya* and *Dictyoxiphium*, the leaf-structure suggests condensation, and
coenosori are a marked feature. A vascular commissure accompanies the

linkage, and completes it physiologically. There is reason to believe from its sporadic occurrence, even in individual leaves, that linkage has been initiated repeatedly and independently in individuals, species, and genera of Davallioid Ferns. The facts indicate this for *Diellia, Tapeinidium, Nephrolepis*, and *Lindsaya*. If this is so within the Davallieae, there will be no need to see in such linkage any proof of their relationship with the Pterideae, though it is in them that linkage finds its highest development as seen in Ferns of marginal origin.

These three modifications of the sorus go along with various other structural details already noted as leading to a full Leptosporangiate state. Moreover, they are all physiologically advantageous. The transit from a marginal to a superficial position of the sorus gives protection, especially perhaps from the incidence of intense light. The abortion of the lower indusium may be held as the economic removal of a part that has become superfluous in consequence of that transit. The linkage, with its vascular commissures, gives the advantage of equalising supplies available for any individual leaf or segment. In addition to these changes there is the passage from the gradate to the mixed type of sorus. This secures the spread of the drain of spore-production over a longer time and a larger space. It goes along with the elongation of the sporangial stalk. It is true that this carries the mature sporangium farther from the source of supply: but as it approaches maturity it has little further need of food, while by the elongation it secures the shedding of the spores into the free space above the younger and shorter-stalked sporangia. Thus the structural advances which mark the Davallioid Ferns may be held not merely as morphological features that serve for classification, but as evolutionary advances that have come into existence as sources of physiological advantage They possess in fact "survival value."

With these general considerations as a basis the Davallioid Ferns may be grouped in relation to the Dennstaedtiinae, which form an outlier of phyletic advance towards them from the Dicksonioid source. Three natural sections may be distinguished by their soral condition.

I. **Primitive individuality of the sorus retained (except in some species of Nephrolepis): position more or less marginal. Dermal scales.**

    (i) *Humata* (Cav., 1802)   ...   .   14 species.

        Leaves once or more pinnate, venation free. Sori intramarginal, edges of inner indusium free.

    (ii) *Davallia* (Smith, 1793) (reduced)   ...   67 species.
        Leaves variously pinnate, venation free. Inner indusium with edges fused to leaf-surface.

(iii) *Nephrolepis* (Schott., 1834)   ...   ...   17 species.
Leaves once pinnate, with continued apical growth. Venation free. Sori marginal or variously intra-marginal: marginal coenosori in some species: where intra-marginal the inner indusium kidney-shaped, with free margins.

II. **Individuality of the sori retained: position intra-marginal. Inner indusium partially or completely abortive.**

(iv) *Arthropteris* (J. Smith, 1854)   ...   ...   4 species.
Dermal scales and habit as in *Humata*, of which it may be held as an ex-indusiate type.

III. **Distinguished by relatively condensed leaf-structure, culminating in simple blades. Dermal scales. Venation mostly open, but reticulate in the most condensed types. Sori marginal, fused in varying degree to form coenosori.**

(v) *Tapeinidium* (Presl, 1849) C. Chr. 1906 ...   4 species.
Leaves simply pinnate. Sori slightly intra-marginal, often fused.

(vi) *Diellia* (Brackenridge, 1854)   ...   ...   7 species.
Leaves simply or repeatedly pinnate. Occasional fusion of sori.

(vii) *Odontosoria* (Presl, 1836) Fée, 1850   ...   19 species.
Leaves repeatedly pinnate. Sori variously fused.

(viii) *Lindsaya* (Dryander, 1793) (reduced) ...   90 species.
Leaves variously pinnate, or even simple: pinnae unilateral, veins free or sometimes anastomosing. Sori marginal or sub-marginal, variously fused.

(ix) *Dictyoxiphium* (Hooker, 1840) ...   ...   1 species.
Leaves simple, venation reticulate. Coenosorus marginal, continuous, but with upper indusium obsolete.

Genus incertae sedis.

Oleandra (Cavanilles, 1799)   ...   ...   10 species.

For a discussion of the affinity of *Deparia*, *Cystopteris* and *Prosaptia*, see Chapter XLVIII.

The grouping given above must not be taken as any detailed exposition of the phyletic relationships of the Davallieae. It is based upon the recognition of methods of evolutionary advance, any one of which may be carried out homoplastically in a plurality of phyla. For instance, Section III includes genera which have in common a probable condensation of leaf-structure from some more complicated primitive branching. This may naturally bring with it the formation of coenosori, and the soral fusion is seen as it were tentatively in several distinct genera: this may be held as a homoplastic result produced in a plurality of phyletic lines. The fact that it is seen in certain species of *Nephrolepis* (*N. dicksonioides* and *acutifolia*, Fig. 592), illustrates the point in a genus in which the sori for the most part retain their individuality.

Difficulties are inherent in any classification which aims at being natural, that is evolutionary. In the present instance it may be held that the three sections of the Davallioid derivative Ferns exhibit *states* or *conditions* arrived at probably from a common origin, but along a plurality of individual lines. They would thus rank theoretically with such states or conditions as were expressed by the definitions of the old genera *Acrostichum* or *Polypodium*: these can no longer be maintained in their old classificatory sense, since both states have undoubtedly been attained polyphyletically. The difficulty thus encountered in handling this first large group of advanced Leptosporangiate Ferns is inherent in greater or less degree in all other large groups. It need not oppress the mind unduly so long as the position is clearly understood. A complete artificial classification is always possible, and is indeed necessary for floristic use. A complete phyletic classification will only become possible with complete knowledge of the descent of the organisms classified. The second cannot replace the first under present conditions, owing to the imperfection of present knowledge. But it can lead to a correction and amendment of classification for floristic use, so as to make it run ever more nearly along the lines of probable evolution. This is what appears to result from the suggested grouping of the Davallioid Ferns, as compared with the catalogues of the genera of this group given elsewhere.

BIBLIOGRAPHY FOR CHAPTER XXXVII

593. HOOKER. Genera Filicum, Plate LXII, *Dictyoxiphium*. 1842.
594. HOOKER. Species Filicum, Vol. I, p. 150. 1846.
595. KUHN. Die Gruppe der Chaetopterides. Berlin. 1882.
596. DE BARY. Comparative Anatomy, Oxford, pp. 287, 347. 1884.
597. LACHMAN. *Nephrolepis*, Ann. Soc. Bot. de Lyon. 1888.
598. PRANTL. Das System der Farne. Breslau. 1892.
599. CHRIST. Farnkräuter, p. 289. 1897.
600. BOWER. Studies in Spore-Producing Members, IV, Phil. Trans. p. 75. 1899.
601. DIELS. Natürl. Pflanzenfam. I, 4, p. 204. Also p. 139, where the systematic literature is fully cited up to 1902.
602. GWYNNE-VAUGHAN. Solenostelic Ferns, II, Ann. of Bot. XVII, p. 689. 1903.
603. VAN ROSEBURGH. Malayan Ferns, Batavia, pp. 255, 567. 1909.
604. BOWER. Studies III, Ann. of Bot. XXVII, p. 443. 1913.
605. GWYNNE-VAUGHAN. Climbing Davallias, Ann. of Bot. p. 495. 1916.
606. BOWER. Studies VII, Ann. of Bot. XXXII, p. 1. 1918.
607. VON GOEBEL. Organographie, 2te Aufl. p. 1143. 1918.
608. VON GOEBEL. *Prosaptia*, Ann. du Jard. Bot. de Buit. XXXVI, p. 148. 1926.

CHAPTER XXXVIII

PTEROID FERNS

UNDER the heading VI Pterideae, as adopted by Diels in Engler and Prantl, *Natürl. Pflanzenfam.* I, 4, p. 254, numerous Ferns are arranged of which the collective characterisation is as follows: "Sori länglich bis lineal, längs der Adern, an deren Enden oder einer Queranastomose, Indusium meist fehlend. Blattrand häufig umgeschlagen, oft modificirt, den Sorus überdachend. Blätter ungegliedert dem Rhizome angefügt. Spreite seltener ungeteilt, meist zusammengesetzt. Keine Spicularzellen. Bekleidung Spreuschuppen oder Haare, letztere zuweilen Wachs ausscheidend." Then follows a list of some thirty genera, grouped under the headings, I Gymnogramminae; II Cheilanthinae, III Adiantinae; and IV Pteridinae. In point of fact the Pterideae of Diels includes the bulk of the Pterideae of Hooker, together with his Grammitideae (see *Synopsis Filicum*, 1873, pp. 9, 10).

Diels' grouping is based upon a comprehensive rather than an exact diagnosis. The adult sorus is naturally placed in the forefront; but its objective features only are taken into account, without reference to their evolutionary or even their ontogenetic origin. It is a systematic rather than a morphological diagnosis, in fact convenient rather than scientific. If, however, the question be raised how the soral state that is used as a basis for the grouping has been arrived at, it will be evident that at least two distinct sources are possible for the conditions that are shown by the included genera. One is from a fusion of originally isolated, two-lipped, marginal sori, such as are characteristic of the Dicksonioid Ferns; the other is from some types such as *Todea* or *Plagiogyria*, where the sporangia were already present upon the leaf-surface, and were without any indusium. The immediate question will then be how to discriminate in any individual case between these two or other possible sources; for upon such decisions any more natural grouping must depend. The reply is by comparison of allied forms, based upon developmental study of the sori, and checked by comparison of other characters of form and structure. The grouping, if it is to be natural, should be according to induction founded upon a wide comparison and upon development, rather than upon any arbitrary principle dictated by systematic convenience.

Those genera in which there appears to be the best chance of applying consecutive comparative argument will be taken first. They are clearly those few genera in which the sorus is actually marginal, with a double indusium, such as is familiar in the Dicksonioid-Davallioid series. For reasons to be explained later, the first choice will fall upon the Bracken Fern, *Pteridium*. Then will follow a comparison with such other genera as have always been held to be closely allied with it, even though they may not always possess the double indusium. In fact we shall first examine the Ferns included under Diels' heading of the Pterideae-Pteridinae (*l.c.* p. 287), taking them approximately in the reverse order of their description by him. Those genera are *Pteridium*, *Paesia*, *Lonchitis*, *Histiopteris*, *Pteris*, *Ochropteris*, *Anopteris*, *Amphiblestria*, and with certain more doubtful congeners in *Cassebeera* and *Actiniopteris*. The three first-named genera were grouped by Prantl as his "Lonchitidinae." Of all the Fern-systematists of the 19th century Prantl was the writer who most readily worked phyletic views into his classification, the basis of his comparisons being widened by the results of his excellent laboratory-technique. He himself regarded his Lonchitidinae as the most primitive of the Pterideae, a position which accords with the reasoning to be developed here; and this provisionally gives justification for describing them first (*Arb. Königl. Bot. Gart.* Breslau, 1892, pp. 16–18).

PTERIDIUM

The Common Bracken is one of the most successful of all vascular plants, as shown by its cosmopolitan distribution and its gregarious habit. This has probably been promoted by its underground creeping rhizome, which is thus protected during adverse seasons. The unusual habit is shared by *Stromatopteris*, but without any similar cosmopolitan success (Vol. II, p. 200). The habit is established in the Bracken immediately on germination. After bearing 7 to 9 alternating leaves the primary axis bifurcates, and each shank burrows downwards into the soil bearing alternate leaves, and forking repeatedly. The further development of the branch-system tends to become dichopodial (Fig. 606): see Vol. I, pp. 74–76. As a progressive rotting of the older parts reaches a branching and passes it, two separate individuals may result. It is to this method rather than to sexual propagation that increase of individuals is mostly due. The long-stalked leaf bears a highly-branched deltoid blade, with narrow pinnatifid segments having an open venation. The rachis is marked by lateral pneumatophoric lines, which are continued upwards to the pinnae. Glands secreting nectar are seated on the bases of the lower pinnae, and are visited by ants (Sir F. Darwin, MS. 1876): but their exact use is uncertain (see Vol. I, p. 205, Fig. 195). The rhizome and young leaves are covered by a felt of simple hairs, and scales are absent. The equal dichotomy, open venation, and the absence of protective scales may be held as indications of a relatively primitive state.

This conclusion appears to accord ill with the elaborate vascular structure of the rhizome and leaf-stalk. The stelar system has already been described

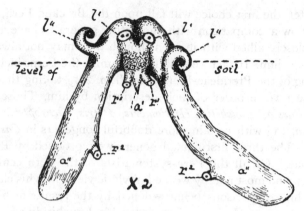

Fig. 606. Young plant of *Pteridium aquilinum* (L.) Kuhn, seen from the convex side (obliquely downward directed) of the curved primary axis, and showing the first dichotomy, with downward directed shanks. a' = primary axis; a'' = shanks of first dichotomy; l' = leaves borne on the primary axis: l'' = leaves borne on shanks of dichotomy; r' = roots attached to the primary axis; r^2 = roots on shanks of dichotomy.

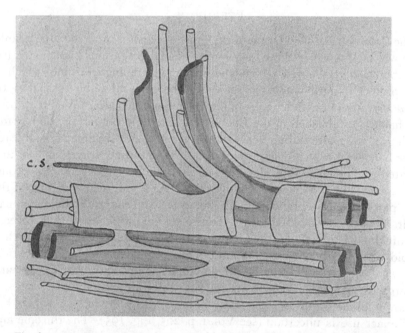

Fig. 607. *Pteridium aquilinum.* Vascular system of a node seen from the dorsal side, somewhat simplified. The internal system and lateral folds of the leaf-trace are darkly shaded. A portion of the dorsal meristele of the outer cylinder is cut away to show branching of the internal strands. $c.s.$ = compensation strand. (After Tansley and Lulham.)

in Vol. I, p. 4 (Fig. 3)[1]. It may be regarded as consisting of a highly perforated solenostele, represented by the outer ring of meristeles, and a medullary system consisting of a variable number of tracts, usually two, disposed as an inner ring. The connections of these at the departure of a leaf-trace are shown somewhat diagrammatically in Fig. 607, from which it is seen that both the inner and the outer systems contribute to the leaf-trace, of which the constituent strands are disposed in the usual horse-shoe, while a compensation strand (*c.s.*) connects the inner system with the outer above the leaf-insertion. This is clearly a complex and apparently an advanced type of construction. It may be interpreted as consequent on relatively large size, a double solenostele being affected by numerous and large perforations. Size will not explain this altogether, but the fact that the dimensions are large is probably a real factor in the comparative problem. Thus the external features seem

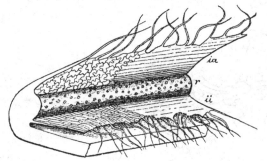

Fig. 608. *Pteridium aquilinum.* Part of the margin of a fertile segment of the second order, greatly enlarged. *r* = receptacle, after removal of the sporangia, whose positions are indicated by scars; *ia* = outer, *ii* = inner indusium, the cellular structure of each being indicated to the left of the drawing. (After Luerssen.)

to mark the Bracken as relatively primitive: the vascular anatomy suggests advance. We shall naturally examine the soral characters with enhanced interest, since these bear a special importance as evidence in such questions.

Of all the Pteroids the Bracken has the most elaborate sorus in that there are two well-formed indusial lips between which lies the receptacle (Fig. 608). The sorus is continuous along the margin of the pinnules, often for consider-able distances; but its continuity is apt to be broken irregularly. Detailed examination shows that the receptacle is traversed by a vascular commissure, which runs beneath the insertion of the sporangia, and links together the endings of the otherwise free veins. In all of these structural facts *Pteridium* corresponds to *Lindsaya.* Speaking of the Section *Paesia* St Hilaire, in which he included *Pteridium,* Sir W. Hooker remarks that "According to

[1] Unfortunately this figure was inverted in the setting: the lower side of the rhizome being placed upwards on the page.

strict technical characters this group of species, which differs from the rest of the genus also in habit of growth, has as good a claim to be placed in Lindsayae as Pterideae (*Syn. Fil.* 1873, p. 162). In both the structure of the sorus suggests the linkage of a number of marginal, two-lipped sori of a Dicksonioid type to form linear coenosori: a view that greatly adds to the phyletic interest of the Pterideae.

This being so, it becomes a matter of importance to know the development as well as the adult structure of the coenosorus of the Bracken. As seen in

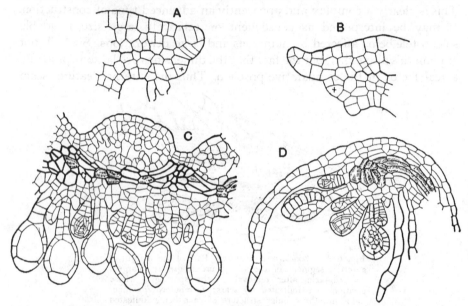

Fig. 609. Development of the coenosorus of *Pteridium aquilinum*. *A, B*=vertical sections through the margin of the pinnule, showing how the receptacle (×) originates directly from the marginal segmentation, while the indusial flaps are of superficial origin (× 180). *C*=a vertical section of an older coenosorus, following the line of the vascular commissure (× 90). *D*=a vertical section of a similar sorus, for comparison with *A* and *B* (× 90). The sporangia are of mixed ages.

vertical section, the regular marginal segmentation of the pinnule continues with its row of wedge-shaped initials directly into the receptacle (Fig. 609, *A, B*). The upper or adaxial indusium is the first to appear, as a superficial outgrowth back from the extreme margin; the lower or abaxial appears slightly later, and it is also smaller and less constant in structure than the upper. These facts for *Pteridium* are in accord with what has been seen in *Lindsaya* (Fig. 603). The order of appearance of the sporangia upon the receptacle shows some variety: sometimes there is a regular basipetal sequence, as in *P. aquilinum* var. *caudatum* from Jamaica, the oldest sporangia being adaxial (Fig. 610). But in the native Bracken the succession is less regular, and the sorus soon acquires the mixed condition (Fig. 609, *D*). A vertical section parallel to the margin of the pinnule, and following the vascular commissure, is seen in Fig. 609, *C.* It shows the vein-endings, their positions being indicated by the indentations of the upper surface; the tracheides elongated in the plane of the section link them together, forming the receptacular commissure. The sporangia of various ages are seen intermixed, and covering the whole length of the commissure. The condition is that seen as a transparency in *Lindsaya* in Fig. 602. The conclusion from such facts is inevitable, that in *Pteridium* as in *Lindsaya* the coenosorus is the result of a lateral linkage of marginal two-lipped sori, originally of gradate type, such as are seen in the Dicksonioid Ferns. It does not necessarily follow that these genera are of common descent, but at least their similarity of soral

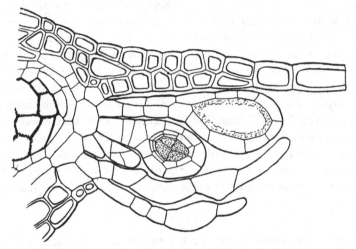

Fig. 610. Sorus of *P. aquilinum* var. *caudatum* cut vertically, showing a regular basipetal succession of the four sporangia. (× 150.)

construction does suggest a common origin, from some Dicksonioid-Dennstaedtioid source. The relation of *Pteridium* to *Dennstaedtia* is specially indicated by the external habit of the creeping rhizome, the elongated rachis, and much branched deltoid blade, with open venation, and dermal hairs, not scales. The polycyclic stele is, however, advanced beyond that of the Dennstaedtiinae in respect of its prevalent perforation, though this is occasionally seen in *Dennstaedtia* itself. The gradate, two-lipped sorus of Dicksonioid type is still traceable in the development of *Pteridium*, but it is disguised by the lateral linkage into coenosori, with their vascular commissures connecting the otherwise free veins. On this view *Pteridium* stands as a synthetic type between the Dicksonioids and the Pteroid Ferns. It will now be shown comparatively within the latter group how, by gradual steps, the condition seen in *Pteris* itself may probably have been attained, and finally even that of *Acrostichum*.

PAESIA

Paesia has always been associated by systematists with *Pteridium*: in
fact the latter has been included in the former genus by Hooker (*Syn. Fil.*
p. 162). It includes seven species, mostly of smaller size than the Bracken.
Like it they have a creeping bifurcating rhizome, bearing ample pinnately
cut leaves, with open venation. The dermal appendages are again hairs, not
scales. Christ notes as diagnostic the undivided and channelled leaf-trace,
with horse-shoe transverse section in the petiole (*Farnkr.* p. 164). This arises
from a typical solenostele(Gwynne-Vaughan,*Ann. Bot.* XVII, p. 691). The sori
are essentially of the same type as those of the Bracken, but more markedly
curved downwards, the upper indusium appearing like a marginal flap. The
coenosori do not extend far without interruptions. The development of the
sorus has been traced in *P. viscosa* and *scaberula* (Studies VII, *Ann. of Bot.*
XXXII, p. 24). It accords with that in *Pteridium* in being basipetal, typically
bi-indusiate, and in having a vascular commissure. But the lower indusium,
which is closely appressed to the lower surface of the pinnule, is never more
than one layer of cells in thickness, and it may frequently be absent altogether (Fig. 611). These features accord with the view that *Paesia* includes
Ferns closely related to *Pteridium*, but of smaller size and simpler construction of the vascular system: while the sorus is of the same type, but more
closely protected, with indications of the obliteration of the lower indusium
that find their explanation in its close apposition to the lower leaf-surface.

LONCHITIS L.

The third genus included by Prantl in his Lonchitidinae is *Lonchitis*
itself. Its eight species are large upright plants with compound leaves of
soft texture, and with pubescent surfaces. The chief distinctive features are
the position of the sori mostly at the sinuses of the pinnatifid leaf-segments,
and the absence of the inner indusium which has already been seen to be
inconstant in *Paesia*. Observations have been made on *L. aurita* L. (incl.
L. Lindeniana Hk.), and on *L. hirsuta* L., a Fern of which Hooker remarks
(*Syn. Fil.* p. 160) that: "Though in technical character a *Pteris*, this is far
more like the two species of *Lonchitis* in habit." The two species above
named are both now included in *Lonchitis*; but they differ in certain details,
and it will be seen that they supply further intermediate states between
the bi-indusiate Pteroids and the true *Pteris* with only a single indusium.

Both species contain in the massive stock a large solenostele, about an
inch in diameter. Its origin was traced in *L. hirsuta* by the late Prof. Gwynne-
Vaughan, from protostely, through a *Lindsaya*-state to solenostely, and
occasional perforations may occur in this species (G.-V. MS. notes). In both
species the leaf-trace is given off as two distinct straps, separate from one

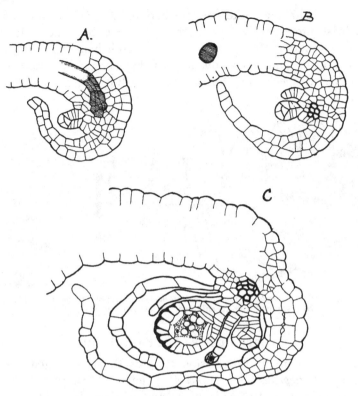

Fig. 611. Vertical sections through the young sorus of *Paesia viscosa* St Hil.
A = a young sorus with upper indusium consisting of two or more layers of
cells: the lower indusium is absent. The section follows a vein till it ends at
the junction with the commissure. Only one sporangium is present. *B* = a
similar section traversing the commissure and sorus, showing one sporangium
covered by a well-developed upper indusium: the lower is much feebler, and
represented in section by a row of cells. *C* = a more advanced sorus, showing
upper and lower indusia overlapping, two sporangia with suggestion of basipetal succession, and two hairs. (× 75.)

another before detachment from the stele, a state seen in *Pteris cretica*.
Both plants are hairy; but whereas *L. aurita* has only hairs, there are
numerous narrow scales on the stock of *L. hirsuta*. On the other hand,
whereas *L. hirsuta* has an open venation, that of *L. aurita* is coarsely
reticulate (Fig. 612). The soral characters are the same for both. The sori
are protected by a thick flap of the upper indusium, and the receptacle bears
sporangia of mixed ages, interspersed with hairs. The position of the lower
indusium is occupied by about two ranks of hairs, suggesting a resolution
of the indusium into independent cell-rows. The absence of a continuous
lower indusium led various authors to place these Ferns under *Pteris*. It is
just these intermediate states which provide the basis for close phyletic
sequences. In vascular characters in investiture, in venation and in soral

characters *Lonchitis* lies intermediate between *Pteridium* and *Pteris*: and not the least interesting feature is that *L. aurita* bears no scales but has a reticulate venation, though *L. hirsuta* has both hairs and scales, but its venation is open. Thus such advances as they show do not march parallel in the different species of the same genus. But still a general trend is maintained.

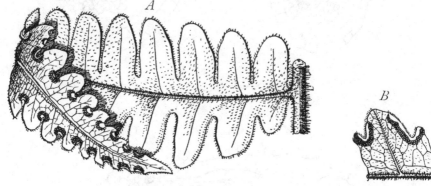

Fig. 612. *Lonchitis aurita* L. (= *L. Lindeniana* Hk.). (*A*) one pinna, natural size, and (*B*) a segment from it, enlarged. (After Hooker, from Christ.)

HISTIOPTERIS (Agardh) J. Smith

A further step in the progression towards the typical genus *Pteris* is supplied by *Histiopteris* (*Pteris*) *incisa* (Thunbg.) J. Smith, a Fern of very wide distribution in both hemispheres, and of variable habit. On account of its occasional reticulate venation it was placed in the section *Litobrochia* of Hooker's comprehensive genus *Pteris*, but it is now (with a Philippine species) restored to generic rank under Agardh's name of *Histiopteris*. The usual type of the species resembles *Pteridium* in its creeping habit, with underground rhizome and deltoid leaves. But its venation is variable, sometimes free but frequently with coarse anastomoses. The surface when young bears simple hairs, but there are also scales upon the rhizome. The adult leaves are glaucous. These vegetative characters point to *Pteridium*, but with a suggestion of advance.

The adult anatomy discloses a solenostele in the rhizome: it is liable to corrugation in large rhizomes, a device that gives an enhanced proportion of surface to bulk in the conducting tracts (Fig. 613). The relation of the stele to the foliar trace has been worked out by Tansley and Lulham, and it is found to correspond essentially to the *Hypolepis*-type, but the vascular tracts are more deeply folded, and the structure is complicated by the fact that each leaf in the specimens examined arises in the angle between the shanks of a dichotomy of the rhizome (see Tansley, *Fil. Vasc. Syst.* Fig. 65). The leaf-trace is at first undivided, but it soon breaks up into separate tracts. The ontogeny of the stele has been traced in several sporelings : it illustrates the usual stages (of protostele, medullated protostele, *Lindsaya*-condition, and solenostele), variously curtailed or lengthened in individual examples. A section above the last leaf-trace which shows no axillary pocket reveals the *Lindsaya*-structure, which is unusually protracted in this Fern

Fig. 613. *Histiopteris incisa* (Thunbg.) J. Sm. Transverse section of an internode of the rhizome, showing corrugation of the solenostele. (× 10.)

(see Studies VII, Fig. 23). This vascular structure accords with a position of advance on *Hypolepis*, but it is simpler than in the highly disintegrated *Pteridium*. Taking all the vegetative characters together, some showing conservatism (vascular structure), others advance (venation and dermal scales), they would suggest that *Histiopteris* is a probable Dennstaedtioid derivative, holding a position parallel with *Pteridium*.

Such comparisons make the question of the soral condition of *Histiopteris incisa* all the more interesting. The detail of two leaf-segments, in a form where the venation is reticulate, is seen in Fig. 614. The coenosori follow the lateral margins, while the upper indusium appearing as a marginal flap covers the receptacle: there is no lower indusium. The immediate comparative question will then be as to the relation of the upper indusium and the receptacle to the marginal segmentation. Examination of a large number of sections has shown some fluctuation in detail. The segmentation of the sterile pinnule is of the usual type : but where the sorus is to be formed the curved outline of the section appears slightly irregular, as seen in transverse section (Fig. 615, *A, B*), and it becomes difficult to trace the fate of the real margin by its segmentation. Comparison of numerous sections suggests that the receptacle (*R*), though it often seems as though it were on the lower

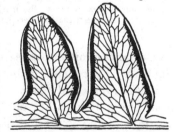

Fig. 614. *Histiopteris incisa* (Thunbg.) J. Sm. Two fertile leaf-segments enlarged. (After Mettenius from Christ.)

surface, is essentially of marginal origin ; meanwhile strong growth of the
adaxial segments produces the indusial flap (F), which thus appears as though
it had been of marginal origin, but was really superficial. There is no indication of a lower indusium. Later, single cells from the apex of the now clearly
superficial receptacular ridge (R, Fig. 615, C) grow out to form sporangia (D);
and others appearing later, with which hairs are associated, indicate a basipetal
succession, which is, however, not strictly maintained. Beneath the receptacle

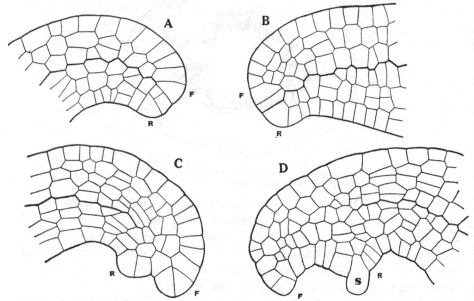

Fig. 615. Sections vertically through the margins of young pinnules of *Histiopteris incisa*, showing
the relation of soral origin to the marginal segmentation. R = receptacle, S = sporangium, F = indusial flap. It is seen that the receptacle is close to the margin in A, B, but definitely superficial
in D. (× 200.)

a vascular commissure appears: thus the structure comes to be that shown in
Fig. 616. It is concluded that the sorus of *H. incisa* is essentially marginal,
as in *Pteridium* and *Paesia*, but that it is more advanced towards a superficial origin than either of these, and that this goes along with complete
abortion of the lower indusium. These observations indicate what is the
probable phyletic history: that *H. incisa* originated from some Dennstaedtioid
source. The steps of advance which it shows are: (i) some degree of
elaboration of the vascular system, (ii) a tendency towards reticulate venation,
(iii) the presence of dermal scales, (iv) loss of the lower indusium, (v) a
tendency for the sorus to slide to the lower surface: this last comes out
more clearly in the mature than in the young state. On all of these grounds
collectively we see in *H. incisa* a nearer approach than in other Ferns to the
condition seen in the genus *Pteris*.

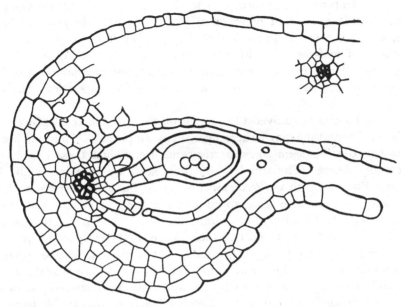

Fig. 616. Vertical section through a sorus of *Histiopteris incisa* well advanced
towards maturity. (× 75.)

PTERIS L.

This large genus includes 157 species as stated in Christensen's Index.
The wide synonymy that surrounds it is a witness to the divergences of view
to which its treatment has been subjected in the past. Some authors have
narrowed the limits of the genus, giving separate generic rank to various
outstanding types, while others have included all which have "marginal
linear continuous sori, occupying a slender filiform receptacle," under the
generic title (*Syn. Fil.* p. 153). From the evolutionary point of view the
Pteroid Ferns as a whole are those which share a certain biologically
successful innovation; this consists in the linking together of numerous
marginal two-lipped sori into a linear series. That series once so constructed
has been subject to certain modifications, such as the slide of the resulting
coenosorus towards a superficial position, the elimination of its lower
indusial lip, and finally the spread of the linear receptacle as a broadening
area over the lower surface of the leaf-blade. Roughly speaking the generic
title of *Pteris* is applied to all those Ferns in which the lateral fusion of the
marginal sori and the elimination of the lower indusium are complete. It is
a large and cosmopolitan genus, including plants of almost every kind of
leaf-form and venation: in fact the institution of the Pteroid-sorus was an
evolutionary success. But compared with the Dicksonioid source, from which
comparison shows these Ferns to have originated, they may be expected to

show other features of advance than in the sorus only. In the vegetative system there is a constant presence of protective scales in place of hairs alone, an advancing disintegration of the vascular tracts, and a frequent integration of leaf-surfaces, with which goes also an advancing reticulation of the veins. These remarks are made in advance of the statement of the facts upon which they are based, so as to clarify the exposition that is to follow.

Flattened scales are universally present on the rhizome and leaf-bases of species of *Pteris*: they may be accompanied by simple hairs. This is in broad contrast to the *Dicksonia-Dennstaedtia* series from which these Ferns were presumably derived. The simple hairs there seen are repeated in the bi-indusiate Pteroids, such as *Pteridium* and *Paesia*; but for some reason which is obscure the protective scale appears in the genera *Lonchitis* and *Histiopteris*, while coincidently the inner indusium is obliterated. The remarkable point is not that there should be a divergence in this respect, as we have seen, between the two species of *Lonchitis*, but that there should be any near relation between the incidence of features so distinct as dermal scales and a lower indusium. Such facts tend to establish both as trustworthy data for broad comparison in a progression which involves so many variable features. They accentuate the recognition of the recurrent protective scales in *Pteris* as a feature of advance in a genus where the lower indusium is always wanting.

In so large a genus as *Pteris*, including ferns sometimes with long creeping rhizomes, as in *P. grandifolia*, sometimes with a compact upright habit, as in *P. longifolia, cretica*, or *podophylla*, some variation in vascular structure is to be expected. A general comparison relates them all as natural derivatives from solenostely with an undivided leaf-trace, such as is characteristic of their probable Dennstaedtioid source. The species investigated range between typically solenostelic structure and a complicated polycyclic dictyostely: while the leaf-trace varies from an uninterrupted horse-shoe to two straps originating separately from the stele of the axis. Typical solenostely is seen in *P. grandifolia*, where there is a long creeping rhizome. A transition to dictyostely with overlapping leaf-gaps appears in many more compact or ascending shoots, such as *P. tremula, cretica, flabellata, heterophylla, pellucida, biaurita*, and *Swartziana*. In most of these the leaf-trace remains undivided; but in some, such as *P. cretica*, it appears as two separate straps, though these may fuse upwards.

A more marked modification is seen in the presence of accessory vascular strands. A relatively simple case has been described at length by Gwynne-Vaughan in *P. (Litobrochia) Kunzeana* Ag. (*P. elata* var. *Karsteniana* Kze.). Here the erect or oblique rhizome contains a perfect solenostele with an internal vascular cylinder connected at each node with the outer by a com-

pensation-strand (Fig. 617). But a much more elaborate structure is seen in
P. (Litobrochia) podophylla Sw. Here the stock is conical enlarging upwards
from the sporeling, and the vascular com-
plexity increases with its size (Fig. 618). At
the base it is protostelic (i); then follow the
states of medullated protostele and solenostele (iii, iv): but soon an encroachment of
vascular tissue appears from the inner surface
of the solenostele, which provides the compensation strand for the next leaf-gap (iv).
From this point onwards there is a continuous
medullary system. The next step is the expansion of the medullary strand into a second
ring, and this is repeated again as the stock
increases rapidly in size (v). The result in

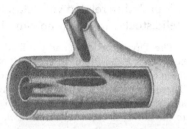

Fig. 617. *Pteris (Litobrochia) Kunzeana*
Ag. (= *P. elata*, var. *Karsteniana* Kze.).
Diagram showing the vascular tissue at the
insertion of a leaf. A piece is supposed to
be cut out of the side of the solenostele,
so as to show the internal vascular system.
(After Gwynne-Vaughan.)

the largest stock examined was as shown in Fig. 618 (vi), where there are
three concentric solonosteles, with a central strand suggesting the initiation
of a fourth. The similarity is obvious between this and what has been seen in
Thyrsopteris (Vol. II, Fig. 528), and in *Saccoloma* (Vol. II, Fig. 538). But the

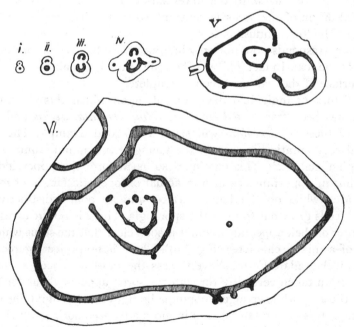

Fig. 618. A series of transverse sections of the stem of *Pteris (Litobrochia)*
podophylla Sw., all drawn to the same scale. They show the great increase
in stelar complexity as the conical stem expands upwards. (× 4.) The stelar
tracts are shaded.

difference in detail of structure, as well as in the degree of aloofness of the Ferns in question indicates, that the similarity is homoplastic. In all of them it is probably related very closely with increase in size of the upright soleno-stelic stocks, in which no cambium provides for secondary thickening.

The Thyrsopterideae and the Dennstaedtiinae are here regarded as relatively primitive congeners of the Pteroid Ferns. All of them appear to be susceptible to increasing elabor ation of their essentially solenostelic or dictyostelic structure, by medullary developments These are probably to be viewed as independent consequences of adjustment to increasing size, rather than as being truly homogenetic. More cogently still does this argument apply to the polycyclic structure of *Matonia* (Vol. II, Figs. 498, 499), for that genus is still further aloof systematically, while the structural details, though similar in principle, differ in detail. Still further afield systematically is the somewhat similar structure that appears in the rhizome of *Selaginella laevigata*. For a discussion of the underlying principle of Size, see Vol. I, Chapter X. What these various plants have in common is the capacity for such developments as they present; and this is what gives its interest to the similar medullary system to be described later in *Acrostichum aureum* (p. 59).

Turning now to the development of the sorus of *Pteris*, it has been seen that in *Histiopteris incisa* the lower indusium is absent, and that the reference of the receptacle to a strictly marginal origin is doubtful. This probably illustrates a step in the "phyletic slide" of the sorus from the actual margin, as in *Lindsaya* or *Paesia*, to the lower surface of the blade. In *Pteris* that superficial origin of the sorus is constant, so that it has become a generic character. The lower indusium is regularly absent, and the persistent upper indusium corresponds now, not only in appearance but also in origin, to the margin of the leaf. In fact in *Pteris* the slide of the originally marginal sorus to the surface of the leaf has become complete.

The history of individual development, upon which this conclusion is founded, has been traced in *Pteris longifolia, serrulata, cretica*, and *quadri-aurita*: of these *P. serrulata* will serve as a good example. The normal marginal segmentation is seen in the young fertile pinnule running out to the marginal cell itself. This continues so as to form the upper indusium, which thus in appearance as also in actual ontogenetic fact is of marginal origin. Meanwhile on the lower surface, and at points distinctly intra-marginal, cells grow out to form the sporangia. There is no great regularity in the order of their appearance, and it is soon clear that the sorus which they form is of a mixed character (Fig. 619). There is no projecting receptacle as in typical gradate Ferns. Nevertheless the most advanced sporangium in any section taken vertically through the sorus appears about the middle of the fertile area, while younger sporangia lie right and left of it (Fig. 619, *B*). Evidence of this may often be seen in more advanced sori, in which the re-ceptacle still remains flat, with a broad vascular commissure underlying it, the whole being very fully covered by the marginal indusium (Fig. 619, *C*). Still it may be noted in this section of *P. cretica* that the oldest sporangium, represented only by its stalk, lies at the centre of the receptacle, while the later sporangia right and left show the "mixed" condition of the sorus fully established.

Anopteris hexagona L., C. Chr. (= *Pteris heterophylla* L.) has been brought into prominence by Mettenius in relation to his theory of connation of hairs (*Verwachsung*) to form the lower indusium. He stated that the "paraphyses" are disposed in a series along the inner limit of the sorus, and are so closely ranged that they seem to form an indusium, which grows out distally into hairs projecting as cilia upon its margin. He concludes that the inner indusium, such as is seen in *Pteridium*, has arisen by fusion of hairs like those of *Anopteris* (*Farngattungen*, Frankfurt, 1858). It has been found in sections cut vertically through the young sorus of this Fern that the hairs are very numerous, but never webbed; the sporangia are relatively few. The first hairs originate in no definite relation to the

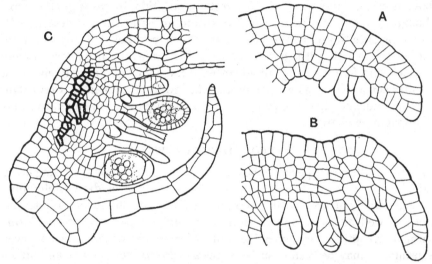

Fig. 619. *A, B* = vertical sections through young sori of *Pteris serrulata* L. fil. (× 150.) The indusium is here a direct continuation of the marginal growth, and the flattened receptacle and sporangia arise from the lower surface. *C* = a similar section through a nearly mature sorus of *Pteris cretica* L. (× 75.)

inner limit of the receptacle, where theoretically the inner indusium should be; later the whole width of the receptacle may be occupied by them. These facts are quite out of harmony with Mettenius' theory, which itself does not accord with the facts and comparisons detailed above for other Pteroid Ferns. Such hairs as are seen in the sorus of *Anopteris* may be held simply as paraphyses scattered over the receptacle, as they are in other species of *Pteris*; though here they are more closely arranged than usual. (See Studies VII, 1918, p. 41, Fig. 30.)

Comparing the mode of origin of the sorus thus described for *Pteris* with that seen in the Dicksonioid Ferns, or at nearer hand with that in bi-indusiate types such as *Lindsaya, Pteridium* or *Paesia*, the differences are (i) that in *Pteris* there is no lower indusium, (ii) that the upper indusium is actually marginal in its origin, appearing as a direct continuation of the marginal segmentation of the blade; (iii) that the sporangia arise superficially from the lower surface of the blade, without any markedly convex receptacle preceding them; and (iv) that the sorus shows only slight signs of a gradate sequence, and is of a "mixed" type from an early stage of development. Its gradate character has departed with the loss of the conical receptacle. These differences have made their appearance within the circle of Ferns which have always been held as closely related by the best systematists. It cannot be doubted that the morphological identity of the sorus is the same throughout the series. Hence the only possible conclusion is that in the course of the evolution of *Pteris* the sorus has passed bodily from a marginal to a superficial position, while at the same time the gradate sequence of the sporangia has been substituted by a "mixed" condition of the sorus. The series of Pteroid Ferns above described illustrates the suggested transition by very gentle steps. Though they may not themselves constitute an actual line of descent, still their existence supports the view that the change has actually occurred. Finally, the fact that it is accompanied by the transition from dermal hairs to scales, and frequently by an advance from a simpler to a more elaborate vascular construction, as well as by the change from an open to a reticulate venation, indicates that the progression has been parallel in vegetative and in propagative characters.

Pteroid Derivatives

A central Pteroid type would be one with a rhizomatous or ascending habit and pinnate leaves with open venation, a solenostelic or dictyostelic axis, leaf-trace undivided or possibly separated into two equal straps, with dermal scales as well as hairs, and a sorus of a mixed type, with narrow intra-marginal receptacle, an upper indusium of marginal origin, and no lower indusium. It may be seen in such a species as *Pteris longifolia* L. and others of the section designated as *Eu-Pteris*. Though the Pteroid type does not show very marked further advance, there are certain lines of specialisation which may be recognised. The most frequent and obvious is the passage to a reticulate venation. When partially carried out this gives the foundation for the sections *Heterophlebium* Fée, exemplified by *P. grandifolia* L.; and *Campteria* Presl, exemplified by *P. biaurita* L. When the anastomoses are copious but without included veinlets, this is the character of the section *Litobrochia* Presl: the extreme condition, where included veinlets are also present, is the character of Hooker's section *Amphiblestria* Presl, which is now distinguished again as a distinct genus. We need not pursue this further than to remark that anastomosis may be held to be a secondary condition, as a change which brings higher physiological efficiency, and that it is apt to be most pronounced where the cutting of the leaf is reduced and the segments expanded, as in *P. splendens* Kaulf., or in *Amphiblestria* Presl.

Along with this, however, a widening of the receptacle may appear, so as to encroach inwards upon the blade. This may be seen in some degree in

P. (*Litobrochia*) *podophylla* Swartz, where the sorus presents in section a broad expanse with a flat receptacle, beneath which lies an extended vascular commissure. The numerous sporangia, with various ages intermixed and numerous paraphyses, are borne upon it, but still with some indication of a gradate sequence of those earliest formed (Fig. 620).

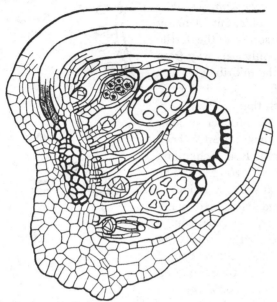

Fig. 620. Vertical section through the mature sorus of *Pteris* (*Litobrochia*) *podophylla* Sw., showing the greatly widened receptacle bearing mixed sporangia, but with some indication of a gradate sequence. (× 85.)

ACROSTICHUM L.

It is but a step from this state as seen in *P. podophylla* to those few Ferns which remain under the greatly reduced designation of *Acrostichum* L. The best known species is *A.* (*Chrysodium*) *aureum* L., which is a large, rather coarse-textured, leathery-leaved Fern, peculiar as growing in brackish water throughout the tropics. The stock is massive and upright, and bears large scales, while smaller scales are found on the leaves which are pinnate, the lower pinnae being sterile, the upper fertile. The thick fleshy roots that project from the stock may be hexarch, but the thin fibrous roots are diarch. The leaves have reticulate venation with small hexagonal areolae, and no free veinlets. The fertile pinnae usually have the whole surface occupied by sporangia together with numerous protective hairs; but there is no indusium, and no indication of distinct sori. Diels, while referring this Fern to the old genus *Acrostichum* L., remarks that little can be made out as to the affinity of this isolated type.

The key to the puzzle is found in the less common species, *A. praestantissimum* Bory, native in the Antilles, figured in Hooker's *Garden Ferns*, 1862, Pl. 58. Its habit is rather coarse, with a thick upright stock, simply pinnate leaves, and reticulate venation of the *Litobrochia*-type. The leaves are dimorphic, but in many specimens the sorus of the fertile pinnae extends only part way from the margin to the midrib. This and the presence of an upper indusium comparable with that of *Pteris* suggested to Sir W. Hooker a comparison which has recently been revived by Frau Eva Schumann (*Flora*, 1915, pp. 220, 243). She indicates specially the relation to *P. splendens*, and has added many facts bearing on the comparison.

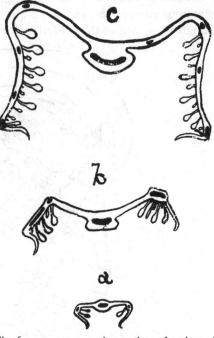

Fig. 621. *a—c* = successive sections of a pinna of *Acrostichum praestantissimum* Bory, from the apex downwards, showing the gradual widening of the Pteroid sorus inwards from the margin. (× 20.)

Structure linking the fully Acrostichoid state with the *Pteris*-type of sorus may be found by examining pinnae of *A. praestantissimum* from the apex downwards. At the distal end a purely *Pteris* structure appears with well-marked indusial flap and a narrow receptacle (Fig. 621, *a*): lower down the receptacle widens, and the sporangia are at first restricted to it (*b*); but lower down still they encroach upon the free surface of the pinna (*c*). Thus the fertile area is not due merely to a widening of the receptacle, as in *P. podophylla*, but to an actual spread of the sporangia on to the leaf-surface (Fig. 622).

Such a comparison is insufficient to determine affinity without structural evidence.

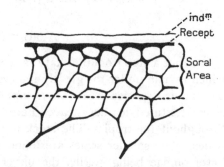

Fig. 622. Margin of a fertile pinna of *Acrostichum praestantissimum*, showing in surface view the indusium, receptacular commissure, and soral area. (× 10.)

Structurally *Acrostichum aureum* is a remarkable plant in all its parts: this may probably be related to a gross habit of growth in its brackish habitat. The thick strut-roots have a bulky lacunar cortex surrounding a

stele with as many as six protoxylems, a feature related to size; for the
small roots of the sporeling are diarch. The stem has a complex vascular
system extending upwards into the highly segre-
gated leaf-traces (Thomas, *New Phyt.* IV, 1905,
p. 175; Frau Eva Schumann, *Flora*, 1915, Bd. 108,
p. 211). The adult axis is traversed by a three-
angled dictyostele with a complex medullary
system. This has lately been brought into relation
with a like structure in the larger Pteroid stems
by the observations of Prof. J. McL. Thompson,
who has reconstructed the vascular skeleton up
to the insertion of the 12th leaf, as shown in
Fig. 623, and supplied the description which
follows. Protostelic at its base the xylem becomes
medullated, and phloem may appear centrally:
then follows as we ascend the stem a spindle-
shaped island of parenchyma bounded by endodermis: but this again fades out upwards. The
stele now widening more rapidly, the central
parenchymatous tract becomes permanently
established, shut off from the vascular tissue by
a continuous endodermis, by which the whole
stelar system is completely invested. The
medullary tract appears above the 6th leaf as
a vascular rod traversing the cavity of the dictyostele obliquely. Thence upwards the medullary
system becomes continuous, but with irregularities that often repeat the islands of parenchyma
and phloem, enclosed each by its own endodermis. Thus the whole is a dictyostele with
an irregular medullary system added, cognate
with what is seen in *Pteris* (*Litobrochia*) *podo-*
phylla (Fig. 618).

Fig. 623. A reconstruction from
sections of the vascular skeleton of
Acrostichum aureum, from its protostelic base upwards: showing
leaf-traces and leaf-gaps, numerous
root-traces and a medullary system.
(After Prof. J. McL. Thompson.)

Each leaf-trace arises from an angle of the
dictyostele: it soon breaks up into distinct strands,
while the leaf-gap closes laterally, a medullary
strand actually completing the closure, as in *Litobrochia podophylla*: but the
medullary system takes no direct part in the leaf-supply. According to the
account of Frau Eva Schumann (*l.c.* p. 212) the vascular supply to the petiole
appears to be a very complex derivative, with high segregation, from the
usual horse-shoe trace, with added fusions of a type similar to those which
appear in the leaf of *Pteris* (*Litobrochia*) *podophylla* (Studies VII, p. 37,

Fig. 26). The leaf-blade is coarsely reticulate, with hexagonal areolae, after the manner of *Litobrochia*, bearing very numerous stomata on the lower surface: it is invested with scaly hairs when young. The Acrostichoid soral areas may cover the whole lower surface of the fertile pinna, or only parts of it: the sporangia are large, having about 20 indurated cells of the annulus, while the stomium consists of about six. Abnormalities of the sporangia are not uncommon, particularly about the limits of the fertile area. Protection is given to the sporangia while young by branched hairs (Schumann, *l.c.* p. 220, Figs. 10, 11). The spores are tetrahedral, and are without any perispore: a fact that distinguishes them from those of other Acrostichoid Ferns in which a perispore is present (Hannig, *Flora*, 1911, p. 338).

It is interesting to note that the gametophyte of *A. aureum*, after developing first a spathulate form, establishes a lateral meristem, in the manner described by Von Goebel for *Pteris longifolia* (Vol. I, p. 276, Fig. 267): but later it may become many-lobed.

The facts thus summarised for *Acrostichum aureum* L., and *A. praestantissimum* Bory, fully substantiate the claim that they are Pteroid Ferns. The steps in soral origin of such a type may be traced from the bi-indusiate sorus as it appears in the Dennstaedtioid Ferns (Vol. II, Fig. 540). These already show a bias of the sorus towards the lower surface of the blade, which is accentuated in *Pteridium* (Fig. 609). A definite transit to the lower surface together with a loss of the lower indusium resulted in the state seen in *Pteris* (Fig. 619): the receptacle is seen flattened and widened out in *Litobrochia* (Fig. 620), and still more fully spread over the lower surface in *Acrostichum praestantissimum* (Fig. 621); finally it covers the whole of the lower surface of the blade as in *A. aureum*. The two species last named, together with *A. fasciculatum* C. Chr., are now the sole representatives of that great mass of Ferns that were congregated, by systematists who followed a single leading character, under the comprehensive name of *Acrostichum*. It is now clear that the Acrostichoid sorus is not the prerogative of one natural group alone, but of many. It cannot be held as a generic character from the phyletic point of view: it connotes rather a state or condition of the sorus that has been achieved along many distinct evolutionary lines. It is well that the steps of its origin can be so conclusively demonstrated in the Fern to which Linnaeus gave the name in 1753: so that in it we return, after the lapse of nearly two centuries, and great systematic confusion, to a generic designation which can be upheld in its original purity.

Returning now to the list of genera included by Diels under his heading of Pterideae-Pteridinae, here designated the Pteroid Ferns, those which have been accurately examined from the phyletic point of view may be disposed according to the results obtained, as follows:—

Pteridium Gleditch (1760) ... ... 1 species.
Hairs only are borne: vascular system highly disintegrated: venation open: bi-indusiate: receptacle marginal: sporangial succession at first gradate, later mixed.

Paesia St Hilaire (1833) ... ... 7 species.
Hairs only: solenostelic: leaf-trace undivided: venation open: bi-indusiate, but lower indusium inconstant: receptacle marginal (?): sporangial succession gradate to mixed.

Lonchitis L. (1753) ... ... 8 species.
L. aurita L. Hairs only: venation coarsely reticulate: solenostelic, with occasional perforations: leaf-trace binary: lower indusium replaced by hairs: sporangia mixed.

L. hirsuta L. Hairs and scales: venation open: large solenostele with occasional perforations: leaf-trace binary: lower indusium replaced by hairs: sporangia mixed.

Histiopteris (Agardh) J. Smith (1875) ... ... 2 species.
Hairs and scales: venation occasionally reticulate: solenostelic (corrugated): leaf-trace undivided at departure: lower indusium absent: receptacle intermediate between marginal and superficial: sporangia basipetal at first, later mixed.

Pteris L. (1753) ... ... 157 species.
(i) *Eu-Pteris* Newman. Scales: leaf-trace divided: venation open: lower indusium absent: receptacle superficial: sporangia mixed.
(ii) *Litobrochia* Presl. Scales: reticulate venation, with no free veinlets: solenostelic, with medullary system, and undivided leaf-trace: lower indusium absent: receptacle superficial and widened: sporangia mixed (*L. podophylla* Sw.).

Acrostichum L. (1753). Scales: reticulate venation: solenostelic to dictyostelic, with medullary system: leaf-trace soon divided: indusium absent: receptacle superficial, covering part (*A. praestantissimum* Bory) or the whole of the space between margin and midrib (*A. aureum* L.).

The sequence of genera thus quoted may not actually represent a simple phyletic line. The fact that the changes of detail do not all march parallel, one with another, makes such simplicity appear improbable. But the progressive sweep from solenostely to disintegration: from hairs to scales: from open venation to reticulate: from marginal sori to superficial: from bi-indusiate to uni-indusiate: from gradate sori to mixed: from a convex receptacle to one flattened and extended over the leaf-surface, forms a body of evidence that leads finally to the state of *Acrostichum aureum*. The stelar evidence appears less conclusive if *Pteridium* be held as a starting-point: but it is an exceptional plant in many ways. If, however, the smaller species of *Paesia* be held as such, then there is progression from typical solenostely with undivided leaf-trace to higher states of disintegration seen in passing along the series. Looking back to the source from which this whole sequence probably sprang, there can hardly be any two opinions. The series of Pteroid Ferns are Dennstaedtioid derivatives, and they lead that phyletic sequence on to the common goal of the Acrostichoid state.

In conclusion a brief reference may be made to the other genera grouped by Diels under his Pterideae-Pteridinae, which have not been included in the above list. Of these *Cassebeera* Kaulfuss is probably a specialised Cheilanthoid; *Actiniopteris* Link must be held in doubt as a "genus incertae sedis." The reference of the other three genera—viz. *Amphiblestria* Presl, *Anopteris* Prantl, and *Ochropteris* J. Smith—to a Pteroid affinity must also be left in suspense, pending further enquiry. But such uncertainty, regarding any or all of these, leaves unimpaired the argument based on the comparison of those Ferns which clearly find their place in the Pteroid series.

BIBLIOGRAPHY FOR CHAPTER XXXVIII

609. METTENIUS. Filices Horti Lipsiensis. 1856.
610. METTENIUS. Farngattungen, III. Frankfurt. 1858.
611. HOOKER. Species Filicum, II, p. 154. 1858.
612. HOOKER. Synopsis Filicum, p. 153. 1873.
613. BURCK. Indusium der Varens. Haarlem. 1874.
614. LUERSSEN. Rab. Krypt. Flora, III, p. 100. 1889.
615. PRANTL. Arb. K. Bot. Gart. pp. 16–18. Breslau. 1892.
616. CHRIST. Farnkräuter d. Erde, p. 161. 1897.
617. DIELS. E. & P. Natürl. Pflanzenfam. I, 4, p. 254. 1902.
618. TANSLEY & LULHAM. New Phyt. III. 1904.
619. THOMAS. *Acrostichum aureum*, New Phyt. IV, p. 175. 1905.
620. TANSLEY. Filicinean Vasc. Syst., New Phyt. Reprint, No. 2. 1908
621. HANNIG. Flora, p. 338. 1911.
622. SCHUMANN. Flora, Bd. 108, p. 201. 1915.
623. VON GOEBEL. Organographie, II, p. 1137, etc. 1918.
624. BOWER. Studies VII, Ann. of Bot. XXXII, p. 1. 1918.

CHAPTER XXXIX

GYMNOGRAMMOID FERNS

IN the opening paragraphs of Chapter XXXVIII it was pointed out that the "Pterideae" of Diels (Engler and Prantl, I, 4, p. 254) comprise the bulk of the Pterideae of Hooker, together with his Grammitideae. It was further noted that this association is based upon a diagnosis that is comprehensive rather than exact, and that it was applied without reference to evolutionary or ontogenetic origin. As a consequence it places together groups which may have had a quite distinct evolutionary story: this is probably the fact on the one hand for the Ferns styled by Diels the Pterideae-Pteridinae, which have been dealt with under the name of the Pteroid Ferns in the preceding chapter; and on the other those which he designates the Pterideae-Gymnogramminae, -Cheilanthinae, and -Adiantinae, here to be styled the Gymnogrammoid Ferns. On the ground of consecutive comparison of Ferns clearly related to one another it has been shown in Chapter XXXVIII with high probability that the former sprang from a bi-indusiate source, such as is seen in the Dicksonioid-Dennstaedtioid series. A leading feature in the argument has been the gradual disappearance of the lower indusium; for this a high degree of physiological probability rests on the fact that, as the sorus shifts from the marginal to a superficial position and the upper indusium curls over, the lower indusium is no longer essential for protection, which was its original function. In taking up the treatment of the Gymnogrammoid Ferns, the first question that will arise will therefore be whether any such argument will apply for them. Are any vestiges of the lower indusium to be found? If not, there would appear to be no justification drawn from the soral characters themselves for attributing to these Ferns a common origin with the Pteroids, which had a bi-indusiate ancestry. It may be stated at once that no such evidence is forthcoming for the Ferns ranked as Gymnogrammoid. It will therefore be necessary to consider whether any other types of non-indusiate Ferns of primitive type exist, in relation to which the origin of a Gymnogrammoid state may be traced. Such primitive types without any specialised indusium, and with sori already superficial, and more or less extended along the veins, are to be found in the surviving genera *Todea* and *Plagiogyria*, while the Schizaeaceae should also be borne in mind, and particularly *Mohria* and *Anemia*. With these considerations before us we may enter upon the study of the Gymnogrammoid Ferns.

For purposes of convenient description here the Gymnogrammoid Ferns may be grouped into four sections:

I. Those which show a relatively primitive condition, as seen especially in their sporangia. This section would comprise *Llavea*, *Cryptogramme*, *Onychium*, *Ceratopteris*, and *Jamesonia*.

II. A second section would centre round *Gymnogramme* and *Hemionitis*, together with such minor genera as *Pterozonium*, *Syngramme*, *Anogramme*, *Coniogramme*, *Gymnopteris*, *Ceropteris*, and *Trismeria*: all of which have from time to time been included in the old genus *Gymnogramme* Desv., 1811. Though these distinct genera may be upheld as now defined, there is no reason to doubt their near affinity to the central genera named. It is necessary, however, to remark the exclusion of the genera *Ceterach* Adanson, 1763, and *Pleurosorus* Fée, 1850; for they are now regarded as partially or completely non-indusiate Asplenieae: also the position of *Aspleniopsis* Mett. and Kuhn must be held in suspense (see *Natürl. Pflanzenfam.* I, 4, p. 272, Fig. 145, *A*).

III. A third section would consist of the Adiantinae, represented by *Adiantum* itself.

IV. A fourth section includes specialised xerophytic types, such as *Pellaea*, *Doryopteris*, *Cheilanthes*, and *Notholaena*, together with *Saffordia* and *Trachypteris*.

The central features that all these Ferns have in common are that the sorus is superficial in origin, never actually marginal, and that there is not any vestige of an inner indusium. Nevertheless the leaf-margin may be variously developed and recurved, so as to give protection to the sporangia. The receptacle often extends a considerable distance along the underlying veins, while in extreme instances the insertion of the sporangia may not merely extend along them, but also spread laterally from them so as to cover the leaf-surface, in the manner known as Acrostichoid. It will be realised that these features are not in themselves very distinctive, and that they fall short of being clearly diagnostic. If the original position of the sporangia in Ferns generally was, as believed, distal or marginal, a transit to the surface might appear in any phylum where the leaf-area became extended (see Vol. I, p. 225). Thus along a number of distinct phyletic lines the Gymnogrammoid character may have been acquired. The natural inference will then be that Ferns showing that character may belong to several distinct lines of descent, and not necessarily be phyletically related to one another at all. Accordingly we must be prepared for some degree of segregation, and possibly for the recognition of affinity of one of these sections, or even of some single genus, to one phyletic source; while others may be referable to some quite distinct source. These preliminary remarks form a necessary introduction to the description of the more prominent genera grouped as already suggested: they leave the question of the phyletic unity of the Gymnogrammoid Ferns open for discussion after the facts have been disclosed.

I. The Primitive Gymnogrammoid Ferns

LLAVEA Lagasca

Llavea is a monotypic genus. Its single representative is *L. cordifolia* Lagasca, a native of the Mexican uplands, figured by Sir W. Hooker (*Gen. Fil.* Tab. XXXVI). The ascending stock and leaf-bases bear hairs and scales. The leaves are spirally arranged and long-stalked, and they may be tri-pinnate. The lower pinnae are sterile, with segments of the Osmundaceous type, having forked venation. The upper are fertile with inrolled margins, giving an appearance as in *Plagiogyria*. The very numerous sporangia are inserted along the veins, as in *Gymnogramme*, and they are fully protected by the leaf-margin till mature (Fig. 624). The stock is traversed by a slightly dictyostelic solenostele, as in *Plagiogyria*: the vascular ring is sometimes complete when seen in transverse section. The leaf-trace departs as a single strap, which shortly after separation is seen to be tetrarch, and the pinna-traces are marginal in origin (see Thompson, 647, Text-figs. 8–11). In these structural points *Llavea* compares with *Plagiogyria*: but there is no basal enlargement of the leaf-stalk, and pneumatophores are absent.

The sporangia are pear-shaped, with a three-rowed stalk, but the annulus and stomium are variable. Sometimes the annulus is vertical and regular, but this is relatively uncommon. Frequently it passes obliquely, being interrupted in its course, and its cells being even bi-seriate at some points. This instability is a feature shared by other related Ferns, and especially by *Cryptogramme*. The spore-counts give figures 46–52, indicating a typical number of 48–64. For further details reference may be made to Dr J. McL. Thompson's *Memoir* (647).

CRYPTOGRAMME R. Brown

Cryptogramme is a wide-spread genus, with 4 species, of which *C. crispa* (L.) R. Br., the Parsley Fern, is native in Britain. It shares many of the features of *Llavea*, but on a smaller scale, and the two genera have habitually been classed together. Here, however, the sori occupy the vein-endings (Fig. 624, *D*). The Parsley Fern has a creeping rhizome bearing dermal scales, and spirally arranged leaves that are 2–4 pinnate (Fig. 625, *A*). It is dictyostelic, a state consequent on the close arrangement of the leaves, and each leaf-trace is an undivided meristele. The venation is open, with anadromic branching: thus in essentials the structure is as in *Llavea*, but smaller. The similarity extends also to the fertile segments (Fig. 625). The chief interest lies in the varying details of the sporangia, which are pear-shaped. They arise superficially, close to the vein-endings, and without any vestige of an indusium: moreover there are indications of a "mixed" condition, though these are only slight (Fig. 626). The mature sporangium has a short stalk of variable structure. It may be three-rowed, or it may consist of four to six rows of cells. The annulus is also variable: the chain of cells sometimes runs obliquely past the insertion of the stalk, though its induration

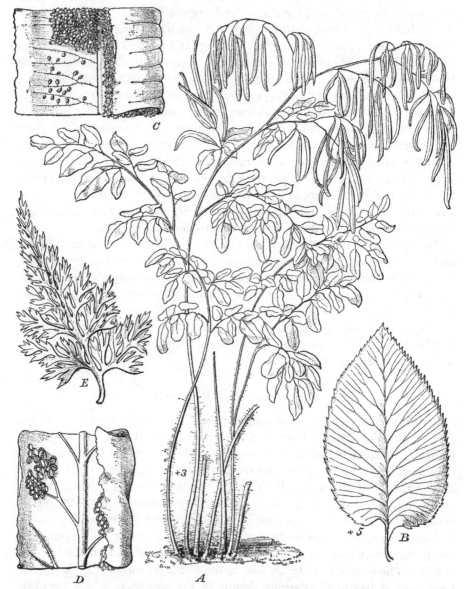

Fig. 624. *A—C = Llavea cordifolia* Lag. *A* = habit. *B* = sterile pinna with venation. *C* = part of a fertile pinna, the left-hand flap flattened out to show the venation. *D, E = Cryptogramme* R. Br. *D* = part of a fertile pinna with venation and sori: on the left the margin is flattened out. *E* = *C. japonica* (Thunb.) Prantl, sterile pinna. (*D*, after Luerssen, the rest after Diels.)

is interrupted: occasionally the series of cells may be doubled (Fig. 627). Such instability of structure is held to be an indication of a primitive state, since it occurs in sporangia that are short-stalked and pear-shaped, and the variants point towards an uninterrupted oblique annulus, such as is actually present in *Plagiogyria*, and the more complex stalk characteristic of relatively primitive Ferns. Spore-countings give numbers varying from 45 to 52.

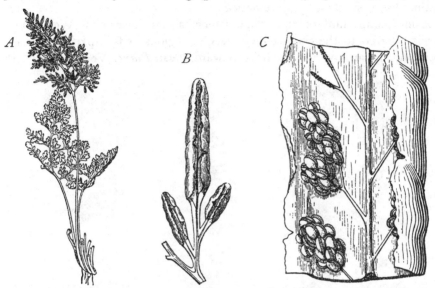

Fig. 625. *Cryptogramme crispa* R. Br. *A* = part of a strong plant, with one fertile and two sterile leaves, ¼ natural size. *B* = segment of the second order with inrolled margins still covering the sori (× 4). *C* = part of a segment of the third order, enlarged: the left flap reflexed to expose the sori: those of the uppermost nerve-fork have been removed. (After Luerssen.)

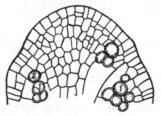

Fig. 626. *Cryptogramme* (*Allosorus*) *crispa* (L.) R. Br. Tip of a sporophyll-segment, seen from below. The sporangia are shaded. Magnified. (After Von Goebel.)

ONYCHIUM Kaulf.

Onychium, which also bears its sporangia upon the vein-endings, was separated as a distinct genus by Kaulfuss, and it is maintained as such in Christensen's Index, though it was included in *Cryptogramme* by Prantl (634, p. 413). The difference lies in the fact that the veins which bear the sori are connected marginally by vascular commissures. The result may appear similar to what is seen in *Pteris, Lindsaya, Diellia,* or *Nephrolepis.* But the innovation of a connecting commissure may be held as homoplastic in them all, rather than as an evidence of kinship. It ranks in the propagative region with the adoption of a reticulate venation in the sterile: and this innovation is seen to arise independently in various groups of Ferns quite distinct from one another systematically.

A more important point is the existence as far back as the Wealden of fossils referred to the genus *Onychiopsis*, in regions as far apart as Britain, Germany, Japan, and South Africa (Seward, *Fossil Plants*, Vol. II, p. 377). If

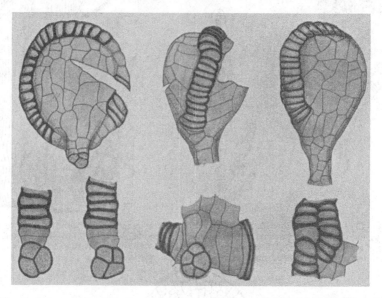

Fig. 627. Sporangia of *Cryptogramme crispa*, showing instability in the obliquity of the annulus, in its induration, and in the structure of the stalk.

the reference to this affinity be correct, the early occurrence of a Fern so close in kinship to *Cryptogramme* would enhance the interest which the indications of antiquity seen in the living representatives of that genus already possess.

JAMESONIA Hook. and Grev.

Jamesonia is a genus comprising some 14 species: they live in the higher Andes, and are characterised by a reduced habit, but particularly by a continued apical growth of the simply pinnate leaves: they live on exposed boggy ground together with *Sphagnum*. The genus has usually been grouped systematically with *Gymnogramme*, from which it is distinguished by habit. This is excellently shown by Karsten's drawings (Fig. 628). The thin creeping rhizome forks: it bears alternate leaves with closely ranged pinnae of circular outline, while the whole when young is covered by dense golden-brown hairs that are simple and multicellular. There are no scales, but small mound-

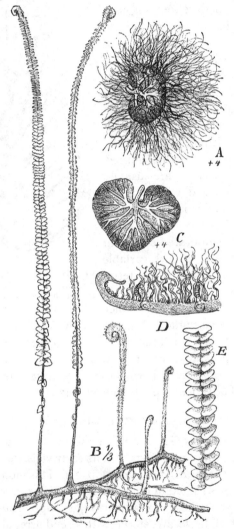

Fig. 628. *Jamesonia* Hook. and Grev. *A* = pinna of *J. nivea* Karst. *B—E* = *J. canescens* (Klotzsch) Kze. *B* = habit. *C* = a pinna. *D* = part of a transverse section through a fertile pinna, the lower surface uppermost. *E* = part of a leaf. (After Karsten.)

like outgrowths are common, on which the hairs are often seated. The upper surfaces of the adult pinnae are naked, but the woolly hairs remain permanently on the lower surface protecting the sporangia (Fig. 628, *D*). The vascular system of the axis is a slender solenostele, in which the alternate leaf-gaps are elongated, so that in transverse section it often appears as two separate straps. The leaf-trace departs as a simple strand. The venation of the leaves is an open dichopodium that spreads fan-like, its ends extending into the recurved margins of the leaves (see Thompson, 647).

The sporangia are distributed along the veins: but some are inserted on the general surface of the pinna between them: this may be held as an advance upon the usual Gymnogrammoid type towards that which is described as Acrostichoid (Fig. 629). The various ages of sporangia are intermixed. The sporangia themselves are usually pear-shaped but lop-sided, with a three-rowed stalk. The annulus is vertical, and interrupted at the insertion of the stalk: it has about twenty indurated cells, but these and the stomium are both variable as regards detail. The spores are tetrahedral. Spore-counts range from fifty-six to seventy-two. The latter figure, which

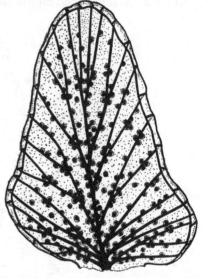

Fig. 629. Fertile pinna of *Jamesonia verticalis* Kze., showing the venation, and insertion of the sporangia, the latter is "Acrostichoid." (After Dr J. McL. Thompson.)

was found in large sporangia with an irregular annulus, is most unusual in Ferns with a mixed sorus. All these features together suggest a transitional state, where the sporangial structure has not become fully standardised. They point out *Jamesonia* as a Fern combining characters that are primitive with indications of advance (see Thompson, 647).

CERATOPTERIS Brongn.

Ceratopteris thalictroides (L.) Brongn. is usually placed as the sole representative of the Parkeriaceae. The story of its early vicissitudes of classification is told in Hooker's *Species Filicum*, Vol. II, p. 236. This monotypic Fern grows throughout the tropics rooted in mud or shallow water, or sometimes floating. The short upright shoot bears spirally arranged sappy leaves with reticulate venation. The first of these are sterile with broad irregular lobes, the later leaves are fertile with narrower lobes, their recurved margins protecting the lower surface on which the large isolated sporangia are borne. Scales are present on the axis and leaf-bases.

The aquatic habit and the very sappy nature of its tissues have made the study of the vascular system difficult. The account of it here given is based upon the MS. notes of the late Prof. D. T. Gwynne-Vaughan. The axis after initial protostely widens to produce a typical radial dictyostele, with added strands that cross the leaf-gaps. At first there are no medullary strands, though these arise later. The earlier leaves show a leaf-trace of two pairs of strands: the lower pair springs from a median point at the base of the leaf-gap, and they either fuse at once laterally, or are connected later by a cross-strand. They supply the abaxial curve of the rachis. The second pair of strands arise high up on the leaf-gap, and they are connected by one or two cross-strands. They supply the adaxial curve of the rachis. Higher up still two additional leaf-strands may arise, one at each side of the base of the leaf-gap. At first the leaves receive no strands from the medullary system, for it does not exist. But as the axis increases medullary strands arise, and the later leaves receive strands direct from them, which branch and anastomose in the rachis, but finally disappear by fusion with the bundles of the leaf-trace itself. The supply to the pinnae springs from the margins of the outer series of the rachis, the connections being with both the abaxial and the adaxial curves of the highly disintegrated system (compare Vol. I, p. 173, Figs. 169, 170).

It thus appears that the vascular system of *Ceratopteris* is cognate with what is seen elsewhere in large Ferns of various affinity. The axis is dictyo-stelic with a medullary system. The petiole shows high disintegration of the primary trace, which is reinforced in the older leaves by medullary connections: and the origin of the pinna-traces is extra-marginal. All these conditions are probable for a sappy water plant, though they suggest an advanced state of adaptation. This is also indicated by the reticulate vena-tion.

A greater interest attaches to the soral conditions, which present strangely conflicting features. The fertile leaf-segment is lanceolate in outline: the sporangia that it bears are not associated into sori, but are seated solitary upon the anastomosing veins, so that they appear in irregular rows parallel to the margins of the segment: these overarching protect the young sporangia in the absence of any more definite indusium (see Kny, 631). The sporangia themselves are large and short-stalked, the stalk consisting of a rosette of five or six cells. Each sporangium arises from the leaf-surface by the out-growth of a single cell, in which the first segment-wall impinges directly on the basal wall, or sometimes on one of the oblique lateral walls (Fig. 630), after the manner of Ferns with sporangia larger than those of the typical Leptosporangiates (see Fig. 238, Vol. I, p. 244). The sporangial head is spherical, and very variable in structure and contents (Fig. 631). The annulus is usually vertical and interrupted at the insertion of the stalk. It may consist of very numerous cells: as many as seventy have been counted by Benedict (642), but forty is usual, though the numbers may be much smaller, and it is stated that sometimes the induration may be absent altogether (Hooker, *l.c.* p. 236). While the annulus may thus be sometimes incomplete, at others it may be doubled. Such conditions show that here the sporangium is far from being standardised. The same holds for its spore-output. This is often stated

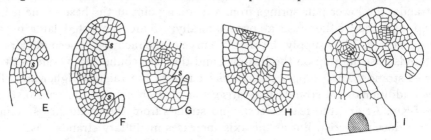

Fig. 630. *Ceratopteris thalictroides*: vertical sections of young sporophylls to show the superficial origin of the sporangia (*s*). (× 100.)

as sixteen, but Benedict found it to be definitely thirty-two for many plants, while Thompson (*l.c.* p. 389) found the spore-count in his material to be anything between twenty-four and twelve. This again shows that the sporangium of *Ceratopteris* is not a constant entity.

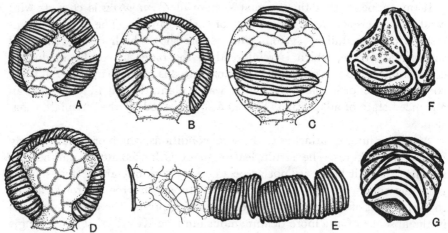

Fig. 631. *A—E* = sporangia of *Ceratopteris thalictroides*, selected to show the variability of the annulus. *F*, *G* = spores. (After Dr J. McL. Thompson.)

COMPARISON OF THE PRIMITIVE GYMNOGRAMMOID FERNS

Llavea has always been grouped with *Cryptogramme*, while J. Smith's name for it, *Ceratodactylis osmundioides*, points the comparison also with *Osmunda*. Sir W. Hooker (*Sp. Fil.* Vol. II, p. 125) remarks, "This is in every respect a very striking plant, closely allied in its fructifications to *Cryptogramme* Br. but with a very different habit, in some respects approaching *Osmunda*." None of the early writers compare it with *Plagiogyria*: but Diels (E. and P. I, 4, p. 255) places these three genera *Plagiogyria*, *Llavea*, and *Cryptogramme* in close relation, which is probably correct. It has already been pointed out

(Vol. II, p. 280) that a relation exists between *Plagiogyria* and *Osmunda*, so that the comparison of Sir W. Hooker of *Llavea* with *Osmunda* now receives a tardy though indirect corroboration. These genera appear to form a nexus of Ferns that suggests an origin of a primitive Gymnogrammoid type, with superficial sori throughout, quite independent of the Dicksonioid origin of the Pteroids, in which the phyletic origin of the sorus has been shown to be marginal.

The relatively primitive character of those Gymnogrammoids above named appears in features that are distributed rather arbitrarily among them. The shoots are creeping or ascending: mostly with scales on the rhizomes and leaf-bases (*Llavea, Cryptogramme, Ceratopteris*), but sometimes with hairs only (*Jamesonia*). The vascular system of the adult is based upon a type of solenostele which is not far removed from a typical one in *Llavea* and *Jamesonia*, though in *Ceratopteris* there is an elaborate dictyostele, with medullary strands also in the sappy axis. The leaf-trace of *Llavea* and of *Cryptogramme* comes off as a single strap, but in the latter it shows an indication of a median xylic break that probably corresponds to the median fission seen in *Jamesonia* and in *Gymnogramme*. But *Ceratopteris* has a highly disintegrated leaf-trace, with medullary complications: this accords further with its reticulate venation, while the others have forked veins with free vein-endings. Thus anatomically *Llavea, Cryptogramme* and *Jamesonia*, and in particular *Llavea*, are relatively primitive, while *Ceratopteris* is advanced. In respect of dermal appendages *Jamesonia* is peculiar in this whole affinity in having hairs only, thus comparing with *Plagiogyria* with which anatomically both *Llavea* and *Jamesonia* have much in common.

It is in respect of the sori and sporangia that the most important comparisons appear. In all these Ferns various ages of the sporangia are intermixed, and they share the absence of any true indusium, while their sporangia, seated individually rather than grouped in definite sori, are distributed along the veins, and are large and pear-shaped or approximately spherical, but with short stalks. In *Ceratopteris* a section of the stalk may present a rosette of five or six cells, comparable with that of *Plagiogyria* (Vol. I, Fig. 243, *m*), and with other Ferns that are relatively primitive. But in *Llavea* and *Jamesonia* there is the usual three-rowed stalk. *Cryptogramme crispa*, however, gives varying results between a three-rowed stalk and one with six rows. The most striking sporangial feature is the inconstancy of the annulus, which all these genera share. This is most marked in *Ceratopteris*, where normally the annulus consists of an unusually large number of indurated cells. It is as a rule interrupted at the insertion of the stalk, but *Cryptogramme* gives indications of obliquity of position, even with continuity of the series of cells past the stalk-insertion. In all the four genera there is this inconstancy of structure of the annulus, it is not merely exceptional: in *Jamesonia* Thompson found 150 irregular out of

500 sporangia observed. The annulus may be interrupted, incomplete, or even doubled in varying degree. The induration may be very limited, and it is stated to be absent sometimes in *Ceratopteris*. Such signs of instability, not in one genus only but in all, are held to mark a transitional state between large sporangia with an oblique annulus, and smaller sporangia with a vertical annulus. A parallel is found in *Platyzoma* (Vol. II, p. 209, Fig. 492), which is also a transitional type. It is also seen in *Acrostichum aureum* (Schumann, *l.c.* p. 220) at the region where fertility of the leaf-surface begins. Lastly, the spore-enumerations give very varied results. It will be remembered that the spore-output of *Plagiogyria* is only 48, though that of the Osmundaceae and Schizaeaceae is relatively high. It is then interesting to note that in *Jamesonia*, which like them has no scales but hairs, the spore-counts show the exception of exceeding 64, so rare among Leptosporangiate Ferns with mixed sori: the numbers recorded range between 48 and 72. On the other hand, the typical numbers in the other genera vary widely. In *Llavea* and *Cryptogramme crispa* they are 48 to 64, but in *Ceratopteris* Benedict found 16 to 32 for *C. thalictroides*, but only 16 in his *C. deltoides*. Thompson says the number may be anything between 12 and 24. These facts appear again to point to a transitional state of the sporangium. The spores in all of these Ferns—including also *Plagiogyria* and *Osmunda*, and most of the Schizaeaceae on the one hand, and the Gymnogrammoid Ferns on the other—are spherical-tetrahedral, often large, with very bold markings of the outer wall (Fig. 631). According to Hannig's record (*Flora*, Bd. 103, p. 338-9), there is a prevailing absence of a perispore in these Ferns, and this is found to be so also in the Osmundaceae and Schizaeaceae (*l.c.* p. 342).

The general result of these comparisons appears to be that the four genera under discussion are probably related to one another, and present many primitive features that point towards such antique types as the Osmundaceae, Plagiogyriaceae, and in a less degree the Schizaeaceae. Further, they suggest that they occupy a transitional position: but inasmuch as the archaic features that they show are not concentrated in any one of them, being in fact distributed irregularly between them, it is difficult to place any one genus as the most primitive of them all. If any such indication can be held to have emerged, it would point to a special relation of *Llavea* to *Plagiogyria* or to *Osmunda*. The general result is, however, to suggest the probability that a Gymnogrammoid state, such as these Ferns actually show, has originated directly from types already having superficial sporangia, such as *Todea* or *Plagiogyria*. There is no evidence in any of them of obliteration or of transformation of either or both of the indusia of a two-lipped marginal sorus, such as is seen in the Pteroids. This being so, these primitive Gymnogrammoids must be held as standing phyletically on their own feet, distinct by descent from the Pteroid Ferns. Even the linkage of the sori of *Onychium* into coenosori does not shake this conclusion, since such fusions occur homoplastically in a plurality of phyletic lines.

II. THE CENTRAL GROUP OF GYMNOGRAMMOID FERNS

Under this head are included *Gymnogramme* Desv., together with the allied genera *Pterozonium* Fée, *Syngramme* J. Sm., *Anogramme* Link, *Coniogramme* Fée, *Hemionitis* L., *Ceropteris* Link, and *Trismeria* Fée. The genus *Gymnopteris* Bernh. is also included in Christensen's Index (p. xxxviii). These are Ferns of moderate size and wide distribution, varying in habit, and particularly in the outline of the leaf. They all bear dermal scales, especially upon the rhizome and leaf-base. They have for the most part a relatively primitive stelar structure, but the leaf-trace is commonly divided into two equal straps. Their most marked common feature is that the sorus, which is without any true indusium, is spread in varying degree along the length of the veins, but without extending to the actual margin of the blade. Since the venation is in many of these Ferns open, and in others closed, the sori are often linear as in *Gymnogramme*, but sometimes reticulate as in *Syngramme* or *Hemionitis*. The latter state appears as a natural sequel to the expansion of the blade, and may be held as derivative. The sporangia are of the ordinary Leptosporangiate type, with a vertical annulus, but without specially long or thin stalks. The spore-output, so far as ascertained, appears to be the normal 48 to 64. The spores are often large, and spherical-tetrahedral in form, and are without perispore.

The peculiar perennation of the gametophyte in *Anogramme*, combined as it is in *A. leptophylla* with the unusual condition of an annual sporophyte, has already been described (Vol. I, p. 276, Fig. 267). The facts point to a special seasonal adaptation rather than to any line of comparative argument.

The stelar system of them all is based upon the solenostele, and is without medullary strands. It is marked by perforations, the most frequent of these being a long slit which divides the leaf-trace down to its base into two straps. This is seen in a relatively simple form in *Gymnogramme* (*Ceropteris*) *calomelanos* (L.) Und. (Gwynne-Vaughan, *Sol. Ferns*, II, Fig. 9): but it appears also accompanied by numerous other perforations in the more highly disintegrated system of *Gymnogramme* (*Coniogramme*) *japonica* (Thunbg.) Diels (see Vol. I, Fig. 160), which is a species of large size.

There will be no need to enter here into a detailed description of the several genera. But as *Trismeria trifoliata* has recently been examined in some detail by Dr J. McL. Thompson (647), that Fern though somewhat specialised may be taken as an example of the central group. The result of Dr Thompson's enquiry was to conclude that *Trismeria* need not be considered as a genus distinct from *Gymnogramme*, so alike are they in essential features.

TRISMERIA Fée

This genus, as maintained in Christensen's Index, contains only two species, of which the better known is *T. trifoliata* (L.) Diels, a Fern of tropical America. It has been ranked with *Acrostichum* by Linnaeus, and is included under *Gymnogramme* in the *Synopsis Filicum*, as *G. (Ceropteris) trifoliata* Desv. The question of retention of its rank as a separate genus may be left to the systematists, with the remark that while it shows an Acrostichoid tendency, its close relation to *Gymnogramme* is accentuated by recent observations (647).

The Fern is well figured in Hooker's *Garden Ferns*, Plate IV. It has an erect stock bearing leaves arranged alternately, the axis and leaf-bases being covered with brown scales. The leaves bear numerous subcoriaceous pinnae, the lower of which are usually ternate, the upper undivided. The distal pinnae are fertile, the whole lower surface appearing to be covered by sporangia and glandular hairs, with a characteristic white or yellow secretion of wax (De Bary, *Comp. Anat.* p. 99). The venation is always open.

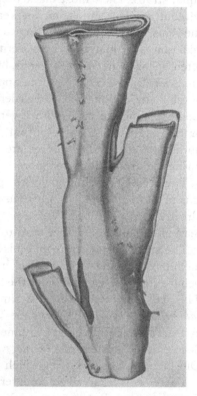

The axis is traversed by a solenostele, which opens to give off successive leaf-traces. These consist each of two vascular straps, which are already separated from one another before they are detached from the stele (Fig. 632). The pinna-traces of the basal pinnae are of extra-marginal origin from the still divided straps of the rachis, but these unite towards the distal region, and the pinna-traces are there marginal in origin. This change within a single leaf shows that there is no essential difference between the two types of origin of the leaf-trace: it appears to depend chiefly upon the size and the degree of curvature of the vascular tract of the rachis. A comparison of the vascular system of *Trismeria* with that of *Gymnogramme japonica* (see Vol. I, Fig. 160) shows that

Fig. 632. Reconstruction of the stele of *Trismeria* showing solenostele, with short leaf-gaps and binary leaf-traces. (After J. McL. Thompson.)

though both have the divided leaf-trace, the former is nearer to a primitive solenostele than the latter, differing from it in the absence of those perforations which are so marked a feature in the latter. This may perhaps be related to the large size and sappy nature of *G. japonica*, for Gwynne-Vaughan found the stelar structure of *G. calomelanos* not far removed from solenostely (*l.c.* p. 698). In this *Trismeria* more closely resembles this Fern than it does *G. japonica*. The general conclusion is that, given a divided leaf-trace, all these Ferns are essentially solenostelic, but with occasional and variable perforations, and with overlappings of the foliar gaps which are dependent upon the leaf-arrangement.

The sporangia of *Trismeria* are mostly inserted on the veins: some, however, are seated on the general leaf-surface. The various ages are intermixed. The sporangial stalk is three-rowed, and the annulus vertical and regular, with a well-defined stomium. The tetrahedral spores give spore-counts between the extremes of 49 and 35. Sorally in fact there is nothing of importance to note beyond the slightly Acrostichoid state. It will be a question for further enquiry how prevalent an Acrostichoid development has been among the Gymnogrammoid Ferns. Evidence of it has been found in *Jamesonia* and in *Trismeria*, but there is little sign of it in other genera though it might occur in any of them. Attempts have been made to trace *Leptochilus* from forms like *Gymnogramme* or *Hemionitis*: but this is held in doubt by Frau Eva Schumann (*Flora*, Bd. 108, p. 249). She finds little suggestion of that affinity in the external habit of *Leptochilus*: moreover, while the Gymnogrammoid spores are tetrahedral and naked, those of *Leptochilus* are, like those of the Dryopteroid *Meniscium*, kidney-shaped, and have a perispore. This makes it appear improbable that such developments as are seen in *Leptochilus* are of Gymnogrammoid origin. No other Acrostichoid type suggests this affinity: in fact the Acrostichoid spread of the sporangia seems in these Ferns to be potential rather than widely realised.

The way has been opened by the comparative study of the relatively primitive Gymnogrammoids for that of the central group. It is true that the latter do not exhibit that variability of structure of the sporangia, nor the short thick stalk and pear-shaped head, which are such marked features in the former. But the general soral condition is here so far similar to theirs that the central Gymnogrammoids may reasonably be regarded as advanced derivatives of some similar type. In fact we may hold that they also originated from a non-indusiate ancestry, in which at an early period the production of the sporangia passed from an originally marginal position to the surface of a widening blade: but that in the course of their further development the sporangium had become progressively stabilised after the usual Leptosporangiate type.

III. ADIANTUM Linné, 1753

Adiantum is a widely spread genus containing 184 species of Ferns usually of shady and moist habit, and with a preponderance of species in tropical America. They are characterised by having sori apparently marginal, but really superficial in origin, and covered by a sharply reflexed leaf-margin which has the appearance of an indusium, since it is either membranous or brown-coloured when mature. As a matter of fact there is no true indusium. The sporangia are inserted upon the distal region of the veins that traverse the fertile lobe, and it is the fertile region of the blade itself which is reflexed at the proximal limit of the fertile zone. It is this flexure that gives the essential difference between *Adiantum* and the central group of the Gymnogrammoid Ferns. Strictly speaking, each such fertile lobe bears a group of sori seated upon parallel veins. The whole structure corresponds to a small but highly specialised leaf-segment of *Gymnogramme* reflexed sharply on itself in relation to protection (Fig. 633).

The genus thus characterised comprises a coherent group of Ferns with upright or creeping scaly rhizome, bearing leaves spirally or alternate. The leaf has as a rule a shining black and brittle petiole: the blade may be entire (*A. reniforme* L. or *Parishii* Hk., Fig. 633, *A*); or simply or repeatedly branched. It is often of delicate texture in relation to the moist and shaded habitat, and is traversed by forking veins, which are usually free: but in some species they anastomose to form a reticulum (*Hewardia* J. Sm.). The branching of the leaf is clearly related to dichotomy (Fig. 633, *G*). The ultimate segments are often deltoid in outline, and when these are borne upon the polished black branchlets of the highly divided rachis they give the characteristic appearance of the Maidenhair Fern (Fig. 633).

Anatomically *Adiantum* is closely related to *Gymnogramme*. Those species which have an elongated rhizome may show actual solenostely, as in *A. pedatum* L. and *hispidulum* Sw. But commonly the stele is gutter-shaped, owing to the great elongation of the leaf-gaps (G.-V. MS. notes). In those with short internodes, and particularly where the rhizome is ascending or upright, the leaf-gaps overlap, and the stele appears in transverse section as several meristeles disposed in a ring. This condition is obviously based upon solenostely, from which it is not far removed (Gwynne-Vaughan, *Ann. of Bot.* XVII, p. 695, *A. trapeziforme* L. and *petiolatum* Desv.). In this there is near correspondence to what is seen in *Pellaea* and *Notholaena*. In *A. capillus-Veneris* L. the number of meristeles in a transverse section may be 5–7, of circular form (Luerssen, *Rab. Krypt. Fl.* III, p. 81). The leaf-base receives two closely related strands, which unite upwards to form a single four-angled meristele, a condition not uncommon in contracted leaf-stalks.

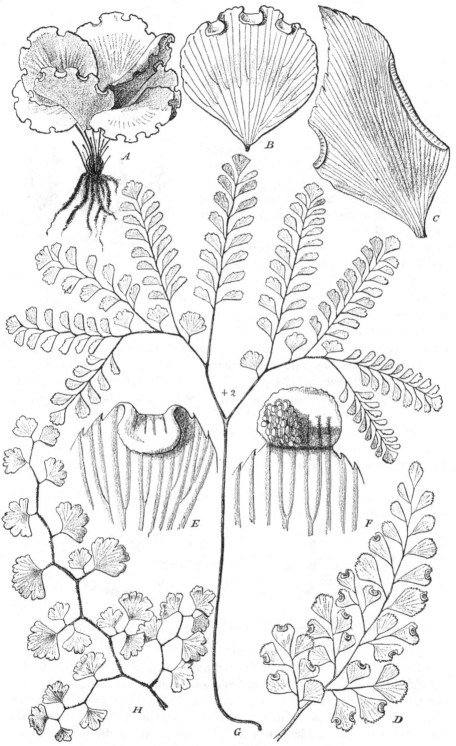

Fig. 633. *Adiantum*. *A, B = A. Parishii* Hook.: *A* = habit, *B* = underside of leaf. *C = A. macro-phyllum* Sw., underside of pinna. *D = A. venustum* Don., underside of primary pinna. *E, F = A. capillus-Veneris* L., parts of a fertile pinna enlarged: in *F* the fertile lobe is raised to show the sori. *G = A. pedatum* L., leaf. *H = A. Féei* Moore, primary pinna. (*A, B* after Hooker: *E, F* after Luerssen: the rest after Diels.)

The origin of the pinna-trace is extra-marginal, as it is also in *Notholaena*. In all these points *Adiantum* fits naturally in with other Gymnogrammoids.

In the epidermis of the blade of *Adiantum* "spicular cells" have been found, the significance of which will appear in relation to the Vittarieae (see Chapter XLVII).

The sori consist of sporangia inserted upon the distal region of the separate veins of the reflexed lobe. This is the usual type, and it has been held as the character of the section *Eu-Adiantum* Kuhn. But sometimes the sporangia may spread also on to the surface of the blade between the veins, giving thus as in *Gymnogramme* and *Jamesonia* an essentially Acrostichoid state. This has been taken as the diagnostic charac- ter of *Adiantellum* Kuhn, and it is seen in *A. macrophyllum* Sw. (Fig. 634).

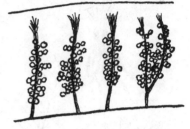

Fig. 634. *Adiantum macrophyllum* Sw., fertile leaf-lobe, with sporangia inserted on the parenchyma between the veins. (After Frau Eva Schumann.)

Dr Horvat has described for old prothalli of *Adiantum* a structural peculiarity which bears some degree of significance. Bauke had many years ago noted a collenchymatous thickening of the walls in the prothalli of *Anemia* (Pringsh. *Jahrb.* 1878, p. 628, Taf. XLI, Fig. 1): and a like structure was found by Heim in *Lygodium* (*Flora*, 1896, p. 368). Thus it appears in representatives of both series of the Schizaeaceae. Dr Horvat (1923) finds that similar collenchymatous thickenings appear in *Adiantum*, and he has since described a like structure in *Cheilanthes* (1925) (compare Fig. 635, 1, 2). The comparative significance of this will be considered again later.

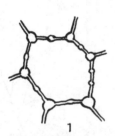 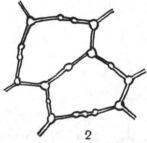

Fig. 635. 1=a cell from the flap or an old prothallus of *Anemia phyllitidis*: the walls show a peculiar bead-like thickening. 2=cells from the prothallus of *Adiantum cuneatum* L. and F. (After Horvat.)

Adiantum appears as a clearly defined and somewhat isolated genus. It is not difficult to see how it may have originated from a source related to other Gymnogrammoids. The effective protection of the sori while young, by the close folding of the distal end of the blade, would in itself help to explain the success of the Maidenhair Ferns, as measured by the large number of their species, and by their wide geographical spread.

IV. The Cheilanthoid Group

This group includes, as in Christensen's Index, the genera *Pellaea* Link, *Doryopteris* J. Smith, *Notholaena* R. Brown, and *Cheilanthes* Swartz. All of these Ferns fall into the "Pterideae-Cheilanthinae" of Diels (E. & P. I, 4, p. 255): but here we exclude *Hypolepis, Llavea, Cryptogramme*, and *Plagiogyria*, as also *Aspleniopsis* Mett. & Kuhn. The last is probably an ex-indusiate *Asplenium*, while the other four genera have been assigned their natural places elsewhere, for reasons already stated at length in Chapters XXXV and XXXVI, and in this Chapter. The difficulty that has been felt in drawing a clear line between the central group of the Gymnogrammoid Ferns and those associated with *Cheilanthes*, and more particularly between *Gymnogramme* and *Cheilanthes*, is in itself a sufficient index of their probable near affinity in descent. The Cheilanthoid group is characterised generally by having superficial sori *seated upon the distal region of the veins*, but more or less extended backwards from the margin, and sometimes appearing to coalesce when mature. Here again the rachis is often black and polished, and the rhizome bears dermal scales. A xerophytic character is prevalent throughout the group.

PELLAEA Link

This large genus may be held as the least specialised of the group, though not necessarily on that account the most primitive. It consists of Ferns with a creeping rhizome, bearing scales, and with leaves that are variously branched, with glossy dark rachis, and segments often of a harsh and rigid character, suggesting xerophytic conditions. The venation is pinnate, but sunk in the opaque mesophyll: it is usually free, but sometimes the veins anastomose, as in *P. Holstii* Hieron (Fig. 636, *F*). The sori are linear following the veins, but near to their endings. When mature they appear confluent, but in reality they are separate, not being united by a commissure. They are covered by a continuous "involucre," which is the reflexed and often membranous margin of the blade. The whole structure corresponds to that in *Adiantum*, except that here the veins and sori stop short of the overarching margin (Fig. 636, *B*).

The stelar structure in this, as in all of the related genera, is based upon the solenostele: some species of *Pellaea* (*P. atropurpurea* (L.) Link, and *falcata* (R. Br.) Fée) are even included in Gwynne-Vaughan's list of typical solenosteles. In other species, however, in which the leaves are placed nearer together, the leaf-gaps overlap, with the result that the ring is twice interrupted, a state explained by the reconstruction in Fig. 637. Each leaf-trace arises as a single vascular strap, with three protoxylems, and sometimes this primitive structure is retained throughout the length of the leaf (*P. andro-*

medifolia). In others, as in *Pteridella* Kuhn, sub-sec. A, the trace is diarch, a feature which goes with a reticulate venation, and an almost *Pteris*-like marginal sorus (*P. Holstii* Hieron, Fig. 636, *F*): these details indicate an advanced state.

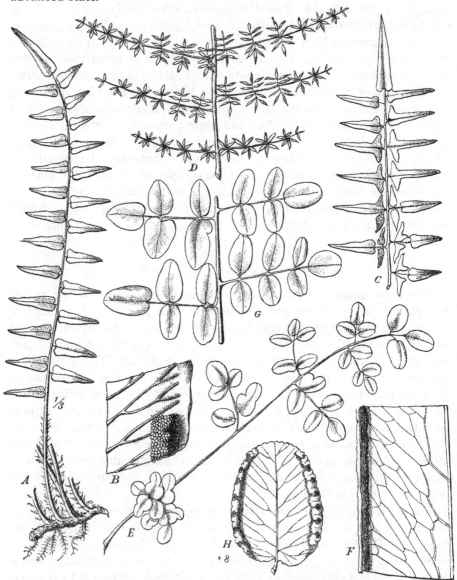

Fig. 636. *Pellaea* Link. *A*, *B* = *P. falcata* (R. Br.) Fée: *A* = habit, *B* = part of a pinna with venation and sori. *C* = *P. atropurpurea* (L.) Link. *D* = *P. ornithopus* Hook., part of a leaf. *E* = *P. nivea* (Lam.) Prantl, part of a leaf. *F* = *P. Holstii* Hieron, part of a pinna with venation and sori. *G* = *P. pteroides* (Thunb.) Prantl, part of a leaf. *H* = a pinnule of the same, with venation and sori. (*B* is after Hooker: all the rest after Diels.)

The sporangia are of superficial origin, arising near to the margin, and opposite to the vein-endings: they show no regular or characteristic sequence in their appearance. The margin itself is reflexed and indusioid (Fig. 636, *B*, *F*); in *P. intramarginalis* (Klfs.) J. Sm., a strong tissue-growth at the point of the greatest curvature forms a flange of tissue, comparable in

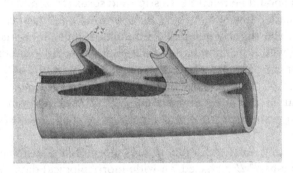

Fig. 637. Vascular system of *Pellaea rotundifolia* (Forst.) Hook., showing the departure of two leaf-traces (*L.T.*), and their relation to the solenostele of the axis. (After Gwynne-Vaughan.)

position to that so largely developed in *Blechnum*: it also finds its parallel in Kaulfuss' old genus *Cassebeera*, now merged in *Cheilanthes*, and apparently also in *C. lendigera* itself. These developments, which are variable in occurrence and extent, may be held as a natural consequence of the region of greatest curvature offering the least resistance to growth from within. The result is the feature which suggested the specific name (Fig. 638). The sporangia themselves are of an ordinary Leptosporangiate type. The annulus is almost vertical, but it does not extend down to the insertion of the stalk. The spore-counts for *P. falcata* (R. Br.) Fée show a typical

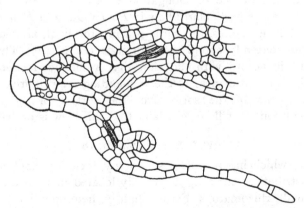

Fig. 638. Vertical section through the young sorus of *Pellaea intramarginalis* (Klfs.) J. Sm., showing the superficial origin of the sporangium, the indusioid margin and the flange. (× 100.)

number of 64: but in *P. intramarginalis* and *andromedifolia* it is 32, and in *P. hastata* it is only 24, and sometimes 16. It will be seen later that a wide variability in spore-counts is frequent in this group.

A new species of *Pellaea* from China has lately been shown by Dr C. Christensen to bear characters of special interest, since it raises a doubt to which genus it should be referred. In stipe and sorus it resembles *Adiantopsis*, the imparipinnate pinnae point to *P. andromedifolia* or *flavescens*, their shape to *P. flexuosa*. The slightly reflexed continuous or lobed margins imperfectly cover the sori, while the sporangia are large and few, or sometimes solitary. Since it appears to offer characters intermediate between *Adiantopsis*, *Pellaea*, *Cheilanthes* and *Notholaena*, Dr Christensen has named it *Pellaea connectens*. As will appear in the later comparisons, Dr Horvat finds in this plant a synthetic type, pointing downwards in the direction of *Mohria* (Christensen 649).

Prantl (*Engler's Jahrb.* III, 1882, p. 416) attempted a systematic arrangement of the genus *Pellaea* based on wide morphological data. He included underit Kaulfuss' genus *Cassebeera*, which is maintained by Christ as a substantive genus (*l.c.* p. 154). He also included *Doryopteris* J. Smith, though this is likewise maintained as a separate genus by Diels (*l.c.* p. 269), and by Christ also (*l.c.* p. 162). Taking thus a comprehensive view of the limits of *Pellaea*, Prantl disposed its constituents in a sequence which in the main was probably natural and phyletic: for it was based upon a comparison of facts of anatomy and soral development not always regarded by systematists. He placed first those with the sori free (*Platyloma*, *Eu-Pellaea*, *Cincinalis*), and proceeded to those with occasional reticulation of the blade, and with sori more or less fused, "anastomosantes" (*Pteridella*, *Cassebeera*, *Dorypteridastrum*, *Doryopteris*, *Pteridellastrum*). He does not attach great importance to the degree of elongation of the sorus from the margin inwards: but he notes particularly how in *Eu-Pellaea* and *Platyloma*, which he places first in his series, the sorus is truly terminal, *i.e.* that the vein-endings do not extend beyond the sporangia (*l.c.* p. 405). This grouping appears to be in accord with the view that a distinct receptacle and a position of the sorus at or near to the vein-ending are primitive, while soral fusion, or a spread of the sorus over the surface of the blade, is a derivative condition. It will be seen later that this view is probably correct.

DORYOPTERIS J. Smith, 1841

This genus, which has a profuse synonymy, includes about 40 species of very wide distribution, but they are chiefly located in the western tropics. They are small rhizomatous Ferns, slightly heterophyllous, with black polished petioles, and with coriaceous blades, often pedate. The sori are sometimes separate, and round: in other species a narrow linear fusion-sorus runs along the entire margin of the fertile blade, which is recurved for protection. The type of this genus as founded by J. Smith was *Pteris pedata* of Linnaeus (Hooker, *Exotic Ferns*, Pl. 34). In the *Synopsis Filicum* (p. 166) it was included with eight other species as a section of *Pteris*: but it was reinstated as a substantive genus by Prantl, on the basis of more searching anatomical and soral examination (*Engler's Jahrb.* III, p. 403). He draws a distinction between typical species of *Pteris* and the species now included in *Pteridella* and *Doryopteris*: while in *Pteris* the formation of sporangia begins on the commissure which underlies the coenosorus, linking the veins together, in the Gymnogrammoids the first sporangia appear upon the vein-endings, and spread from them, "so that a real fusion of the originally separate sori arises." He extends this statement to include also *Onychium* and *Cryptogramme*. Whether or not this view will hold in all cases, it was applied by Prantl to separate the soral anastomoses at the margins in the Gymnogrammoid Ferns from the familiar marginal coenosori of the Pteroid Ferns: two series which are now generally held as distinct though parallel.

The rhizome bears scales, and in *D. ludens* (Wall.) J. Sm. it is typically solenostelic, with an undivided leaf-trace (G.-V. MS.). The venation of the genus as at present defined includes species with open venation, as in *D. concolor* (Langsd. & Fisch.) Kuhn: others are reticulate, as in *D. ludens*: and these have the more definitely continuous sori. An example of the latter type is seen in *Doryopteris pedata* (L.) Fée (Fig. 639), which shows the coarse reticulation of the sterile blade (1), and a portion of the fertile blade (2):

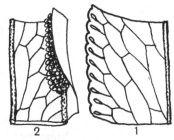

Fig. 639. *Doryopteris pedata* (L.) Fée, after Hooker. 1 = portion of the sterile blade, showing venation. 2 = portion of the fertile blade, with coenosorus. The receptacle is cleared in the lower part of the drawing: (enlarged).

here the venation stops short at the receptacle which is narrow and continuous, and the sporangia are covered over by the marginal flap. The apparent similarity to the *Litobrochia* section of *Pteris* is very marked, and this readily accounts for the earlier inclusion of *Doryopteris* in that genus.

SAFFORDIA Maxon and TRACHYPTERIS André

These two small genera here find their natural place. A small Fern collected by Darwin from the Galapagos Islands, but later found also in Ecuador and in S. Brazil, was first described as *Hemionitis pinnata* Hk. fil. Subsequently it was styled *Acrostichum (Chrysodium) aureo-nitens* Hk., but now it has been given generic rank under the name *Trachypteris* André, as *T. pinnata* (Hk. fil.) C. Chr. Christ (1899) regarded it as closely allied to *Elaphoglossum*. Such steps clearly point to a soral spread over the surface of the sporophyll, which is its marked feature. More recently this isolated type has been compared with *Saffordia*, a new genus described by Maxon from the mountains of Lima, Peru: and he has placed them both with *Doryopteris*.

The nearer of the two to *Doryopteris* is *Saffordia*, which is a small Fern with an ascending stock bearing numerous leaves, the whole being covered by closely appressed imbricate scales. The petiole is purplish brown, and the lamina in general outline deltoid, but pinnately-parted. Its upper surface is smooth and concave, with involute margins, and clearly marked midribs: the lower is densely covered by imbricate scales, which conceal the broad continuous marginal receptacles, with their numerous sporangia. The venation

Fig. 640. *Saffordia induta* Maxon. The lower surface of a basal pinna (pedate), cleared of its scales and sporangia, and showing the continuous marginal receptacle curved convexly upwards, bearing scars of insertion of the sporangial stalks, only the black main veins are visible, slightly enlarged: from a specimen supplied by the Smithsonian Institute.

which is sunk in the opaque mesophyll is areolate, without included veinlets: the areolae are mostly hexagonal, as in *Doryopteris* (Fig. 640). The sorus is of the "mixed" type, the sporangia are globular, with 16 to 18 indurated cells of the annulus, the thickening stopping short of the three-rowed stalk: there is a well-formed, four-celled stomium. The spores are spherical-tetrahedral, and the spore-output is 48.

Trachypteris is well figured in Hooker's *First Century of Ferns*, Plate XXXIII (1854), under the name of *Acrostichum aureo-nitens*. It differs from *Saffordia* in being strongly heterophyllous, the sterile leaf being entire and spathulate: but the fertile leaves are pinnate, with 5–9 pinnae showing a pedate branching. Their undersides are "uniformly clothed with capsules mixed with chaffy scales" (Fig. 641, 1, 2). This is a further difference from *Saffordia*, in which the coenosorus is a relatively narrow band. Here again the venation is reticulate, with hexagonal areolae. Maxon's reference of both of these Ferns to a relation with *Doryopteris* may be held as correct. The three genera appear to form a natural series of progressive Acrostichoid development, from a source related to the Cheilanthoid Ferns. There was first a coenosoral fusion (*Doryopteris*), and then a widening of the receptacle (*Saffordia*), till it occupies the whole lower surface (*Trachypteris*). The series forms a close parallel to that seen in the Pteroids, in *Litobrochia*, *Acrostichum praestantissimum*, and *A. aureum* (see p. 58).

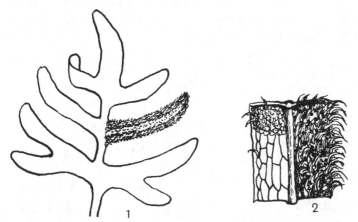

Fig. 641. *Trachypteris pinnata* (Hk. fil.) C. Chr. 1. Blade of fertile frond from Bolivia (Williams, 1177), intermediate between Brazil and Galapagos forms, natural size. 2. Part of a fertile pinna of *T. pinnata*, after Hooker, showing the coenosorus extending to the midrib, with the scales and sporangia in part removed to show the reticulation.

CHEILANTHES Swartz

The relation of *Cheilanthes* to *Pellaea* is a very close one. Christ defines *Cheilanthes*, which comprises more than 100 species, as consisting of small usually xerophytic plants, bearing leaves divided into minute segments, with free veins. The small rounded sori are borne on the swollen tips of the veins, and usually give the appearance of coalescence at maturity. The protection is afforded by the more or less membranous margin, or by a projecting tooth of the segment. It often appears irregular in form, and may be continuous from one sorus to another, but not coherent throughout (Fig. 642). The difference between *Pellaea* and *Cheilanthes* is one of detailed habit rather than of essential construction of the sorus, and this is borne out by the similarity of the anatomical detail. These Ferns are xerophytes of high-lying plateaux, particularly of Mexico and South Africa, and they are almost wholly absent from the moist equatorial zone (Christ, *Farnkräuter*, p. 142).

The stelar structure of the scaly stock shows a range from the simple solenostele of *C. Fendleri* Hk., through *C. gracillima* Eat. with the leaf-gaps overlapping, to *C. lanuginosa* Dav., which rarely shows a complete ring in transverse section: while in *C. persica* (Bory) Mett. there is a radial dictyostele much more highly segregated than in these (Marsh, 645). Parallel with this go changes in the bulk of the xylem, and in the increasing proportion of parenchyma that it contains. The petiolar meristele varies also from the primitive type with three endarch protoxylems (*C. Fendleri*, at base only) to an exarch median protoxylem, which is finally eliminated as in *C. gracillima* and *lanuginosa*. Thus anatomically these species suggest a progression in which *C. Fendleri* is the most primitive: it leads from solenostely to confirmed dictyostely (Marsh, Fig. 11, p. 680).

Fig. 642. *Cheilanthes* Sw. *A*=leaf of *Ch. fragrans* (L.) Webb & Berth. *B*=secondary pinnule or the same with venation and sori. *C*=*Ch. farinosa* Kaulf., habit. *D*=primary pinna of *Ch. Regnelliana* Mett. *E*=part of a pinnule of the same, with sorus and protective margin. *F*=*Ch. myriophylla* Desv., part of the blade. (*A, B* after Luerssen: *D E* after Baker, *Fl. Brasil*: *C, F* after Diels.)

The sori contain each a relatively small number of sporangia, which are
long-stalked at maturity, but with other features similar to those of *Pellaea*
or *Notholaena*. There is again some diversity in spore-numbers: the highest
counts are in *C. vestita* Sw. and *Fendleri* Hk., which point to a typical
number of 48 to 64. But in a number of other species, including the
anatomically advanced *C. gracillima* and *uliginosa*, the counts indicate
numbers ranging from 24 to 32 (Marsh). The parallelism of progression
between anatomy and spore-number is worthy of note. There is no perispore.

The question of actual origin of the sori in the Cheilanthoid Ferns has
been left aside by most systematic writers: and yet such facts appear
essential to the determination of their affinity. The main questions are
whether the sporangia are superficial or marginal in origin, and whether

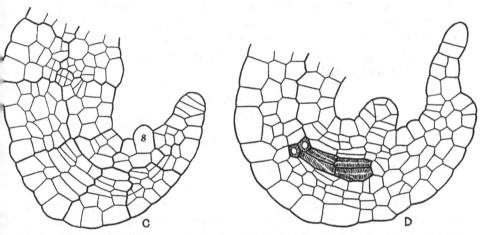

Fig.643. Vertical sections of the young fertile pinnule of *Cheilanthes microphylla* Sw., showing that the
indusioid flap is of marginal origin, and the sporangia (*s*) definitely superficial. (× 200.) *C*=a younger,
D=a more advanced state.

there is any sign of an inner or lower indusium. The facts have been
ascertained for *C. microphylla* Sw., a species from the section *Eu-Cheilanthes*,
with finely divided leaves, roundish sori, and a narrow pale "involucre."
Since here the adult sori are seated close to the margin, their development
should give evidence of value as to the relation of the sorus to the marginal
segmentation. Sections of the very young pinnule show clearly that the
segmentation leads continuously to the marginal flap: thus the "involucre"
or "indusium" represents the actual margin itself. It becomes strongly
reflexed as it develops, corresponding in this to the reflexed flap of *Adiantum*
or *Pellaea*. The first sporangia arise as superficial outgrowths of single cells
at points well within the margin, which at once overarches them (Fig. 643).
In this species there is no definite sequence in the production of the few
sporangia: but Horvat (1925) finds in *C. pteridioides* (Reich.) C. Chr.

(= *C. odora* Webb & Berth.) an indication of a gradate state. This is probably exceptional for the genus and is not continued, as the flat receptacle would imply. The first segmentation of the sporangium follows the Schizaeoid type (Vol. II, Fig. 455), and the stalk is at first massive and short.

The prothallus is of the cordate type: its vegetative cells show those collenchymatous swellings of the walls seen in *Adiantum*, and recurring in the Schizaeaceae. The antheridia have an undivided cap-cell. At first the young sporophyte bears hairs, but later scales are also produced (Horvat, 1925).

Adiantopsis Fée, a genus of shady habit in Brazilian woods, was included in *Cheilanthes* in the *Synopsis Filicum* (p. 131), and that is probably its proper place. It is characterised by peculiarly radiating pinnae, readily referable in origin through comparison of the juvenile leaves to a catadromic helicoid branching, as in *Matonia*: this appears also in *Adiantum*, *Saffordia*, and *Trachypteris* (Von Goebel, *Organographie*, II, p. 1046, Fig. 1131). Its round sori with few sporangia are, as in *Cheilanthes*, terminal on the veins, each protected by a marginal lappet.

NOTHOLAENA R. Br.

This is again a genus of xerophytic Ferns of dry, warm lands, particularly of Western America and South Africa. The tufted leaves are usually once-pinnate, with free vein-endings, which bear the more or less elongated sori. These consist as a rule of few sporangia, sometimes of a single one, and they are not unfrequently unprotected, or in varying degree covered by the slightly altered leaf-margin. But efficient protection may also be given by hairs and scales which abound on the lower surface of the leaf. The close relation of the genus to *Cheilanthes* is shown by the difficulty in diagnosis of the two genera, and by the frequent synonymy.

As follows naturally from the closer grouping of the leaves, the stelar structure, though based upon the solenostele, shows the leaf-gaps overlapping, so that the ring is interrupted more than once in the transverse section, as in *Pellaea* and others (compare Fig. 637). The petiolar trace is as in *Pellaea* and *Cheilanthes*: thus the anatomy supports the affinity of these genera.

The sporangia are essentially of the same structure as theirs, with short three-rowed stalks, and an approximately vertical annulus that stops short of the stalk-insertion. There is, however, a high degree of variety in the size and number of the spores. The counts vary from 12 to 64. The largest numbers are made up of the smallest spores, and the smallest of larger spores. They are all of tetrahedral type. Whether or not the difference in size of these spores is related to heterospory is uncertain in the absence of culture experiments: but at least the small numbers seen in *N. distans* R. Br.

(12), and in *N. sinuata* (Lag.) Klf. (16–32), suggest that these species are advanced in this feature, while the numbers in *N. affinis* (Mett.) Moore (16–64) present a specially high degree of variability (see Thompson, *Trans. Roy. Soc. Edin.* Vol. LII, Pt. II, p. 386).

Since in certain species of *Notholaena* the relatively large and even solitary sporangia appear very close to the leaf-margin when mature, a particular interest attaches to the question of the exact point of their origin relatively to the marginal cells. The excellent figure of Schkuhr (*Farnkräuter* Tab. 99) shows how in *N. trichomanoides* the mature sporangia appear as though marginal, and the sori apparently confluent. It would therefore seem to be a suitable species for testing the question. Sections of the young

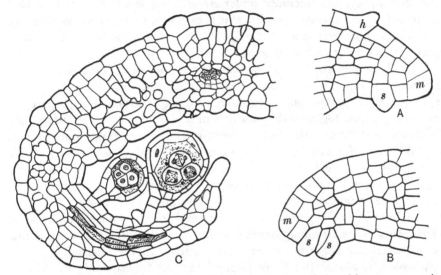

Fig. 644. *Notholaena trichomanoides* (L.) R. Br., showing the close relation of the young sporangia to the leaf-margin. *s* = sporangia: *h* = hair: *m* = marginal cell. (*A, B* × 200. *C* × 100.)

pinnae show clearly that the sporangia are intra-marginal: they arise from segments back from the actual marginal series, though in near proximity to it. Moreover the first cleavage of the sporangial cell is by a wall that impinges upon the basal wall (Vol. II, Fig. 455). These facts suggest that the sporangium itself is of a relatively primitive type, with its first segmentation after the manner of the Schizaeaceae: but while in the latter the sporangia actually originate from the marginal cells, here they are superficial, though springing from some of the latest segments cut off from the marginal series (Fig. 644).

The prothallus bears glandular hairs of exactly the same nature as those so profusely present on the young leaves of the sporophyte (Vol. I, Fig. 186).

COMPARISON

The designation "Gymnogrammoid Ferns" covers a large number of genera and species, if it be taken as connoting those Ferns which show the general features of the old genus *Gymnogramme*: *i.e.* superficial sori, more or less extended along the veins, but without any specialised indusium. We may now add to this from the more modern point of view the *proviso*, that there shall not be any comparative ground for believing that an indusium existed in their ancestry. This definition would cover the bulk but not all of those genera which were grouped by Diels under his designation of Gymno-gramminae, Cheilanthinae, and Adiantinae, together with *Cassebeera*: while some Ferns previously included under *Acrostichum* would have to be added on grounds of comparative study. But it excludes such genera as *Ceterach*, *Pleurosorus*, *Hypolepis*, and *Aspleniopsis*; while *Plagiogyria* would be held apart as a specially archaic type.

With characters so diverse as some of the included Ferns show, and a common character so comprehensive, it would be against reason to assume at the outset any single phyletic source. We ought rather to be prepared to find here grouped together the products of a plurality of evolutionary lines. By comparison among themselves, and by reference to types more archaic than they, or of earlier geological occurrence, it may be found possible to segregate some at least of them, and to assign them to their probable phyletic sources. The old conception of *Gymnogramme* as a genus stood next in indefiniteness to those of *Polypodium* and *Acrostichum*. Like them it probably did not represent a single natural group. One general feature which holds in the Gymnogrammoid Ferns more than in most others is that its characters are negative rather than positive. In the Gleichenioids and Cyatheoids, as also in the Hymenophylloids and Dicksonioids, the sorus is a clearly circumscribed entity, with a definite receptacle round which the sporangia are grouped, and in many of them more or less constant protective indusial growths are present. It is true that the identity of the sorus may be blurred or even obliterated by soral fusion, with or without subsequent fissions: or by the adoption of an Acrostichoid spread in those types which are held as the most advanced. But in the Gymnogrammoid Ferns the sorus is not a definite entity in the same sense as in other Ferns: it is without the circumscribing influence of an indusium, and the sporangia appear from the first to have been liable to spread indefinitely along the veins, or even into the areas between them. This indefiniteness, while it makes the whole group more comprehensive, removes from the sorus those distinctive features which have aided its comparison elsewhere. The effect of this is to accentuate the importance of the sporangia themselves, and to concentrate comparison upon the details of their position, structure, and spore-output.

The facts relating to the sporangia together with a rather primitive vascular construction have led to the segregation of *Llavea*, *Cryptogramme Jamesonia*, and *Ceratopteris* as Primitive Gymnogrammoid Ferns. Of these *Llavea* and *Cryptogramme* are confirmed in the position near to *Plagiogyria* assigned to them by Diels (*l.c.* p. 279). With them goes also *Onychium* Kaulf., which is ranked as a section of *Cryptogramme*, differing chiefly in the formation of coenosori. The occurrence of fossils described as *Onychiopsis* from so early an horizon as the Wealden appears to support this reference. For reasons stated at length above (pp. 72–74), it appears probable that these genera represent derivative lines which originated in relation to such antique types as the Osmundaceae, and particularly *Todea*: nevertheless some Schizaeoid relation for them is not to be excluded.

It is more difficult to place *Jamesonia* with any confidence, for its characters are so mixed. In its hairs, and the absence of scales, as well as in its high spore-counts it possesses primitive features: in the Acrostichoid spread of the sporangia it shows advance, while the instability of sporangial structure indicates transition. Its anatomy is of a primitive type. It may be held to represent an isolated genus, perhaps the sole representative of an independent line in the plexus of the Gymnogrammoid Ferns.

The same appears also for *Ceratopteris*. The vascular anatomy is of a more advanced type than the rest, with its disintegrated leaf-trace, and in large stems a medullary system within the dictyostele: these details together with the presence of dermal scales place it structurally in advance of *Jamesonia*. But the non-soral sporophylls, with their superficial thick-stalked sporangia, and the wide latitude of the annular variations point to a primitive and transitional state: nevertheless, the low but very variable spore-output appears to suggest advance. Again the phyletic decision must be left in the air. *Ceratopteris* may owe some of its peculiarities to its aquatic habit: apart from this, which may account for its apparent isolation, it may also be best regarded as the single representative of a transitional line from relatively primitive Simplices.

The salient interest of all these relatively primitive genera reflects downwards rather than upwards. It would be a mistake, however, to say that none of them suggest linkage with higher and more definitely Leptosporangiate types: but the most distinctive features which they have in common lie in the sporangia: these are all relatively massive and short-stalked. The numerical output of spores is low and variable, but the spores themselves are large and constantly tetrahedral, while the segmentation of their sporangia is that associated with transition from the more massive types of the Simplices, and more particularly with those of the Schizaeoid and Osmundaceous Ferns.

The central group of the Gymnogrammoid Ferns presents less interest than the rest, for in them the sporangia are standardised after the usual Leptosporangiate model, and are disposed for the most part in linear sori along the veins, but stopping short of their actual endings. These Ferns possess dermal scales, and a vascular system derived from solenostely, but with an occasional tendency to perforation, with a binary leaf-trace (Vol. I, Fig. 160). There is also reticulate venation in some of them, with other signs of advance including a tendency to Acrostichoid development. It would be easy to see in them the results of specialised advance from such types as have been above held as primitive: but the phyletic lines have not been exactly traced, so as to give such a suggestion any degree of certainty.

The Maidenhair Ferns (*Adiantum*) may with greater confidence be referred to some central Gymnogrammoid or Cheilanthoid source : but here the linear sori are borne upon the distal region of the veins, and the marginal lappet that bears them has become sharply reflexed. This arrangement was clearly successful, as shown by the existence of nearly 200 species. But the genus offers little sign of advancing specialisation beyond occasional vein-fusions (*Hewardia*), and a tendency to an Acrostichoid state (*Adiantella*). There is, however, a structural detail of the sporophyte that acquires importance in relation to the Vittarieae; that is, the existence of "spicular cells" in the epidermis. In the gametophyte also the collenchymatous thickening of the walls of the prothallial cells presents a feature that has a special comparative importance in relation to the Schizaeaceae.

Of all the Gymnogrammoid Ferns it is the Cheilanthoid group that presents the highest phyletic interest. Robert Brown, in founding the genus *Notholaena*, speaks of "sori marginales" (*Prod. Fl. Nov. Holl.* p. 145, 1810): the very name *Cheilanthes* introduced by Swartz (1806) conveys the same for that genus. Thus the earliest writers on these genera held their sori as actually marginal. On the other hand, Hooker (*Syn. Fil.* p. 436) speaks of *Mohria caffrorum* as combining the capsules of the sub-order (Schizaeaceae) with the habit of *Cheilanthes*. Thus early the habit-similarity was recognised between the Schizaeaceae and the Cheilanthoid Ferns, while a marginal sorus was attributed to the latter. These historical notes form a natural preface to their comparison in the light of details later acquired.

First comes the question of fact, as to the relation of the sporangia to the margin. Prantl (*Schizaeaceen*, 1881) demonstrated most fully that the sporangia of *Mohria* arise from marginal cells (Fig. 645), but that they were early displaced to the lower surface by the growth of an indusial false margin: the later result of this is seen in Vol. II, Fig. 450. But observations on *Notholaena*, in which of all Cheilanthoid Ferns the sporangia are most nearly marginal, have shown that in point of fact they arise superficially, though the sporangium nearest to the marginal cell may be only one segment removed from it (Fig. 644, *A, B*). On the other hand, in *Mohria* the indusium is a new formation initiated superficially (*i*, Fig. 645), while in *Notholaena* the indusioid margin continues as the direct outgrowth of the marginal cells. The same argument may be applied here as in the sori of the Pteroid Ferns (pp. 54, 55), and the conclusion drawn that in both cases there has been a phyletic slide of sporangial production from the original position on the margin of the blade to its surface. The ontogenetic comparisons accord with this, and the physiological probability of the better protection of the young sporangia supports it. In other Gymnogrammoid Ferns than *Notholaena* the sporangia arise farther from the margin, and they are liable to be spread inwards along the veins, giving the usual character of the

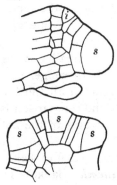

Fig. 645. *Mohria caffrorum* (L.) Desv., drawings after Prantl, showing the relation of the sporangia (*s*) to the marginal series, and to the indusium (*i*), as seen in surface view, and in section. (× 145.)

group. All steps in this invasion of the lower surface can be illustrated by comparison of the species among themselves. In all such questions it is to be remembered that the sporangium is the most constant and important part; the indusia or leaf-margins, and even the flattened blade itself, are ancillary when a broad view of ultimate origin is taken.

Against any close alliance between the living Schizaeaceae and the living Cheilanthoid Ferns there is the fact of the sporangia of the former being themselves relatively archaic, while those of the latter are definitely of the Leptosporangiate type. This fact raises again the general question of phyletic change of the sporangia themselves. The broad lines of comparison now before us for Ferns at large make such a change appear more than ever probable: indeed the conclusion that a reduction in size and spore-output has happened repeatedly seems to be inevitable. Here the extraordinary variability in the individual sporangium seen in certain Gymnogrammoid Ferns has its intimate bearing on the general problem (*Llavea, Ceratopteris, Jamesonia, Cryptogramme*). In all of those genera the sporangia appear variable as well as relatively archaic: but they all bear their sporangia upon the leaf-surface, far removed from the margin. What may be concluded from that is, that in these Gymnogrammoid Ferns changes in position and in structure of the sporangium have not progressed along strictly parallel lines, though still they have affected them all.

Anatomically the Schizaeaceae are as varied as any family of the whole Class of Ferns (Vol. II, p. 157). The most advanced in this are *Anemia* and *Mohria*, those genera with which the Gymnogrammoids may best be compared as regards their sporangia. On both sides of the comparison the structure fluctuates between solenostely and dictyostely, while the leaf-trace is undivided and saddle-shaped in transverse section. Even the median attenuation of the meristele of *Anemia* finds its counterpart in *Cheilanthes* and *Pellaea*. Some of the larger Gymnogrammoids have more highly disintegrated vascular tracts, as in *Ceratopteris* or *Gymnogramme japonica*: but those smaller types which compare sorally with *Anemia* and *Mohria* approach these genera also anatomically. In *Mohria* dermal scales are present, as in all the Gymnogrammoids excepting *Jamesonia*. Thus structurally the comparison is upheld.

The details of the collenchymatous thickenings of the cell-walls in the prothalli of *Adiantum* and of *Cheilanthes* and *Notholaena* (Horvat) have been shown to find their parallel in those of *Lygodium* (Heim), and of *Mohria* and *Anemia* (Bauke). This in itself may appear a slender ground for wider comparisons: but when taken in conjunction with the similarities of these Ferns in other respects, such facts acquire increased importance as indications of affinity. It should be frankly realised that the argument for a phyletic relation between the living Schizaeaceae and the Cheilanthoid Ferns falls short of demonstration. But the variety and cogency of the comparisons may be held as indicating a reasonable probability of its truth.

A general review of the Gymnogrammoid Ferns suggests, and at times may indicate with some degree of clearness, their evolutionary place. They are a rather varied group, but inherent truth may often be learned from studying the general facies of a large body of organisms even if somewhat loosely akin, as these Ferns appear to be. Their kinship is based partly on positive characters of external form and internal anatomy: partly upon the rather negative soral characters, and particularly upon the absence of a true indusium, with its corollary, an indefinite soral construction. They may be held to represent a plexus of phyletic lines, all traceable back with probability to Ferns with marginal sporangia of larger size than theirs, such as the present-day Osmundaceae and Schizaeaceae. A widespread transfer of those sporangia from the margin to the surface of the blade appears to have happened early: this is exemplified in *Todea* and *Plagiogyria* on the one hand, and on the other it is suggested by *Anemia* and *Mohria*. It is reflected as an accomplished fact in *Notholaena*. In none of these is there any specialised indusial protection, other than a recurved leaf-margin. We may picture how, after that change from the margin to the surface had been effected, the production of more numerous sporangia, diminishing in size as their number increased, has spread first along the lines of venation from the margin inwards: subsequently in some instances the areas between the veins may also have produced sporangia, giving an Acrostichoid state, as seen slightly in *Adiantum*, and more clearly in *Jamesonia, Trismeria*, and others. Occasionally a lateral linkage of sori resulted in intra-marginal coenosori, as in *Onychium* and *Doryopteris*: while in *Saffordia* and *Trachypteris* a spread of the fertile tract of the coenosorus inwards from the margin led again to an Acrostichoid state, as in *Acrostichum praestantissimum* and *aureum*. But these several changes, following sometimes on linkage, appear to have arisen in the Gymnogrammoid series quite independently of the similar linkage and Acrostichoid development in the Pteroid Ferns. The two series may be held to have progressed along parallel lines, though with many features in common. This has led to those systematic and terminological difficulties which have always beset the Pteroid and Gymnogrammoid phyla. These can best be resolved by approaching their classification from the evolutionary point of view. In the present case we shall trace the Pteroid Ferns from a bi-indusiate Dicksonioid origin: but the Gymnogrammoid Ferns from an Osmundioid-Schizaeoid source, without any specialised indusium.

No attempt will here be made to group the Gymnogrammoid Ferns according to any exact phyletic relations of the genera one to another, or to ancestral forms. This must be left over till further enquiry shall have supplied the necessarily detailed knowledge. Consequently the table of genera given below is more in the nature of a suggestive catalogue than of a phyletic scheme. But speaking quite generally those earlier on the list link with *Plagiogyria* and the Osmundaceous stock: those named later show rather a Schizaeoid affinity.

THE GYMNOGRAMMOID FERNS

I. Relatively Primitive Genera.
(i) *Llavea* Lagasca, 1816 ... ... 1 species.
(ii) *Cryptogramme* R. Brown, 1823 ... ... 4 species.
(iii) *Onychium* Kaulfuss, 1820 ... ... 6 species.
(iv) *Jamesonia* Hook. and Grev., 1830 ... ... 14 species.
(v) *Ceratopteris* Brongniart, 1821 ... ... 1 species.

II. Central Group of Genera.
(vi) *Pterozonium* Fée, 1850 ... ... 2 species.
[(vii) *Syngramme* J. Smith, 1845 (see Chapter XLVI) 16 species.]
(viii) *Anogramme* Link, 1841 ... ... 9 species.
(ix) *Gymnogramme* Desvaux, 1811 ... ... 50 species.
(x) *Coniogramme* Fée, 1850 ... ... 2 species.
(xi) *Hemionitis* Linné, 1753 ... ... 8 species.
(xii) *Gymnopteris* Bernhardi, 1799 ... ... 12 species.
(xiii) *Ceropteris* Link, 1841 ... ... 7 species.
(xiv) *Trismeria* Fée, 1850 ... ... 2 species.

III. Adiantoid Ferns.
(xv) *Adiantum* Linné, 1753 ... ... 184 species.
IV. Cheilanthoid Ferns.
(xvi) *Pellaea* Link, 1841 ... ... 68 species.
(xvii) *Doryopteris* J. Smith, 1841 ... ... 39 species.
(xviii) *Saffordia* Maxon, 1913 ... ... 1 species.
(xix) *Trachypteris* André, 1899 ... ... 1 species.
(xx) *Cheilanthes* Swartz, 1810 ... ... 101 species.
(xxi) *Notholaena* R. Brown, 1810 ... ... 47 species.

BIBLIOGRAPHY FOR CHAPTER XXXIX

625. SCHKUHR. Farnkräuter, Wittenburg, Tab. 99. 1809.
626. R. BROWN. Prod. Fl., Nov. Holl. 1810.
627. HOOKER. First Century of Ferns, Pl. XXXIII. 1854.
628. METTENIUS. Farngattungen, 1–6. Frankfurt. 1857–1859.
629. HOOKER. Species Filicum, Vol. II, 1858; Vol. V, 1864.
630. HOOKER. Exotic Ferns, Pl. XXXIV. 1859.
631. KNY. Parkeriaceen. Dresden. 1875.
632. BAUKE. Keimungsgeschichte der Schizaeaceen, Pringsh. Jahrb. Bd. IX. 1878.
633. PRANTL. Die Schizaeaceen. Leipzig. 1881.
634. PRANTL. Die Farngattungen Cryptogramme u. Pellaea, Engler's Jahrb. p. 403. 1882.
635. PRANTL. Arb. K. Bot. Garten zu Breslau, I, i, 1892.
636. HEIM. Flora, Bd. 82, p. 368. 1896. Unters. u. Farnprothallien.
637. DUNZIGER. In. Diss. Münich. 1901. *Hemionitis, Gymnogramme* and *Jamesonia.*
638. FORD. *Ceratopteris*, Ann. of Bot. XVI, p. 95. 1902.
639. DIELS. Nat. Pflanzenfam. I, 4, p. 254, etc. 1902.
640. GWYNNE-VAUGHAN. Solenostelic Ferns, II, Ann. of Bot. XVII, p. 689. 1903.
641. GWYNNE-VAUGHAN. MS. Notes, preserved in the Botanical Dept. Univ. Glasgow.
642. BENEDICT. Genus *Ceratopteris*, Contr. N. Y. Bot. Gard. 126. 1909.
643. HANNIG. Flora, Bd. 103, p. 338. 1911.
644. MAXON. *Saffordia*, Smithsonian Collections, LXI, 4. Washington. 1913.
645. MARSH. Anat. of sp. of *Cheilanthes* and *Pellaea*, Ann. of Bot. XXVIII, p. 671. 1914.
646. Frau EVA SCHUMANN. Die Achrosticheen, Flora, 108. 1915.
647. THOMPSON, J. McL. Trans. Roy. Soc. Edin. LII, p. 363. 1918.
648. BOWER. Studies VII, Ann. of Bot. XXXII, p. 1. 1918.
649. C. CHRISTENSEN. Plantae Sinenses, Medd. Från. Göt. Trädgård, p. 84. 1924.
650. HORVAT. Glasnik d. Kroat. Naturwiss. Ges. Zagreb. XXXIII, p. 137. 1921.
651. HORVAT. Beitr. z. Kenntn. d. Marg. Filicineen, Oester. Bot. Zeitschr. p. 335. 1923.
652. HORVAT. Development and affinities of *Cheilanthes*, Act. Bot. Inst. Univ. Zagrebensis, p. 15. 1925.
653. HORVAT. Ursprung der Cheilanthineen, Act. Bot. Inst. Univ. Zagrebensis, II, p. 85. 1927.

CHAPTER XL

DRYOPTEROID FERNS (I. Woodsieae)

IT has already been suggested that the Cyatheoid type of Ferns, with its constantly superficial sori, has given rise to two phyletic sequences, viz. the Dryopteroids which centre round the genus *Dryopteris*, and the Blechnoids which centre round the genus *Blechnum*: and that they differ from one another in the fact that in the former the superficial sori characteristically maintain their individuality, while in the latter they fuse to form longitudinal coenosori (Vol. II, p. 333). It will be convenient to take the former first, and to see how the individual sorus may undergo modifications that result in that characteristic of the large body of Leptosporangiate Ferns which are associated with *Dryopteris*. It will be seen how gentle are the steps of the comparison which leads from the Cyatheoid to the Dryopteroid Ferns themselves, and how on a basis of further comparison these may be held to have advanced to still other derivatives.

The Ferns in question comprise for the most part, though not wholly, those grouped by Diels under the Woodsieae, and the Aspidieae (*Natürl. Pflanzenfam.* I, 4, p. 159). The Onocleinae were included by Diels under the Woodsieae, but for the moment we may hold them over and apart from the present comparisons. The Woodsieae thus limited (excl. Onocleinae) form a natural link with the Cyatheaceae (excl. Dicksonieae). As accepted by Diels they include *Diacalpe* Blume (1 sp.), *Peranema* Don (1 sp.), *Woodsia* R. Brown (25 sp.), *Hypoderris* R. Brown (3 sp.), *Cystopteris* Bernhardi (13 sp.), and *Acrophorus* Presl (1 sp.). The existence of three monotypic genera among these Ferns itself indicates a probability that they are transitory survivals. This applies especially to *Diacalpe* and *Peranema*. The genera thus named will now be examined from a generally morphological point of view but *Cystopteris* and *Acrophorus* will be held over for later discussion (see Chapter XLVIII, on *Genera Incertae Sedis*).

Woodsia R. Brown

The genus *Woodsia* was established by Robert Brown in 1810. It includes 25 species as enumerated in Christensen's Index, and comprises small mountain and rock-dwelling types. The rhizome is upright with nested leaves of relatively simple form, and with open venation. The round sorus has a circular basal indusium, laciniate in varying degree at its margin, and the sequence of the sporangia is basipetal. The habit and the characters of the sorus are in fact those of the smaller Cyatheaceae, of which the genus *Woodsia* may be regarded as including arctic and mountain representatives.

Fortunately the prothalli and sexual organs are particularly well known, and the details confirm this suggestion. The prothalli were found by Schlumberger to be highly variable under special conditions of culture (*Flora*, 1911, p. 384): particularly under weak illumination they may take a filamentous form (Vol. I, Fig. 266). Accordingly he fixed upon features other than mere form that appeared of greater value for comparison, viz. the occurrence of certain hairs, and the structure of the antheridium. Bristle-shaped pluricellular hairs on the surface and margin of the prothallus were held by Heim to be characteristic for the Cyatheaceae (*Flora*, 1896, p. 355, etc.). Hairs of like form, with or without a glandular terminal cell, were found in *W. obtusata*, and also in *Diacalpe aspidioides*, thus providing a common feature. More cogent evidence, however, may be drawn from the structure of the antheridia, and particularly from the behaviour of the lid-cells (*l.c.* p. 387). It has been seen in Vol. I (p. 293, Fig. 283) how, with the diminishing number of the spermatocytes seen in passing from the Eusporangiate to the Leptosporangiate Ferns, there is a simplification of structure of the antheridial wall. In particular this appears in the divisions of the cap-cell. In the Cyatheaceae this is as a rule divided into two or more cells, while in the "Polypodiaceae" it is undivided. Now in *Diacalpe* it is found that the division is as in the Cyatheaceae, viz. usually into two, but occasionally into three cells. But *Woodsia obtusa* shows only two cells, and occasionally the cap-cell remains undivided. In *W. ilvensis* this undivided state is the rule. Schlumberger states that *Diacalpe* and *Woodsia* differ from all the "Polypodiaceae" in having a divided cap-cell: and Von Goebel adopts this so far as investigation has extended (*Organographie*, II, p. 925). The rarity of such exceptions together with their comparative bearings confer a special interest upon these facts relating to the antheridia of *Diacalpe* and *Woodsia*.

The interest extends equally to the sporophyte, and in particular to the sorus. The axis in *Woodsia* is short, and ascending or upright. Both stem and leaves bear superficial scales, which are specially prominent in *W. polystichoides*. The character of the relatively simple leaves of the small mountain species, together with the sori which they bear, is seen in Fig. 646, *A–E*, but in certain larger American species such as *W. mollis* (Klf.) J. Sm., *obtusa* (Spreng.) Torr, *scopulina* Eat., and *mexicana* Fée the leaves may be doubly pinnate, recalling the more elaborate Cyatheaceae in habit. Following on a protostelic and solenostelic stage in the young plant, the vascular system of *Woodsia* widens out upwards into a simple dictyostele. From near the base of each foliar gap arise two separate leaf-trace strands, which enter the petiole, and beneath their point of origin passes out the supply to one or

more roots. There are no accessory strands. Such characters accord with the
suggested relation to the Cyatheaceae (see Schlumberger, *l.c.* Fig. 14, 1–3).
But more distinctive features are yielded by the sorus, the mature characters
of which are seen in Fig. 646, *B*, *C*, *E*. Each sorus of *Woodsia* is seated on

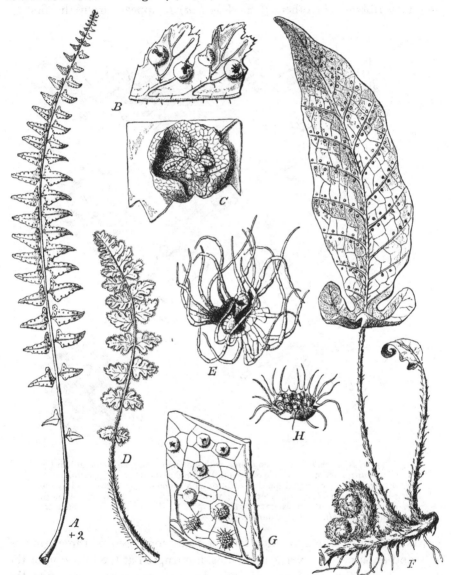

Fig. 646. *A—E = Woodsia* R. Br. *A = W. polystichoides* Eat., leaf : *B = W. elongata* Hook., part of
fertile leaf with sori : *C = W. obtusa* Torr, sorus : *D, E = W. ilvensis* (L.) R.Br.: *D* = leaf : *E* = sorus
with indusium, all sporangia but two removed. *F—H = Hypoderris Brownii* J. Sm.: *F* = habit :
G = part of a leaf, with venation and sori : *H* = sorus and indusium seen from the side. (*B, G, H*
after Hooker : *C* after Bauer : *D, E* after Luerssen : *A, F* after Diels : from *Natürl. Pflanzenfam.*)

a vein near to its ending, and consists of a circular receptacle bearing sporangia distally, but not many of them. Its base is surrounded by an indusium in the form of a more or less complete ring. In some species (§ *Physematium*) the indusium appears as a complete covering, separating later into ribbons: in others (§ *Eu-Woodsia*) it appears from the first as

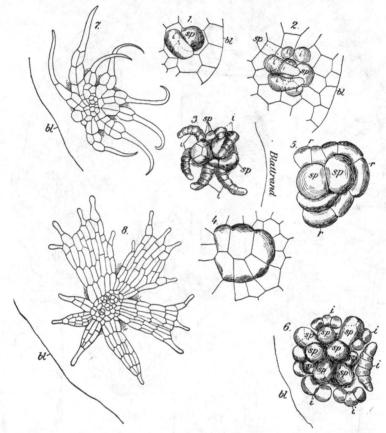

Fig. 647. 1, *W. ilvensis*, quite young sorus, with sporangium (*sp*), and indusial hair (*i*). 2, *W. ilvensis*, later stage, with sporangia (*sp*) and indusial hairs (*i, i*). 3, ditto, older. 4, *W. obtusa*, very young receptacle. 5, ditto, with sporangia (*sp*) surrounded by the ring-like indusium. 6, ditto, older stage, with sporangia (*sp*) and indusium (*i, i*). 7, *W. ilvensis*, mature indusium: *bl*=leaf-margin. 8, *W. obtusa*, mature indusium and leaf-margin (*bl*). (After Schlumberger.)

narrow segments. The sporangia have a stalk composed of three rows of cells, and an apparently vertical annulus, interrupted at the insertion of the stalk, and with the stomium horizontal. In fact they are essentially of the same type as in *Dryopteris filix-mas*.

A special interest lies in the development of the sorus, which has been described by Schlumberger (*l.c.* p. 407; Fig. 647). He tells how in *W. obtusa* (5, 6, 8) a group of epidermal cells surrounding the nascent receptacle arises as a single-layered ring, but open on the side next the leaf-margin: later some "indusial hairs" fill the gap, and these and the ring form lobes fringed with glandular hairs (8). Thus the indusium is here lobed from the first. In *W. ilvensis* the segregation ("auflösung," from his point of view) into single hairs is more advanced (7): the sorus is initiated by the upgrowth of one or two epidermal cells which form sporangia (*sp*): later small upgrowths from the leaf-surface (*i, i*), on the side away from the leaf-margin, give rise to the indusial hairs: others are later formed on the marginal side, and thus the sporangia are surrounded by a ring: but it is open at first on the marginal

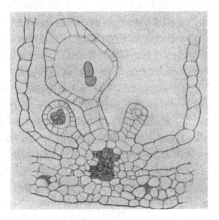

Fig. 648. Vertical section through a sorus of *Woodsia obtusa*, showing the vascular supply into the receptacle, the basal indusium, and indications of a basipetal succession of the sporangia. (× 100.)

side (1, 2, 7). The receptacle in *Woodsia* remains flat, and there is only a slight indication of a gradate sequence of the sporangia: this is natural enough where the sporangia are few. But a basipetal sequence is clearly seen in a vertical section of a sorus approaching maturity (Fig. 648). No evidence of interpolation of sporangia has been obtained in *Woodsia*, which thus takes its place as one of the gradate, but with a very short sequence of sporangia.

The account of the sorus here given follows the verbal description of S. Schlumberger, who regarded the indusium of *Woodsia* as in course of dissolution ("auflösung") into constituent hairs. In the comparison which follows later the converse view is taken, viz. that *Woodsia* illustrates the upgrade origin of a cup-like indusium from constituent hairs.

HYPODERRIS R. Brown

This West Indian genus now includes three species. That first to be discovered was *H. Brownii* J. Sm., which was related by Brown to *Woodsia* on account of its indusial character. Notwithstanding the difference of habit, and as we shall see certain features of soral detail, there has never been reason to reconsider this relationship. *Hypoderris* may be held as a low-country representative of the family, with leaves adapted to a shady habit, and with sorus advanced to a mixed condition.

The creeping rhizome bears solitary leaves right and left, the surface of the shoot being invested by soft brown scales. The leaf is broadly lanceolate, often with two lateral lobes at its base, of variable proportion. But occasionally the blade may be five-lobed, resembling that of *Christensenia* (B.M. specimen from Trinidad H. Prestoc): on the other hand it is usually simply lanceolate. The margin is frequently sinuous. The richly reticulate venation is of the Drynarioid type, and the numerous sori with basal indusium are inserted in lines or series parallel with the primary veins (Fig. 646, *F*). The thick sappy axis is traversed by a dictyostele with large leaf-gaps, and there are also wide-meshed perforations, the latter chiefly on the upper and lower faces. The leaf-traces consist of four strands, each pair right and left representing one of the two strands of *Woodsia*. These facts, together with the reticulate venation, indicate a state of advance on that seen in *Woodsia*.

The soral condition of *Hypoderris* confirms this. The position of the sorus is variable: commonly it is seated upon an arch of the main reticulum, but sometimes on a smaller twig, or even on a vein-ending (Fig. 246, *G, H*). At first the flat receptacle is almost continuous with the general leaf-surface, being fenced off by the fimbriated cup-like indusium before the sporangia appear. These arise sporadically over its broad surface, without any obvious order. Fresh sporangia following on those first formed continue thus to arise till the sorus is crowded, while the heads of those earlier matured break away, leaving their stalks to protect those that follow. Thus a section of an old sorus appears similar to that of certain Mixtae of quite different affinity, excepting for the basal indusium which is clearly like that of the Woodsieae. Such facts suggest that sorally *Hypoderris* is a Cyatheoid derivative that has assumed a mixed sorus. The very numerous sporangia are small and long-stalked. The stalks may appear 1, 2 or 3-celled in transverse section (Fig. 649). This condition arises from the initial segmentation, which is not uniform. Young sporangia of three

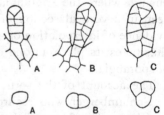

Fig. 649. Young sporangia of *Hypoderris Brownii*, showing three types of segmentation according to size (*A, B, C*): below are the corresponding stalks, *A* consisting of one, *C* of three rows of cells. (× 175.)

types may be found, the largest providing the three-rowed, and the smallest the one-rowed stalk (Fig. 649, *A, B, C*). The annulus of the mature sporangium is indurated on one side, down to the insertion of the stalk where it is interrupted. On the other side is the stomium composed of four cells. Attached to the stalk of many of the sporangia is a hair, often glandular: these hairs are very like those of *Dryopteris filix-mas*, a fact of some comparative importance.

DIACALPE Blume

The single species *D. aspidioides* Bl., from tropical Asia, is still maintained under the name by which it was first defined by Blume in 1828. It has, it is true, borne several other synonyms from time to time, and even been included in *Cyathea* (Moritz, 1845). This stamps a relationship that is probably a real one, though the unanimity of later writers is a witness to the distinctness of type. On the other hand, various species of *Woodsia* have from time to time been included under *Diacalpe*, a fact which points to a near relationship with that genus. Thus the systematic treatment of *Diacalpe* suggests that it is an interesting intermediate though distinct survival. Its vicissitudes of systematic place and designation are fully abstracted by Davie, and need not be repeated here (*Ann. of Bot.* XXVI, 1912, pp. 255–6).

The adult stock is upright, on which are inserted leaves 2–4 feet long, densely scaly below, with a tripinnate blade bearing long hair-like scales and hairs when young. The venation is open, and the veinlets undivided (Fig. 650, *A, B*). The sori are usually seated on the lower anterior veinlet of each segment, below its termination: they are sessile, and protected by an enclosing indusium. Each contains about 100 sporangia, the ages of which are intermixed.

The stock is traversed by an advanced type of dictyostele, but without perforations (Davie, *l.c.* p. 258). The leaf-trace consists of three main strands with a varying number of subsidiary bundles, and it is inserted on the lower half of the leaf-gap. Thus in essentials the vascular system corresponds to those of *Peranema* and *Dryopteris filix-mas*.

The adult sorus is entirely covered in by the basally-attached, coriaceous indusium, as in *Cyathea* (Fig. 650, *B, C*). The convex receptacle, which contains a button-shaped vascular supply, bears sporangia of mixed ages; but there is no evidence of their basipetal sequence. The sporangium itself is borne upon a long three-rowed stalk: its head is encircled by a slightly oblique annulus, interrupted at the insertion of the stalk, and with a lateral well-organised stomium (Fig. 651). The slight obliquity of the annulus makes it possible to distinguish between the proximal and distal faces, as in *Dryopteris* and others (Vol. I, pp. 254–5). Hairs are sometimes present on the sporangial stalk. The spore-output is 48, and each spore has exospore and perispore. In these features *Diacalpe* compares closely with *Dryopteris filix-mas*. The prothallus bears stalked glandular hairs on its lower surface, while the antheridia have a segmented lid-cell (Schlumberger, *l.c.*). These features point to a relationship with the Cyatheaceae, between which and *Dryopteris* the genus appears to form a natural link.

Fig. 650. *A—D=Diacalpe aspidioides* Bl. *A*=pinna from the upper part of the leaf, natural size. *B*=a pair of pinnules, enlarged. *C*=sorus, cut vertically, enlarged. *D*=sporangium, highly magnified. *E—H=Peranema cyatheoides* Don. *E*=pinna from the upper part of the leaf, natural size. *F*=a pair of pinnules, enlarged. *G*=sorus, cut vertically, enlarged. *H*=sporangium, highly magnified. (*B—D*, *F—H* after Bauer: *A*, *E* after Diels, from *Natürl. Pflanzenfam.*)

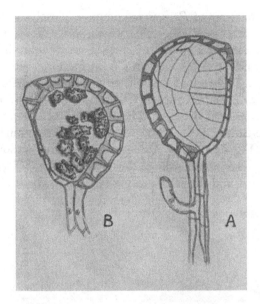

Fig. 651. *A* = mature sporangium of *Diacalpe*, showing a hair borne on the three-rowed stalk. *B* = vertical section through a sporangium, showing the rugged spores, the annulus and stomium. (After Davie. × 150.)

PERANEMA Don

The genus *Peranema* includes a single Himalayan species, *P. cyatheoides* D. Don. This Fern was first collected by Wallich, and described by Don in 1825 under the above name. But Wallich also described it in 1830 under the name *Sphaeropteris barbata* Wall.: naturally the older name must stand. Its further vicissitudes of name and classification have been summarised by Davie (*l.c.* p. 245): here it will suffice to say that it has now definitely taken its place in the Woodsieae under its original designation.

Peranema has an upright stock some 4–6 inches long, bearing bright green leaves 1–3 feet in length, both stock and rachis being covered with red chaffy scales. The leaves are tri-pinnate, with glandular hairs: the venation is open. The stalked sori arise superficially at some distance from the vein-endings. Each sorus is spherical, and is covered by an indusium that is entire, except for a small gap on the side next the leaf margin, close to the insertion of the stalk. At maturity the indusium splits across, its two sides becoming reflexed and finally flattened. The very numerous sporangia are long-stalked, and crowded upon the convex receptacle, while the annulus is slightly oblique. It is apparent that the interest will centre in the peculiar soral condition (Fig. 650, *E*, *H*).

The vascular anatomy of *Peranema* compares closely with that of *Dryopteris filix-mas*. There is a dictyostele without perforations, while the highly divided petiolar-trace, composed of 3 to 5 main strands together with some smaller strands, springs outwards from the lower side of the leaf-gap (Fig. 652, I, II, III). The ground tissue is crowded with nests of sclerenchyma of irregular size and distribution, a feature not uncommon in Ferns which need to resist occasional drought. The structure of the petiole resembles that of the Male Shield Fern, while the origin of the pinna trace is as in Cyatheoid types (Fig. 652, IV, V). A single-stalked sorus is shown in section in Fig. 650, *G*. It arises above one of the lateral veins, but always at some distance from its tip, and the stalk attains a length of about 1 mm. A median

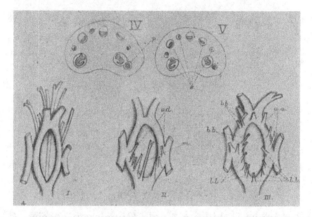

Fig. 652. *Peranema cyatheoides*. I = part of dictyostele seen from within, showing the leaf-gap, and strands of leaf-trace passing outwards from it. II = the same leaf-gap seen from without, *m* = median strand, *ad* = adaxial strands. III = another leaf-gap with its own leaf-trace strands (*l.l.*), together with parts of the leaf-traces above, right and left (*a.a.* and *b.b.*). IV, V = transverse sections of the petiole, showing the origin of the trace of one of the basal pinnae (*p*); *s* = subsidiary strands. (After Davie. Magnified.)

section of a sorus still in an immature state is shown in Fig. 653. The stalk is traversed by a vascular strand which dilates in the receptacle into a loose mass of tracheides, as in *Dryopteris* (Vol. I, Fig. 14). The convex surface of the mature receptacle is covered by a dense mass of sporangia, and the whole is protected by a single-layered indusium, excepting for the narrow opening already mentioned, which lies at the insertion of the stalk, on the side towards the leaf-margin. Here the indusial lip may be curved inwards so as to leave only a narrow well-guarded channel leading to the mass of sporangia within. We have here one of the most accurately protected sori of the whole group. The origin of the sporangia is at first almost simultaneous, though comparison of numerous sections led Davie to conclude that the succession is at the first essentially basipetal: in any case a mixed condition of the sporangia is attained early, and there is no regularity of their orientation in the sorus such as is usual in gradate Ferns.

The development of the sorus as a whole was traced by the late Dr Davie (*Ann. of Bot.* XXVI, p. 245 and XXX, p. 102), from whose work Fig. 653 is taken. But a re-examination of his preparations, and particularly that from

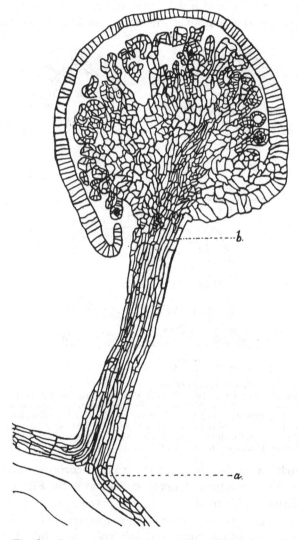

Fig. 653. Immature sorus of *Peranema* in vertical section, with stalk fully developed. The indusium covers the spherical receptacle to which a vascular strand passes up the stalk. A narrow slit is seen between the stalk and the margin of the indusium, on the left side, which is directed towards the leaf-margin. The sorus is "mixed," but with traces of a basipetal sequence at first. *a, b*=intercalated stalk. (After Davie. ×65.)

which the drawing was made, has shown that he was in error in attributing a fully cup-shaped indusium to *Peranema*. The structure which he described as the "laggard portion" of the indusium in Fig. 653 is really a young sporangium, and the indusium is not continuous all round the receptacle. This fact is demonstrated for the young sorus by Fig. 654, *A*, in which the section follows the course of a vein, and traverses the margin: the hemispherical receptacle thus cut in a median plane is seen overarched by the

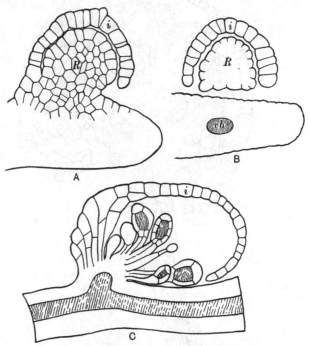

Fig. 654. *A*, *B*=sections of young sori of *Peranema* from preparations made by Dr Davie. *A*=vertical section in plane cutting the leaf-margin. *B*=vertical section in plane parallel to the leaf-margin. *C*=a section of a young sorus of *Dryopteris filix-mas*, in similar plane to *A*, for comparison. (*A*, *B* × 165. *C* × 63.) *C* is rather older than *A* and *B*, and it shows signs of a basipetal sequence of sporangia. *R*=receptacle. *i*=indusium. *vb*=vascular bundle.

indusium, which has a free lip, and there is no corresponding lip on the marginal side. The structure is made still clearer by Fig. 654, *B*, which represents a sorus similar to *A*, but cut in a plane at right-angles to it, and hitting the vein transversely. The indusium appears then as a close-fitting arch, the margins of which right and left are free, while the receptacle appears in this plane of section detached from it, as well as from the leaf-surface. But as the sorus develops the margins of the indusium grow forwards, coming into contact, and even folding inwards so as to form the narrow slit above described. These details are important for comparison with *Dryopteris*; Fig. 654, *C* shows its sorus cut in the same direction as that of *Peranema* in *A*. On the other hand, the lop-sidedness of the sorus finds its equivalent slightly indicated among the Woodsieae, and in *Hemitelia* among the Cyatheaceae, in both of which the indusium is deficient or even absent on the side next to the leaf-margin. Lastly, the origin of the stalk of the sorus in *Peranema* is due to late intercalary growth: apart from this the sorus corresponds in all essentials to that of *Dryopteris*.

The sporangia of *Peranema* are inconstant in their first segmentations: sometimes the first cleavage is transverse, but usually it is oblique. The adult sporangium has a three-rowed stalk, and the margin of the flattened capsule is surrounded by a slightly oblique annulus with 12 to 14 indurated cells. Sometimes the induration is continued past the insertion of the stalk, but more commonly it stops in the cell just at its centre, though the series of cells of the annulus is always continued past the stalk. The stomium has two strongly indurated cells, between which the break occurs. The number of spores is typically 64. These characters indicate an intermediate condition between a gradate and a mixed type. The spores are "Aspidioid" in their character (see Hannig, *l.c.* p. 340).

Davie found the prothallus of *Peranema* to bear unicellular glandular hairs both on the wings and on the cushion. The lid-cell of the antheridia was seen, in every case observed but one, to be undivided. These details accord with an intermediate state between the Cyatheoid and Dryopteroid types, but with a leaning towards the latter. The general conclusion will be that this position commonly assigned to *Peranema* is justified.

COMPARISON

In Volume II a line of comparison among Ferns with consistently superficial sori has been traced, starting from an ancient ex-indusiate type which was frequent in Mesozoic times, and existed even in the Palaeozoic age, as *Oligocarpia* (Fig. 493). It is represented at the present day by the Gleicheniaceae. In Chapter XXIV it was shown how the sub-genus *Eu-Dicranopteris* stands apart from other living Gleicheniaceae in its more advanced vascular condition, and in its dermal hairs; but more particularly in the more numerous sporangia in the unprotected superficial sorus (Fig. 488), in the smaller size of the individual sporangium (Fig. 489), in the smaller number of the spore-mother-cells (Fig. 491), and consequently in the smaller output of spores from each sporangium. The comparison was continued in Chapter XXXII. Here the monotypic genera, *Lophosoria* and *Metaxya*, still ex-indusiate, and retaining certain other relatively primitive features, were regarded as synthetic genera, linking up the superficial Simplices with the Gradatae of this whole phylum, and in particular the Gleicheniaceae with the Cyatheaceae. On that account they were designated the Protocyatheaceae.

In Chapter XXXIII the Cyatheaceae have been examined comparatively, and their place assigned as further steps in the progression of Ferns still retaining consistently the superficial position of the sori; but showing an advance to a gradate state, thereby resolving the difficulty of a mechanical dead-lock of a crowded sorus. A basal indusium, partial in *Hemitelia* and circular in *Cyathea*, afforded the protection that was lacking in *Gleichenia*, *Lophosoria* and *Alsophila*: but in none of these was the step to a mixed sorus taken. It is here that the indusiate Woodsieae come in by supplying this further sign of soral advance, viz. a mixed state: and it is found to go along with other features that indicate a progression towards a Dryopteroid type, where a mixed and indusiate sorus of superficial position is a leading character.

Of the four genera of the Woodsieae above described that which shows the most primitive characters, indicating for it a near relation to the Cyatheaceae, is *Woodsia*. This genus may be regarded as comprising small arctic and alpine representatives of the type of *Cyathea* itself. The upright habit, chaffy scales, the pinnation of the leaves—doubly pinnate in some Mexican and Andean species—the open venation, and the basal sometimes continuously ring-like indusium, together with the signs of a gradate sequence of the sporangia, all support the comparison. The vascular system is, it is true, relatively simple: but its dictyostelic structure and divided leaf-trace are just such features as might be expected in a Cyatheoid type of small size. The gametophyte also lends evidence, the most cogent fact being the divided cap-cell of the antheridium in *W. obtusa*; while a division has occasionally been seen also in *W. ilvensis*. These are points of comparative importance in view of the absence of such divisions in advanced Lepto-sporangiate Ferns, and their constant presence in the Cyatheaceae. The convergence of these lines of comparison gives *Woodsia* a confirmed position in relation to *Cyathea*, a relationship suggested by earlier writers, and now supported by detailed facts to which they could not have had access.

Hypoderris shows advance on *Woodsia* in various features: but there is no reason to doubt the real relationship. It was suggested by Robert Brown thus (*Misc. Bot. Works*, Vol. II, p. 543, 1830): "Among the genera of Polypodiaceae having an indusium one remarkable example occurs in a genus as yet undescribed (*Hypoderris*), which with an indusium not materially different from that of *Woodsia* has exactly the habit of *Aspidium trifoliatum*." The habit may be held as adaptive to growth in moist tropical shade: the real interest lies rather in the vascular and soral advance. The lax dictyostele of *Hypoderris* shows perforations as well as foliar gaps, while the leaf-trace is more highly divided than that of *Woodsia*. These changes are probably concomitant on the habit, and on the fleshy nature of the rhizome. But in the sorus, while the character of the basal indusium is retained, the receptacle is flattened and wide, and there is a mixed succession of the sporangia. The latter are inconstant in the structure of the stalk and in segmentation, suggesting steps in reduction towards an advanced Leptosporangiate type. Incidentally the frequent presence of a hair, glandular or not, upon the sporangial stalk is distinctly reminiscent of *Dryopteris filix-mas*. In fact *Hypoderris* appears to be synthetic between the Cyatheoid and Dryopteroid Ferns, while it affords another instance of the widely homoplastic assumption of a mixed sorus, associated with flattening of the receptacle, and diminution in size of the long-stalked sporangia in which the annulus is definitely interrupted at the insertion, and the stalk may be reduced to a single row of cells.

Other advances not less interesting are seen in *Diacalpe* and *Peranema*, and particularly in relation to their sori. In both of these monotypic genera the stock is traversed by a dictyostele of very similar character to that seen in *Dryopteris filix-mas*, and the leaf-trace is highly divided: but there are no perforations. The sorus of *Diacalpe* conforms nearly to the *Woodsia* type, but with mixed ages of the more numerous sporangia, while these resemble in structure and in spore-output what is seen in *Dryopteris*. On the other hand, by its glandular hairs and the segmented lid-cell in the antheridia the gametophyte points rather to *Woodsia* and *Cyathea*. *Peranema*, with the very Dryopteroid habit of its sporophyte, is particularly suggestive in its soral features. At first sight its stalked sorus recalls that of *Marattia* (*Eupodium*) *Kaulfussii* J. Sm. But such comparisons are only superficial: the stalk of the sorus may be attributed in either case to an intercalary extension below the receptacle: and this being an isolated phenomenon, occurring in two very distinct families of Ferns, must needs be homoplastic. The real interest in the sorus of *Peranema* lies in its one-sidedness combined with the mixed condition.

(*a*) ZYGOMORPHY OF THE SORUS

For the first time in the upward sequence of Ferns with superficial sori the sorus in *Peranema* departs definitely from that radial construction which characterises all the earlier types where sporangia are associated sorally. This remark refers not only to the families and genera above named, but also to the Marattiaceae and Matoniaceae. Indications of zygomorphy of the sorus appear, it is true, in the half-moon-shaped indusium of *Hemitelia*, and in the incomplete ring of the indusial cup of *Woodsia obtusa*: in both of these the indusium is incomplete on the side next the margin of the pinnule. The same holds also for *Peranema*: but there the gap is combined with a lop-sided development of the whole receptacle, which becomes still more pronounced in *Dryopteris* and related Ferns, as will be seen later. This progression appears to be analogous to the secondary adoption of zygomorphy in the shoots and flowers of Angiosperms. It is held to be a derivative state acquired in relation to circumstances not uniform the zygomorphy appearing in relation to the form of the part that bears the sori.

Among Ferns with marginal sori such zygomorphy is naturally more common, for in them the sorus has a more direct relation to the distinct upper and lower surfaces of a flattened sporophyll, and to the unequal incidence of light. An original type of sorus, as it may be seen in *Etapteris* (Vol. II, Fig. 333) or *Corynepteris* (Fig. 334), appears to have been like a radial tassel, distal or marginal upon an organ not markedly bilateral. On the general theory advanced in these volumes from palaeontological data, there was first a flattening and frequently also a webbing of the supporting part, thus giving a broad leaf-surface. It appears that certain Ferns slipped their marginal or distal sori early in descent on to the lower surface of the flattened organ, giving rise to the great group of the Superficiales : but that others styled the Marginales retained the marginal position, though with occasional transitions at later times to the lower surface. The marginal sori nevertheless retained the radial construction of their sorus in some families with pertinacity : this is conspicuously so in the Hymenophyllaceae. But from the Schizaeaceae to the Dicksoniaceae, and especially in the Dennstaedtiinae and their derivatives, the dorsiventral development of the sorus became ever more and more pronounced. Naturally it asserted itself earlier in the descent of the marginal than of the superficial series. Consequently in these parallel but distinct series, while the sori of the Cyatheaceae remained typically radial, those of the Dicksoniaceae became typically lop-sided. Nevertheless lop-sidedness, or rather zygomorphy, does make its appearance eventually among the Cyatheaceae also: but it is only in the later derivatives that it becomes a marked feature, and *Peranema* thus acquires a special interest as combining Cyatheoid characters with a lop-sided sorus. The importance of this will be seen in relation to the Dryopteroid Ferns.

(b) MORPHOLOGY OF THE *"INDUSIUM INFERUM"* IN THE SUPERFICIALES

A more interesting innovation, and one still more open to controversy, is the appearance of an indusial protection in the superficial series. A considerable volume of facts relating to the Cyatheaceae and Woodsieae necessary for a discussion of the morphology of the indusium in Ferns with superficial sori is now before us: and conclusions derived from them will naturally carry on their application to the Dryopteroid Ferns. The question of the origin and nature of any special type of indusium cannot be decided by reference to the indusia alone, for the indusium is a part which has made its appearance after the main categories of parts were established for Ferns: it is a mere accessory to the more ancient sporangia. It appeared later in descent than the establishment of those main evolutionary phyla, the distinction of which is recognised on a basis of comparison, checked by the positive data of palaeontology. It is moreover evident that indusial structures have arisen independently in more than one of those phyla. It cannot then be assumed that all those parts which we call indusia were of the same nature or origin throughout: nor that the evolutionary history was the same, however closely those of different phyla may resemble one another structurally and functionally. Each separate phylum will then present its own evolutionary problem, and it will only be by analogy that the facts relating to one phylum may be held as contributing to the elucidation of those seen in another. For instance, he would be a bold man who would assert that the overarching indusium of *Matonia* is the evolutionary equivalent of that in *Polystichum*. However nearly the one may resemble the other in form and function, they are parts of Ferns as distinct in structure as they are in recorded time. The simple uniseriate sorus of the one is a type characteristic of the Palaeozoic age, the mixed sorus of the other characteristically modern. The indusium of each, notwithstanding a superficial similarity of form and of function, must be held as presenting a distinct evolutionary problem.

Similarly the problem of the indusium in the superficial series may be held as distinct from that of the marginal series. The two are more nearly parallel than are those of *Matonia* and *Polystichum*: owing to convergent evolution the resemblances are sometimes very near. But those who have followed the comparative steps illustrated by *Gleichenia, Lophosoria, Alsophila, Hemitelia*, and *Cyathea* will have recognised a natural sequence, the gradual steps of which have led, in Ferns with consistently superficial sori, from Mesozoic and perhaps from Palaeozoic times to the present day. The indusium imposed upon it is a relatively late step. On the other hand, the marginal series from the Schizaeaceae to the Dicksoniaceae and Dennstaedtiinae showed an earlier adoption of indusial protection. Both palaeontological and comparative evidence indicates that these two sequences, however similar in certain details, have pursued separate courses since Mesozoic times. Hence the comparatively late appearance of indusial developments in one of them must be held as a separate evolutionary problem from its earlier appearance in the other. However far homoplastic resemblance may extend, arguments will not appear convincing which involve any assumption of strict homogeny. And so the origin of the indusium in the Cyatheaceae and Woodsieae must be considered as an evolutionary problem to be solved according to the facts available from the superficial series only.

Professor Von Goebel has interpreted the indusium of *Cyathea* in terms of a Dicksonioid origin (*Organographie*, II, 2, p. 1149, 1918). He derives it from the lower indusial lip of *Dicksonia*, closed round the receptacle, which had itself passed to the lower leaf-surface. The condition seen in *Hemitelia* is held as evidence of this origin. His argument, thus expanded from relatively advanced marginal to relatively advanced superficial types, is founded on comparison of the indusia themselves, which are parts of relatively late origin compared with the sporangia they protect, or the leaves and axes that bear them. It leaves out of account those intermediate types with naked superficial sori, which would lead, as we shall see, to a different conclusion. Moreover, no reference is made to palaeontological evidence, which demonstrates that the superficial series of Ferns has existed as such, with constantly naked sori, since early Mesozoic times. The whole question must be held to be still open for discussion till a wider sweep of evidence is adduced than the mere form of the indusium as it is seen to-day in the Ferns compared.

Historically the interpretation of the Cyatheoid indusium stands thus. Robert Brown, in founding the genus *Hemitelia* in 1810 (*Prod. Fl. Nov. Holl.* p. 158), regarded the semilunar basal scale as a "true involucre." Sir Wm. Hooker (*Sp. Fil.* I, p. 37) says that to him it appears to be of the nature and texture of the bullate scales common on this and other Cyatheaceae, and embodies this in his specific description of *H. capense*: "rachis and costa with small bullate deciduous scales, and one lax laciniated one at the inferior base of each sorus." Mettenius, however, resists this idea (*Fil. Hort. Lips.* p. 110, Taf. XXIX), on the ground that the base of the scale-like indusium surrounds the inner half of the receptacle, and persists even in the oldest sori, while the neighbouring paleae are inserted in the well-known way. He makes no allowance for specialisation, and demands that the indusium shall be exactly like the paleae in position and endurance, otherwise the comparison is invalid. It is strange that so thin an argument should have persisted so long, and be quoted as a demonstration in 1918.

A most valuable contribution to the question was made in recent times by Christ (*Farnkräuter*, 1897, p. 323). In describing the genus *Alsophila* he remarks on its complete identity with *Cyathea* ("volle Uebereinstimmung"), with the sole exception of the indusium. He notes the hairiness of the receptacle, and that here and there a scale is found at the base of the sorus, fringed, short, and caducous. He further remarks how this populous genus has exactly the same appearance, structure, and distribution as *Cyathea*, and that the soral distinction is sometimes so insignificant that in certain forms the suspicion asserts itself that specimens with and without indusia may be included under both genera. Such observations offer a trenchant commentary upon the supposed constancy of the indusium in these genera[1]. They open the door to a theory of upgrade origination of an indusium, as a new and independent but fluctuating protective organ among the Cyatheaceae. Let us see how such an idea would accord with a wider circle of facts.

The fundamental position from which any argument as to the origin and nature of the "*indusium inferum*" of the Superficiales must start is, that these Ferns formed a consecutive phylum from early Mesozoic times. A modern representative sequence for comparative purposes (not necessarily members of an actual line of descent) has been indicated as consisting of *Gleichenia* (and in particular *Eu-Dicranopteris pectinata*), *Lophosoria* and *Metaxya*, *Alsophila*, *Hemitelia*, and *Cyathea*. The Woodsieae will be considered separately later. The *Gleichenia*-type, dating from the Jurassic period, has a simple sorus without hairs or indusium (*Phil. Trans.* B, Vol. 192, Plate 2. Also *Ann. of Bot.* XXVI, Plate XXXIV). *Lophosoria* has also a simple sorus and no indusium, but hairs are present scattered among the sporangia: they are particularly numerous below them, forming a fringe round the base of the receptacle (*Ann. of Bot.* XXVI, Plate XXXV, Figs. 28, 29, 33. Sometimes they are fused laterally, as in Fig. 25). They have basal intercalary growth, which is common for these soral hairs. *Metaxya* has a simple but flat and crowded sorus, and no indusium: but hairs are profusely scattered among the sporangia, particularly below those that are marginal (*Ann. of Bot.* XXVII, Plate XXXII, Figs. 5–9). *Alsophila corcovadense* (Raddi) C. Chr. (=*A. Taenitis* Kze.) has its sporangia "mixed with long copious hairs" (Hooker, *Gen. Fil.* Tab. XXXIV). Hairs are present sparingly in the young sorus of *A. atrovirens* (Vol. II, Fig. 564), but they are present in large numbers and specially at the base of the receptacle in *A. excelsa* (Hooker, *Gen. Fil.* Plate IX, Fig. 6). In this connection it is to be remembered that Christ also noted the hairiness of the receptacle of *Alsophila*, and that here and there a scale is found at the base of the sorus, fringed, short, and caducous. *Hemitelia* bears a more definite and constant scale, similarly placed, and with regular orientation. In *H. capense* it is shown, fringed and short, by Hooker (*Gen. Fil.* Plate XLII, *A*). Hairs are present also scattered among the sporangia (Fig. 566). In *H. horrida* the scale is larger, and semi-lunar, with an irregular margin (Hooker, *Gen. Fil.* Plate IV). The last step is the completion of the cup-like indusium, occasionally seen in *H. horrida* (Mettenius, *l.c.* Pl. XXIX, Fig. 5), but typically present in *Cyathea* (Figs. 577, *C*, and 565). Nevertheless the inconstancy of the indusium in this genus, noted by Christ, is to be remembered as against the statements commonly embodied in the literature, where it is usually assumed to be constant.

If these facts be considered, with a mind as free from bias as if no phylum of Ferns existed other than those under discussion, the natural inference would be that they illustrate the origin of a new and efficient protective organ, viz. the "*indusium inferum*." When the further facts are recalled that in *Gleichenia pectinata*, *Lophosoria* and *Metaxya* no dermal scales are present on the vegetative organs but only hairs, while in *Alsophila*, *Hemitelia*, and *Cyathea* profuse protective scales cover the young stem and leaf—(presumably the result of lateral extension or possibly webbing of the more primitive hairs)—the further inference would be drawn that the hairs, already seen to be present round the base of the receptacle in *Lophosoria*, *Metaxya*, and *Alsophila*, had sometimes in

---

[1] Van Rosenburgh (*Malayan Ferns*, p. 29) specifies in his diagnosis of *Alsophila*, "Indusium wanting or spurious and minute, and then consisting of a semiorbicular, inferior, lateral scale placed at the inner side, or of a circular, entire or lobed scale, or a whorl of fibres with the receptacle in the centre." He specifies this particularly for *A. glauca* J. Sm. Perhaps it was such facts that led Copeland to include all Asiatic species of *Alsophila* and *Hemitelia* under *Cyathea*.

Alsophila itself, and more constantly in *Hemitelia* and *Cyathea*, by lateral extension or possibly by webbing, or both, produced that flattened organ which has been styled in those Ferns the *"indusium inferum."* The whole question requires to be examined afresh, not in one species or only in a few, but very widely throughout the Cyatheaceae, before any final decision can be taken. But meanwhile the facts stated above appear to justify the working hypothesis that the *"indusium inferum"* is in origin of the nature of a dermal appendage : and that it arose by transformation of such dermal appendages as are borne by the vegetative parts.

This is in fact a slight modification of the view first suggested by Sir William Hooker for the Cyatheaceae. It was met with opposition on somewhat slender grounds by Mettenius; this opposition is, however, supported by a general unwillingness that exists to entertain the idea of polyphyletic origins. There is a natural tendency in evolutionary morphology to read all apparently similar organs in terms of that which is the most familiar, and so to reduce the categories of parts to their simplest terms. Instead of yielding to this we may in the present instance contemplate how a protective indusium has arisen not only from a Schizaeoid-Dicksonioid source (which has apparently supplied the type general for the Marginales), but also from a Gleichenioid-Cyatheoid source (which apparently supplied the type general for most of the Superficiales) : another source has been the Matonioid, but it has had only a restricted application : another again gave those distal lobes of the sporangiophore of *Helminthostachys*, which are functionally protective as a rudimentary indusium (Vol. II, Fig. 364, *G*). A fifth type of protection is that by hairs with a stellate or discoid head interspersed among the sporangia, as in *Platycerium* (Hooker, *Gen. Fil.* Plate LXXX, *B*), or *Polypodium lineare* Thbg. (*Gen. Fil.* Plate XVIII). The variety of adaptive methods such as these last points a warning against too rigid a search for uniformity of origin of indusia, however similar in form or structure.

Turning now to the Woodsieae, the development of the indusium in them has interesting features bearing on its probable dermal origin. It has been suggested for the Cyatheaceae that lateral extension and possible webbing may have been involved in the formation of a membranous indusium from the more primitive hairs seen in the sori of non-indusiate related Ferns. Schlumberger's excellent analysis of the details in the developing sorus of *Woodsia ilvensis* and *obtusa* is reproduced as Fig. 647 (p. 102). In his discussion of the facts he refers the indusium of the Woodsieae to the type seen in *Cyathea*, of which he holds it to be a reduced form (*l.c.* p. 406). He contemplates the dissolution ("auflösung") of the continuous cup into individual hairs. But in view of the general argument of the preceding paragraphs, for us his thesis must be inverted. We shall see in these drawings from *Woodsia* early stages in an upward construction of a cup-like indusium from the primitive constituent hairs, such as were already present in the non-indusial sori of *Lophosoria*, or *Alsophila*. In Fig. 647, 1, 2 of *W. ilvensis* the simple hairs (*i, i*) appear isolated : in Fig. 647, 6 of *W. obtusa* some of them are extended laterally, and segmented so as to form flattened scales : in Fig. 647, 5 the receptacle is surrounded by an upgrowth forming a ring-like wall, lateral fusion of the constituent hairs being prevalent ; the ring is open on the side towards the leaf-margin, but later it may be completed (Fig. 647, 7, 8). The final consequence of such steps would be a more or less perfectly cup-like indusium, with a fringed lip, the laciniae indicating the constituent hairs or scales. The fringed indusium of *Hypoderris* will bear a like interpretation (Hooker, *Gen. Fil.* Plate I). When it is remembered that many species of *Cyathea* are lofty Tree-Ferns, while *Woodsia* and *Hypoderris* are small low-growing Ferns, the more complete indusial protection of the sori in the former finds its biological rationale : while the latter, subjected to less drastic demands, may be held to have retained a more rudimentary state of their indusia. On the other hand, species of *Alsophila* are often tall Tree-Ferns, but apparently they have shown less ready adaptation of the sorus.

The result of this whole discussion, as to the origin of the "*indusium inferum*" of the Woodsieae and Cyatheaceae, is to show a reasonable chain of argument, favouring its being of the same nature as the dermal appendages borne superficially upon the vegetative organs. The actual source from which it is held to have sprung is from such basal hairs as are a marked feature in the sori of *Lophosoria, Metaxya*, and *Alsophila*. The steps suggested by the more rudimentary types of indusium seen in the Woodsieae are: (i) widening of the hairs as flattened scales; (ii) coalescence of them laterally after the mode of development of a gamopetalous corolla. Thereafter a thickening by periclinal division of the cells to form more than a single layer would produce the state seen in *Cyathea*. This last is a feature not uncommon in dermal scales. A significant circumstance is that the membranous indusium does not originate in Ferns which bear only dermal hairs (e.g. *Lophosoria, Metaxya*), but in those in which dermal scales are present (Cyatheaceae, Woodsieae). But the appearance of an indusium seems to have lagged in the evolutionary history behind that of the dermal scales of the vegetative region: in particular, *Alsophila* may have dermal scales without any indusium being present. This circumstance seems to support rather than to oppose its suggested origin as the equivalent of those scales.

From this discussion the final conclusion may be drawn that, however closely the "*indusium inferum*" of the Cyatheoids and their derivatives may resemble the indusium of the Dicksonioids and their derivatives, the two appear to have been distinct in phyletic origin. They are consequently homoplastic, not homogenetic parts.

The whole question has been treated here at length, and the argument explicitly stated because it gives a good example of the working of the phyletic method in Comparative Morphology. It is quite clear, from the way in which certain results following from its use have been received, that the method itself has not been fully realised, or the convergent lines of argument followed to their legitimate conclusion.

The general result of the facts here detailed or alluded to is, in the first place, to demonstrate that the Woodsieae (excl. *Cystopteris* and *Acrophorus*) constitute a very natural family, though the genera show some independence of detail. Secondly, they confirm the relation of that family to the Cyatheaceae, both on the ground of the characters of the gametophyte and of the sporophyte. They point also further to the Gleicheniaceae, and thus consolidate on grounds of detailed comparison the sequence of Ferns bearing superficial sori as independent, and progressing by gentle steps from a simple to a gradate, and finally to a mixed sorus. Moreover, this sequence is coherent within itself, and takes its course independently of the marginal series, as seen in the Dicksoniaceae and its derivatives; though both advance along similar but parallel lines.

If the attempt be made to arrange the Woodsieae in sequence, there will be no doubt that *Woodsia* itself stands nearest to the Cyatheaceae, as is indicated by its gradate sorus, and its prothallial characters. *Diacalpe*, with its mixed sorus and basal indusium, together with certain prothallial characters, takes a middle position; while *Peranema*, with its lop-sided or zygomophic sorus, mixed sequence of sporangia, curiously specialised basal indusium, and Nephrodioid prothallus, may be held as the most advanced of the three. The genera thus placed suggest that an evolutionary line has moved from a Gleichenioid source, through types like the Cyatheaceae with a radial sorus, towards types like those found in the Dryopteroid Ferns with a lop-sided sorus, though none of the living genera may themselves have been the actual phyletic links. *Hypoderris*, with its reticulate venation, is clearly a specialised collateral type: but it may very well be held as indicating a line leading in the direction of certain types of Dryopteroids and possibly by complete abortion of the basal indusium may have foreshadowed some Ferns that have been included under the comprehensive title of *Polypodium*.

WOODSIEAE

A. SORUS GRADATE.

   (1) *Woodsia* R. Brown, 1810   ...   .   25 species.

B. SORUS MIXED.

   (2) *Diacalpe* Blume, 1828   ...   .   1 species.

   (3) *Peranema* Don, 1825   ...   .   1 species.

C. VENATION RETICULATE.

   (4) *Hypoderris* R. Brown, 1830 ...   .   3 species.

GENERA INCERTAE SEDIS

   (5) *Cystopteris* Bernhardi, 1806 ...   .   13 species.

   (6) *Acrophorus* Presl, 1836   ...   .   1 species.

BIBLIOGRAPHY FOR CHAPTER XL

654. R. BROWN. Prodromus Florae Novae Hollandiae. 1810.

655. R. BROWN. On *Woodsia*, Trans. Linn. Soc. II, p. 170. 1812.

656. R. BROWN. Notes and observations on Indian Plants, Miscellaneous Works, II, p. 543. 1830.

657. Sir W. HOOKER. Genera Filicum, Tab. CXIX, etc. 1842.

658. METTENIUS. Filices Horti Lipsiensis, p. 110. Leipzig. 1856.

659. DIELS. Engler u. Prantl, Nat. Pflanzenfam. I, 4, p. 159.

660. CHRIST. Farnkräuter, pp. 282, 323. 1897.

661. HEIM. Flora, Bd. LXXXII, p. 355. 1896.

662. SCHLUMBERGER. Flora, Bd. CII, p. 383. 1911.

663. R. C. DAVIE. Ann. of Bot. XXVI, p. 245, 1912, where the literature is fully quoted.

664. R. C. DAVIE. Ann. of Bot. XXX, p. 101. 1916.

665. BOWER. Ann. of Bot. XXVI, 1912; Studies II, p. 306.

666. VON GOEBEL. Organographie, Bd. II, p. 1148, etc. 1918.

CHAPTER XLI

DRYOPTEROID FERNS (II. Aspidieae)

UNDER this heading very numerous Leptosporangiate Ferns are grouped, of various habit, many of them creeping with forked branching of the stock (Vol. I, Fig. 32); but many others have the basket habit with ascending stock, as in *Dryopteris filix-mas* (L.) Schott, which may be taken as a central representative of the group (Vol. I, Fig. 31). The dermal appendages of these Dryopteroid Ferns are chaffy scales, often very plentiful and prominent: but associated with them there may be simple or glandular hairs. The leaf-stalk is as a rule not articulate, and the venation may be open or reticulate. The sori are usually of a roundish form, but they commonly show a more or less obvious lop-sidedness, or zygomorphy. They are superficial, being sometimes placed terminally but for the most part dorsally upon a vein, or seated at a point where veins meet. An indusium is usually present protecting the mixed sorus: it is of variable form and position in the different types: but often it is feebly developed or even absent. Glandular hairs are frequently associated with the sporangia, which have a vertical, interrupted annulus, and usually a three-rowed stalk. The spores are bilateral, and possess a perispore (Hannig, *Flora*, Bd. 103, p. 340).

The adult axis of these Ferns is traversed by a simple dictyostele, typically represented for the larger upright stocks by the well-known reticulum of the Male Shield Fern (Vol. I, Fig. 140). The highly disintegrated leaf-trace here possesses two larger adaxial marginal strands, marking the heel of the horse-shoe, and a varying number of smaller strands, forming the usual abaxial curve between them. But in the smaller creeping forms, where the leaves are alternate and remote, the dictyostele of the axis may appear simplified down to three or even two meristeles, placed respectively above and below, with the alternate and greatly elongated leaf-gaps separating them. This is found in the Oak Fern, and the Beech Fern (Luerssen, *Rab. Krypt. Flora*, III, pp. 297, 301). It is plainly an adjustment according to habit and actual size.

The leaf-trace is variable in its complexity along lines parallel to those of the axis: the results of the observations of Luerssen may be extracted for certain familiar Ferns of this affinity, and tabulated as follows:

A. *Relatively simple*: two strands pass separately into the rachis, but fuse upwards to form a gutter-shaped meristele, as in *Athyrium filix-foemina* (Luerssen, *l.c.* p. 130, Fig. 90).

      Dryopteris Linnaeana C. Chr. Oak Fern.
      Dryopteris Phegopteris (L.) C. Chr. Beech Fern.
      Dryopteris thelypteris (L.) A. Gray. Marsh Buckler Fern.
      Dryopteris oreopteris (Ehrh.) Maxon. Mountain Buckler Fern.

B. *Intermediate*: the petiole receives 2–3 strands, two stronger are adaxial, but fissions upwards give finally 4–6 strands.

    Polystichum lonchitis (L.) Roth. Holly Fern.

    Polystichum aculeatum (L.) Schott. Prickly Shield Fern.

C. *Relatively complex* : the petiole receives 5–11 strands, the two adaxial being larger.

    Dryopteris filix-mas (L.) Schott. Male Shield Fern.

    Dryopteris spinulosa (Mull.) O. Ktze. Prickly Shield Fern.

Such data have been elaborately tabulated by Luerssen for the German Ferns of this affinity, as a help to diagnosis of the species. It may, however, be a question how far this can be reliable, since all Ferns pass in their ontogeny onwards from the simple protostele and undivided leaf-trace, and so the anatomical diagnosis can only apply to the adult state. What does appear from such results is, that all may be regarded as elaborations of structure based upon the simplest type, A: and that the circumstances which have been operative in producing the results observed are the habit—creeping and lax or upright and compact —and the actual size of the part examined.

Of the genera which are associated by Diels in this very comprehensive group, a number bearing kidney-shaped indusia are clearly related to *Dryopteris* (*Nephrodium*). This remark applies particularly to such genera as *Luerssenia* Kuhn, *Fadyenia* Hook., *Mesochlaena* R. Br., and *Didymochlaena* Desv. (Vol. I, Fig. 236, *G*, *M*). Others have a fully circular form of indusium, as in *Aspidium* and *Polystichum* (Fig. 236, *N*): others again, owing to absence of an indusium, have been ranked as *Polypodium*, for instance the native Beech Fern (styled *P. Phegopteris* L.), and the Oak Fern (styled *P. Dryopteris* L.), both now included in the genus *Dryopteris*. Others again are held to be of this same affinity, though their sporophylls appear to be Acrostichoid, with the sori spreading over the originally vacant leaf-surface. As a substantive genus of this nature, but recognised as of Dryopteroid affinity is the widely spread *Polybotrya* Humb. & Bonpl. It thus appears that the sorus of those Ferns which are held as Dryopteroid is variable in its character, but comparison indicates that they may all be traced in origin to a central type, with superficial sori, such as that represented in essential features by *Dryopteris filix-mas*.

There are, however, other Ferns which have been grouped by Diels with the Aspidieae, with which the relationship may be held as open to doubt: the most notable of these is *Dipteris*. This genus has been already assigned a distinct position of its own in Vol. II, Chapter XXXIV, on the basis of its palaeontological history, which traces the Dipterids back as such to Mesozoic times. They constitute a distinct phylum of their own, with which should probably be ranked also *Neocheiropteris*; this genus also will accordingly be removed from the position among the Aspidieae assigned to it by Diels. There is again considerable doubt as to the propriety of placing *Deparia* among the Aspidieae: this question will be taken up later. The remaining genera referred to the Aspidieae by Diels may be held as forming a coherent group of Ferns, related to one another though subject to modifications as to detail. The first step will be to consider the probable connections downwards of that central type which is represented by the Common Male Shield Fern[1].

DRYOPTERIS FILIX-MAS (L.) Schott

The essential features of the Male Shield Fern have been described already in Vol. I, Chapter I. The basket-like habit with ascending stock is similar to that seen in dwarf Tree-Ferns (Fig. 31), and it is very closely shared by *Peranema* and *Diacalpe*, which also have similar pinnate leaves and open venation: their chaffy scales too correspond, while the superficial position of the sori of them all and the insertion of each upon a vein in its course towards the margin are as in the Cyatheaceae and Gleicheniaceae.

Most of the leaves of the Male Shield Fern bear each a bud laterally, some distance above the leaf-base (Vol. I, Fig. I, *B*, *C*). These buds raise the question whether each is to be regarded as a product of unequal dichotomy of the axis, the bud representing the weaker shank carried up upon the leaf-base: or whether it is an adventitious development (Vol. I, p. 78). The fact that dichotomy is prevalent in the creeping species of *Dryopteris* lends some degree of probability to the former alternative (Fig. 32). Such difficulties of interpretation in terms of dichotomy appear to be inherent in the development of a compact shoot by condensation from a laxly forking rhizome. Similar basal buds are seen in *Lophosoria*, *Metaxya*, and *Alsophila*, though these lie nearer to the leaf-base and are median. Collectively such facts suggest frequent progressions, probably polyphyletic, from a lax creeping rhizome to a compact, and ultimately it may be to a dendroid habit.

The simple dictyostele of *Dryopteris* (Fig. I, *E*, *F*), and the insertion of the highly disintegrated leaf-trace upon the lower region of each foliar gap closely resembles what has been seen in *Peranema*. The general conclusion naturally follows that in external form and in internal structure the two genera are closely related: as also is *Diacalpe*. This will make the comparison of their sori all the more interesting.

[1] The Male Shield Fern is so familiar an object, even to elementary students, that it is liable to be treated simply as a "type," while comparison of its features from an evolutionary point of view is apt to be left in abeyance, chiefly perhaps from deficiency of knowledge of the necessary facts. A like treatment is often meted out to the Bracken. It is seldom realised, even by teachers, and very rarely indeed is it suggested by them to students, that these two common and almost cosmopolitan Ferns occupy distinct and very important positions in the phyletic system. They are alike in the fact that both are intermediate types, and that they have interesting relations both upwards and downwards in the evolutionary scale. They take their distinct places: the Bracken, with its dermal hairs in the marginal series, being related downwards to the Dicksonioid Ferns in which scales are absent: the Male Shield Fern, with its chaffy scales so prominent a feature, being related downwards to the scaly Cyatheoid Ferns.

Two pinnules of the Male Shield Fern bearing sori are shown in Fig. 655.
Each sorus is inserted singly upon the anadromic branch of a forked vein,
which continues its course beyond the recep-
tacle. This position corresponds to that in the
Gleicheniaceae (Vol. II, Figs. 486, 488), and
in *Lophosoria* (Fig. 548), while in *Cyathea* all
the branch-veins may bear sori (Fig. 557). In
Diacalpe and *Peranema* also the insertion is
upon the back of a vein, and it is usually so in
Woodsia, though there it is sometimes almost
terminal. In all of these except in *Peranema*
the sorus is radially constructed: but the
well-known kidney-shaped indusium of *Dryopteris* at once reveals the lop-sidedness of its
sorus, which is best elucidated by comparison
with those of *Peranema* and *Hemitelia.* The
structure of the sorus in the Male Shield
Fern is often misunderstood owing to the
circumstance that sections are more easily

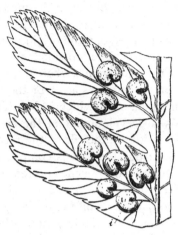

Fig. 655. *Dryopteris filix-mas* (L.) Schott.
Two secondary segments, bearing sori (*i*).
(× 4. After Luerssen.)

cut transversely to the vein than longitudinally. The former sections appear
as in Vol. I, Fig. 14, with the indusium inserted by an apparently central stalk
which overarches the receptacle equally on either side. A much better conception is gained by sections following the course of the vein, and cutting the sorus

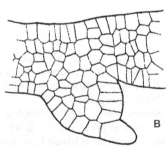
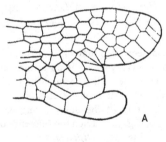

Fig. 656. Vertical sections of young sori of *Dryopteris filix-mas*, cut in the direction
following the vein. *A* is the younger, and it includes the leaf-margin. The indusium
precedes the appearance of the sporangia. (× 250.)

in the median plane of its dorsiventrality. If such a section be taken from
a very young sorus, it will be seen that the receptacle which arises not far
from the leaf-margin is distinctly lop-sided from the first, and begins to be
overarched by the young indusium before the first traces of any sporangia
appear (Fig. 656, *A, B*). Later the sporangia arise upon the slightly convex
receptacle, and there may be clear indications of a basipetal sequence of
those first formed (Fig. 654, *C*): but this is soon superseded by a mixed

origin of the later sporangia. The whole receptacle is at an early stage completely covered in by the arched indusium, which is massively thickened at its base. But even the median section does not give a full conception of the form of the receptacle as a whole. It is horse-shoe shaped and, as Davie has pointed out, the origin of this is prefigured by the sori of *Peranema* and *Hemitelia*. The lop-sidedness may first be traced in the semi-lunar indusium of the latter genus, where it is attached on the side away from the leaf-margin. This encourages a tilt of the previously radial receptacle over towards the marginal side, which is free. The tilt is accentuated in *Peranema* (Figs. 653, 654). Development of the sporangia will consequently be restricted to the free side by the tilt of the receptacle, and this tends to diminish the fertile area: but the loss is compensated by an extension of the receptacle right and left into pouches protected by the indusium, as is seen in Davie's figure (*l.c.* 1916, Pl. III, Fig. 6). Such conditions already indicated in *Peranema* are more fully realised in *Dryopteris*, where the receptacle has accordingly assumed a definitely horse-shoe shape, with the heel directed away from the leaf-margin, and almost encircling the indusial stalk. By such steps illustrated in closely related Ferns, the origin of the lop-sided sorus of *Dryopteris*, with its kidney-shaped indusium and semi-lunar receptacle, can be traced from a radial type such as is seen in the superficial series, starting from the Gleicheniaceae. *Hemitelia*, *Diacalpe* and *Peranema*, but especially the last of these, are the synthetic types that link the sori of the superficial series together in their increasing lop-sidedness. As Dr Davie said, between them *Peranema* and *Diacalpe* appear to unite all the intermediate soral steps between the Cyatheaceae and the Aspidieae.

It has been seen that with transient traces of a gradate sequence the sorus of *Dryopteris* is of mixed type, which is as a rule associated with a vertical annulus interrupted at the insertion of the stalk. Examples have been seen in several distinct phyla of the gradual steps of obliteration of the obliquity of the annulus. First the induration is seen to be incomplete, though the oblique ring of cells is still there (*Thyrsopteris*, Vol. II, Fig. 529; *Loxsoma*, Fig. 521). Then the sequence of cells is almost interrupted (*Dennstaedtia*, Fig. 539; *Metaxya*, Fig. 553; *Dipteris*, Fig. 576). Where the interruption is complete, as it is in most advanced Leptosporangiate Ferns, and as it is seen to be in the Male Shield Fern, it might be assumed that all structural trace of the original obliquity was lost. But it is not so: and the very accurate drawings of Müller, made for Kny's *Wandtafeln*, show the point (Fig. 657). The distinctness of the two faces of the sporangium has already been discussed in Vol. I, p. 254 (Figs. 250–252). The distal face is that which lies encircled by the ring where it is complete: the proximal represents the part of the sporangial wall that lies between the ring and the stalk in primitive types. The two drawings, 4 *a* and 4 *b*, represent two opposed sides of the biconvex head. Two rows of cells of the stalk run up to that shown in 4 *a*, only one row runs up to that in 4 *b*. The former represents the distal or peripheral face of the sporangium, the latter the proximal or central. Moreover, the two sides of the sporangium are not of exactly the same curve or extent. Fig. 657, 3 shows this correctly, the distal side to the left having fewer tabular cells and a slightly different curve from the proximal, which has more numerous narrower cells, and lies to the right. A comparison of Fig. 563 of sporangia of the Cyatheaceae, or of Fig. 576, of sporangia of *Dipteris*, with the drawings of *Dryopteris* (Fig. 657), in regard to the points mentioned, will show that *Dryopteris* still retains traces of the oblique annulus, though in the adult sporangium the ring is interrupted, and the sporangium has apparently, though not exactly, symmetrical sides.

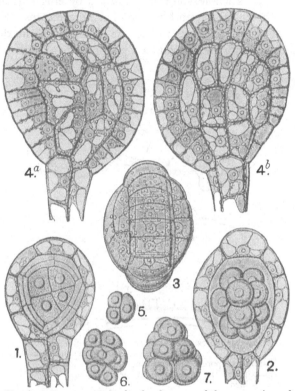

Fig. 657. Later stages in the development of the sporangium of *Dryopteris filix-mas* (after Kny). 1 = the primary segmentation completed. 2 = the spore-mother cells rounded off. 3, 4 = sporangia of about the same age as 2, seen from without. 3 = apical view. 4 *a* = presenting the distal or peripheral face. 4 *b* = presenting the proximal or central face. 5, 6, 7 = the sporogenous group.

On the spore-output, which is normal for the Male Shield Fern, there is nothing to remark. But the presence of a perispore is important (Fig. 658). This deposit on the spore-wall is seen in the Male Shield Fern, and it is present also in the Woodsieae, though it is absent in the Cyatheaceae. Hannig (*l.c.* p. 340), in summing up the results of his observations on the perispore, shows that it is characteristic of certain relatively advanced groups of Ferns, and in particular of the Asplenieae and Aspidieae. It is, however, absent in the great majority of Ferns, particularly in all those that are held as primitive. It is also wanting in those Ferns which I have designated the Marginales. It seems to be the fact that it came into existence in that very phylum which, according to the argument advanced here, is a coherent derivative of the Superficiales, springing from the Cyatheaceae, viz. the Aspidieae. It is well represented in the genera *Diacalpe* and *Peranema*, which link them with the Cyatheaceae. Though the lines of occurrence of the perispore are as a whole somewhat blurred by enigmatical exceptions, the facts give substantial support to the view here expressed, viz. that the Aspidieae having a perispore are a coherent phylum with no direct relation by descent to the Marginales, in which a perispore is absent.

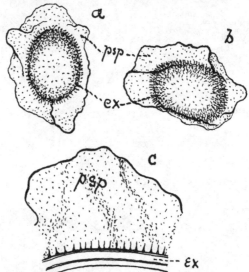

Fig. 658. Spores of *Aspidium trifoliatum*, after Hannig.
a, *b* = ripe spores with prickly exospore (*ex*) and transparent perispore (*psp*), appearing like a loose sac (× 500).
c = part of the exospore and perispore more highly
magnified.

The gametophyte of the Male Shield Fern presents no features of importance excepting the presence of short unicellular hairs on the margin and
lower surface (Fig. 19, Vol. I), such as are present on the prothalli of *Woodsia*
and *Diacalpe*: these are held by Von Goebel as transitional to those of the
Cyatheaceae (*Organographie*, II, 2, p. 959, Fig. 952).

The lid-cell of the antheridium is undivided, according to the ordinary
Leptosporangiate type.

The facts and comparisons thus brought forward confirm the position of
the Male Shield Fern as a derivative from a Cyatheoid ancestry, with the
Woodsioid genera *Diacalpe* and *Peranema* as synthetic intermediate types.
These together appear as later steps in the sequence of Ferns which bear
consistently superficial sori. That sequence leads upwards from the Gleicheniaceae, with their simple radiate, non-indusiate sori of a type prevalent in the
Mesozoic period, and represented even in Carboniferous times. The steps have

been traced through the Gleicheniaceae (Chapter XXIV), the Protocyatheaceae (Chapter XXXII), to the Cyatheaceae (Chapter XXXIII), these last supplying the innovation of a basal indusium. Taking up the sequence, probably from some type of the Cyatheaceae with semi-lunar basal indusium as it is seen in *Hemitelia*, the innovation of lop-sidedness or zygomorphy makes its first appearance in the indusium covering the previously radial receptacle. This is itself tilted towards the leaf-margin in the stalked sorus of *Peranema*: it becomes markedly lop-sided and even horse-shoe shaped in the Male Shield Fern, which however still retains traces of the basipetal sequence of its radiate and gradate ancestry, though this is soon merged in mixed sporangia borne on the zygomorphic receptacle.

The type of sorus of *Dryopteris filix-mas* thus constructed, and interpreted in phyletic terms, may be held as a centre for further advance in at least three directions—the Aspidioid, the Polypodioid, and the Acrostichoid. These advances are registered most clearly in the sorus, but they may accompany other changes in the sporophyte. Of these the most obvious is the adoption of a reticulate venation, in place of the open venation of the Male Shield Fern and other relatively primitive types, a feature shared by typical forms of *Polystichum*. In varied detail vein-fusion has been a feature of the old genera *Cyrtomium* Presl, *Cyclodium* Presl, and of *Eu-Aspidium* Hook. (*Syn. Fil.* p. 258): particularly of the genus *Meniscium* Schreb., now merged in *Dryopteris* under Section II, *Goniopteris* Presl (Diels, E. and P. I, 4, p. 177). Vein-fusion is also characteristically seen in *Luerssenia* Kuhn, and *Fadyenia* Hook. In all of these the reticulate venation may be held as an indication of advance from the open venation, and it is biologically intelligible where the leaf-area is extended.

THE *"INDUSIUM SUPERUM"* AS SEEN IN *ASPIDIUM* AND *POLYSTICHUM*

There can hardly be any two opinions as to the evolutionary relation of the sorus of *Dryopteris* to that seen in *Aspidium* (Swartz, 1801, 138 species), *Polystichum* (Roth, 1799, 112 species), and *Cyclodium* (Presl, 1836, 2 species). Nor yet can their affinity one with another be doubted; indeed they were all included under *Aspidium* by Sir Wm Hooker (*Sp. Fil.* Vol. IV, p. 6). He describes them as Ferns very various in form, size, and composition, with veins either free or variously anastomosing. But he remarks that in their systematic arrangement entire dependence cannot be placed either on the exact uniformity of their venation, or on the shape of their "involucres." These occasionally vary, being sometimes orbicular, sometimes cordate on the same species, and sometimes the form and the point of insertion appear to be intermediate between the two. The venation typically free in *Polystichum* may in the broader-leaved types of *Aspidium* rise to a high degree of reticulation (Fig. 659). Particularly is this so in *Sagenia* Presl, a group largely represented in the Southern Hemisphere.

The prevalence of the two larger genera is itself a witness to the success of the superficial sorus with its orbicular or peltate covering, the *"indusium superum"* of authors (Fig. 660). This type of sorus is clearly secondary and derivative. The source from which it probably arose was in the first instance a type with a basal indusium, and radial receptacle, as seen in the Cyatheaceae.

Fig. 659. *Aspidium Plumierii* Pr., whole leaf, reduced, showing the broad reticulate blade. A small portion is shown magnified, with sori. (After Christ.)

Already the first step towards lop-sidedness has been recognised in *Hemitelia*: this led to the horse-shoe-shaped receptacle, prefigured in *Paranema* with its lateral pouches, and realised in *Dryopteris*. It is a minor step from the horse-shoe-shaped receptacle to the completion of the circle of the receptacle round the indusial stalk, as it appears in *Aspidium* or *Polystichum*. An equal development of the covering all round would then result in the orbicular indusium. The close similarity in habit and structure of those genera that possess it to that seen in *Dryopteris* is amply shown by the richness of the synonymy of these Ferns. This view has been generally adopted as phyletically probable (Von Goebel, *Organographie*, II, Aufl. 2 Teil, p. 1150).

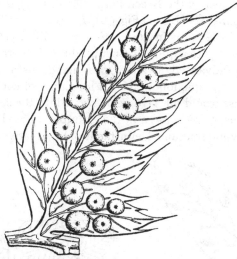

Fig. 660. *Aspidium* (*Polystichum*) *lobatum* Sw. A secondary segment, showing the peltate indusium. (After Luerssen. × 5.)

ABORTION OF THE INDUSIUM

Within the genus *Dryopteris* the indusium is apt to be reduced in size, or to be wholly absent in forms corresponding closely in other details to those which have it normally developed. The result is that the sporangia form a naked hemispherical or slightly elongated mass, attached to the leaf-surface, such as is held to be typical of the old genus *Polypodium*. Here the question will arise whether the indusium has really been aborted, or whether the state actually seen is primitive, there never having been any indusium present in the ancestry. The Oak Fern (*Dryopteris Linnaeana*) and the Beech Fern (*Dryopteris phegopteris*) are examples, and the natural interpretation of them seems to be that they are Dryopteroids that have lost their indusium by abortion. In the genus *Dryopteris* at large, as it is represented in the Monograph of C. Christensen (Copenhagen, 1913, onwards), a very considerable number of species are found to be ex-indusiate, while in others the indusium is noted as "minute" or "fugacious." The effect of this is to suggest that the indusium is an inherited feature no longer of high biological importance, and that it is here seen in course of elimination. Our own native species represent extreme types, on the one hand retaining the indusium in its full development, as in *Dryopteris filix-mas* or *Polystichum aculeatum*; on the other illustrating its complete abortion, as in *Dryopteris Linnaeana* or

D. phegopteris. But the distinction is bridged over by many tropical species leading from the indusiate to the ex-indusiate state. In the latter a tendency commonly appears to extend the sorus along the course of the vein that bears it: this is to be seen in the Beech Fern (Fig. 661, *A*), but more markedly in *Dryopteris decussata* (L.) Urban (Fig. 661, *B*). Such details are reminiscent of a like spread of the sorus in *Hypolepis repens* (Figs. 586, 587, pp. 9, 10), which, however, belongs to a quite distinct phylum. In both we see a feature frequent in the more advanced phyletic types, viz. the discarding of structures protective of the sporangia, followed by a loosening of control of the sorus. In earlier days these would all have been swept into the comprehensive but heterogeneous genus *Polypodium*: now the examples quoted take their places respectively as Dryopteroid and Dennstaedtioid derivatives.

Fig. 661. *A = Dryopteris phegopteris* (L.) C. Chr. A segment of the second order, showing the naked sori elongated along the veins (× 5). After Luerssen. *B = Dryopteris decussata* (L.) Urban, part of a pinna, enlarged, after Mettenius, from Christ.

ACROPHORUS Presl

Presl's monotypic East Indian genus *Acrophorus* appears to find its natural place here. It has been referred to a Davallioid affinity: but its habit, its anatomy, and the presence of plentiful chaffy scales all point towards *Dryopteris*. The upright stock contains a simple dictyostele with highly segregated leaf-traces, after the type of the Male Shield Fern. The sorus is sometimes seated on the end of a vein, hence the generic name. This is seen in Fig. 662, *a*, which also shows the small fimbriated indusium. But often it may be attached laterally on the vein which extends beyond the receptacle, as in Fig. 662, *b*: both types may appear upon the same leaf. The indusium is very variable, and it is often so small as to be hidden by the sporangia when mature. The sorus is of the mixed type. All this points to an affinity of *Acrophorus* with *Dryopteris*. The similarities that led to a reference to the Davallieae are probably homoplastic, and they are negatived by the anatomy, and by the presence of chaffy dermal scales.

PLECOSORUS Fée

It seems probable that the small genus *Plecosorus*, which comprises three Central American species, also finds its best place here. The superficial, apparently non-indusiate sorus, borne on the deeply concave lower surface of the contracted pinnules, together with the chaffy scales and general habit, suggest a highly xerophytic type of Dryopteroid (E. and P. 1, 4, Fig. 101, p. 194).

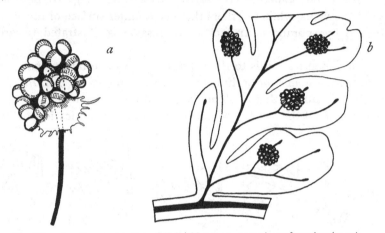

Fig. 662. *Acrophorus stipellatus* (Wall.) Moore. *a* = sorus in surface view (× 35). *b* = pinnule showing the veins continued beyond the sori (× 10).

ACROSTICHOID DERIVATIVES

The developmental study of the Acrostichoid state carried out by Frau Eva Schumann (*Flora*, Bd. 108, 1915, p. 201) has greatly aided the elucidation of the old composite genus *Acrostichum*, and the distribution of its constituents to their several sources of probable origin. A generous share falls to the Dryopteroid source, particularly those Ferns that are associated under the generic name of *Polybotrya*. The relation of the old genus *Meniscium* of Schreber to *Dryopteris* has been so far recognised as to lead to its inclusion within that genus by Diels, and by C. Christensen. It now appears from comparison with *Meniscium* that certain species also of *Leptochilus* are best regarded as being ex-indusiate derivatives of *Dryopteris*.

The absence of an indusium, already seen as a not infrequent feature of *Dryopteris*, gives a greater freedom for the fusion of sori, and for the spread of the sporangia over the general leaf-surface. Various steps of this may be seen leading to that condition which has been described as "Acrostichoid." A very clear series appears to have had its origin in the section *Goniopteris* Presl (incl. *Meniscium* Schreb.). The old genus *Meniscium* comprised tropical ex-indusiate Ferns of Goniopterid-habit, with relatively large leaf-expanses: they differed from *Goniopteris* only by their less highly divided leaves, and by their elongated or confluent sori. There is in fact a consecutive sequence of forms leading onwards from the section *Lastraea* (Diels, *l.c.* p. 177). The characteristic Goniopterid venation is seen in Fig. 663, *A*. The sori are usually roundish, and are seated at the junction of two secondary veins that meet at an acute angle. They may be compared with those of ex-indusiate Dryopteroids. But sometimes the sori are seen to be elongated following the course of the veins, while in others the whole under surface of the leaf may be covered by sporangia. In fact a progression is illustrated towards an Acrostichoid state.

It is but a step from this to the condition seen in certain pinnate species of *Leptochilus*, a genus formerly included in *Acrostichum* (*Syn. Fil.* p. 417). Their general habit is that of *Meniscium*, and comparison of their venation shows that it is of the same type, but with greater complication of its net-

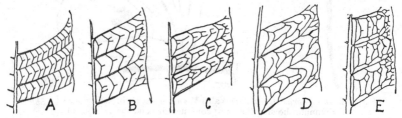

Fig. 663. Venation of *Dryopteris* and *Leptochilus*. *A* = *D. urophylla* (*Meniscium*). *B* = *L. subcrenatus*. *C* = *L. scalpturatus*. *D* = *L. heteroclitus*, sterile. *E* = *L. heteroclitus*, fertile. (× about 2. After Frau Eva Schumann.)

work (Fig. 663, *B–E*). The spores have a perispore, as in *Dryopteris*. Frau Eva Schumann has examined the relation of the sporangia to the veins, both in normal and in half-fertile leaves. In the latter they are seated chiefly along the veins, or in their near proximity (Fig. 664). Where a normal sporophyll has been cut through in an early stage, the oldest sporangia are seated above the veins: but younger sporangia cover the intervening surfaces, showing a completely Acrostichoid state (Fig. 665). The facts are here so cogent that it may be stated definitely that such pinnate species of *Leptochilus* as *L. cuspidatus* (Pr.) C. Chr. are to be held as Dryopteroid derivatives, and that their Acrostichoid condition has been derived from ex-indusiate types such as the old genus *Meniscium*, the sori having spread along the veins and finally over the whole lower surface of the sporophyll.

Sequences of events distinct from this, and from one another, lead to the condition seen in *Polybotrya* Willd. and in *Stenosemia* Presl. We may start in each case from a soral but ex-indusiate and heterophyllous type of Dryopteroid. In *Polybotrya* it has been shown by Frau Eva Schumann how the receptacle of the sorus itself becomes extended to form an enlarged fertile surface (Fig. 666). In i. (*P. articulata*) the receptacle on either side of the mid-rib is narrow: in ii. (*P. scandens*) it is widened, carrying the leaf-margin

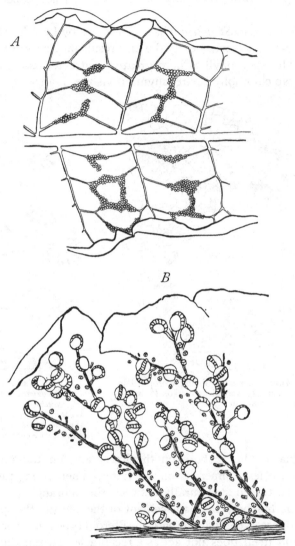

Fig. 664. *Leptochilus cuspidatus*, after Frau Eva Schumann. *A* = middle-form between sterile and fertile, with sporangia still unripe. *B* = middle-form with ripe sporangia. (× about 8.)

backwards: in iii. (*P. osmundacea*) this widening is carried farther, so that the leaf-margin appears to arise from the upper surface: in iv. (*P. cervina*) the receptacle appears to occupy the greater part of the upper surface, while

the margins appear close together, near to the mid-rib. An indication that
this is a correct interpretation is given by a section in the plane including
the course of the vascular supply: such a section (v.) shows the strand as a
recurved loop, following the course which the extension is believed to have
taken.

The development in *Stenosemia* has been shown by Frau Eva Schumann to
be again different, however nearly the end-result may appear to be the same
(Fig. 668). The sporophyll is here greatly contracted compared with the
foliage leaf. No chlorophyll-parenchyma is visible. Transverse sections show

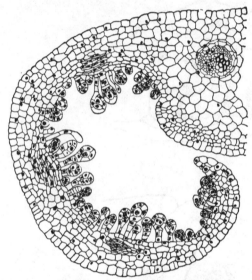

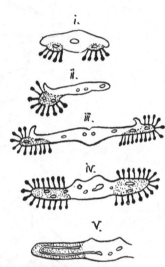

Fig. 665. *Leptochilus cuspidatus.* Young sporophyll in
transverse section, showing Acrostichoid state. (× about
55. After Frau Eva Schumann.)

Fig. 666. Various forms of the sori
of *Polybotrya*. i. = *P. articulata*. ii. =
P. scandens.   iii. = *P. osmundacea.*
iv. = *P. cervina.*   v. = *P. cervina*,
showing course of venation. (After
Frau Eva Schumann.)

that sporangia are borne both on the lower and on the upper surfaces,
though they are less numerous on the latter. When young the actual leaf-
margin can be traced structurally between the sporangia (*R*, Fig. 668), so
that here there has been no displacement of the margin: the sporangia arise
from both surfaces. A comparison with the Pteroid type is not here permissible,
for the margin itself does not act as a receptacle, but remains persistently
unchanged (Schumann, *l.c.* p. 253). The conclusion follows that *Stenosemia*
and *Polybotrya* are rightly placed side by side, but distinct, since they
illustrate distinct methods of arriving at practically the same Acrostichoid
result.

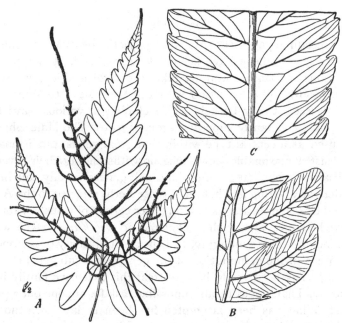

Fig. 667. A, B = *Stenosemia aurita* (Sm.) Presl. A = habit of sterile and fertile leaves. B = part of a sterile segment with venation. C = *Polybotrya Caenopteris* (Kze.) Klotzsch, part of a primary pinna, with venation. (After Diels.)

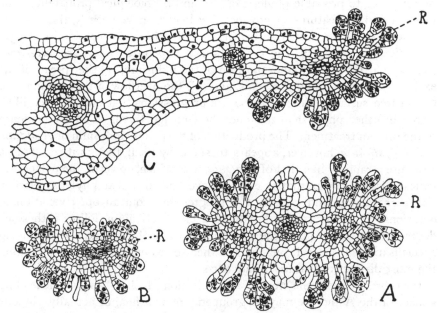

Fig. 668. *Stenosemia aurita*. A = transverse section through a normal sporophyll. B = the same near to the apex. C = transverse section through the blade of a young middle-form between sterile and fertile. R = marginal cell of the blade. (× about 77. After Frau Eva Schumann.)

CONCLUSION

In the preceding pages the probable phyletic relations of the genera included in the Aspidieae have been lightly sketched. It is apparent how the great genus *Dryopteris* naturally takes a central position, which is connected downwards by comparison with the Woodsieae to some probable Cyatheoid source. It may be held as developing in various ways the lopsided type of a constantly superficial, mixed sorus. The Male Shield Fern has been used as a central type within that great genus: provisionally that seems at least a reasonable view, since notwithstanding their several lines of specialisation the related genera and sub-genera readily fall into their places as derivatives from it. That all of these Ferns referred to the Aspidieae by Diels and Christensen are really akin (after certain obviously incompatible genera have been excluded, such as *Dipteris* and *Neocheiropteris*) is witnessed by the richness of their synonymy, and the great diversity of method shown by systematists in the delimitation of the genera and sub-genera.

No attempt has been made here to treat this extensive family in detail, systematically. That would be an impossible task for anyone but a professed systematist. What has been attempted has been to lay down those broad lines which a natural classification must needs follow; and indeed has followed, though not always consistently in the past. The results have been checked by reference to probable phyletic origin, and by biological probability. The most important features of advance are held to have been (i) the advance from a gradate to a mixed sorus, (ii) the adoption of a lop-sided or zygomorphic symmetry, (iii) the production of a peltate "*indusium superum*," (iv) the partial or complete abortion of the indusium, (v) the spread of the ex-indusiate sorus along the veins, and (vi) the overflow of the sporangia on to the free surface. Most of these are progressions that may be seen illustrated in other phyla than this one: the fact that this is so adds confidence to the present treatment. The production of the peltate indusium of *Aspidium* and *Polystichum*, however, appears to stand by itself, though it results in a structure strangely like that of *Matonia*, with which, however, the palaeontological history shows that it cannot have had any near phyletic relation. Other characters, and notably the progression from an open venation to occasional vein-fusions and finally to a close reticulum, follow with some degree of constancy upon the widening of the leaf-area and the advancing specialisation of the sori: but the sori themselves may be held as providing the most distinctive features of advance.

In accordance with the steps of specialisation which they show, the leading genera of the Aspidieae may be grouped, not as a final or actually phyletic arrangement, but so as to bring into prominence the changes which are held to be the most important in leading towards that end. It will be found that the progression thus roughly indicated runs parallel with similar progressions seen in phyla quite distinct from the Aspidioid Ferns. The grouping of the genera of the Aspidieae according to their soral characters will therefore resolve itself as follows:

I. SORI WITH RENIFORM INDUSIUM.

Dryopteris Adanson, 1763	741 species.
Luerssenia Kuhn, 1882	1 species.
Fadyenia Hooker, 1840	1 species.
Mesochlaena R. Brown, 1838	1 species.
Didymochlaena Desvaux, 1811	1 species.

II. SORI WITH PELTATE INDUSIUM (*INDUSIUM SUPERUM*).

Aspidium Swartz, 1801	138 species.
Polystichum Roth, 1799. Schott, 1834	112 species.
Cyclopeltis J. Smith, 1846	2 species.
Adenoderris J. Smith, 1875. Maxon, 1905 ...	2 species.
Phanerophlebia Presl, 1836. Underwood, 1902 ...	10 species.
Cyclodium Presl, 1836	2 species.

III. SORI EX-INDUSIATE, OR SHOWING VARIOUS DEGREES OF ABORTION.
Many species of *Dryopteris*, and particularly the old genera *Meniscium* and *Phegopteris*[1].

Acrophorus Presl, 1836	1 species.
Plecosorus Fée, 1850–52	3 species.

IV. SORI ACROSTICHOID.

Polybotrya Humb. and Bonpl., Willdenow, 1810	30 species.
Stenosemia Presl, 1836	2 species.
Leptochilus Kaulfuss, 1824	the pinnate species.

BIBLIOGRAPHY FOR CHAPTER XLI

667. METTENIUS. Filices Horti Lipsiensis. Leipzig. 1856.
668. METTENIUS. Farnkräuter, IV. Frankfurt. 1858.
669. LUERSSEN. Die Farnpflanzen, Rab. Krypt. Flora, III. 1889.
670. DIELS. Engler and Prantl, I, 4, p. 167, etc. 1902.
671. HANNIG. Vorkommen von Perisporien, Flora, Bd. 103, p. 340. 1911.
672. BOWER. Studies II, Ann. of Bot. XXVI, p. 269. 1912.
673. DAVIE. *Peranema*, Ann. of Bot. XXVI, p. 245. 1912. Also XXX, p. 101. 1916.
674. C. CHRISTENSEN. Monograph of the genus *Dryopteris*. Copenhagen. 1913 onwards.
675. Frau EVA SCHUMANN. Die Acrosticheen, Flora, Bd. 108, p. 201. 1915.
676. VON GOEBEL. Organographie, II, 2, pp. 959, 1148. 1918.
677. BOWER. Studies VII, Ann. of Bot. XXXII, p. 1. 1918.

---

[1] Diels (E. and P. I, 4, p. 167) remarks in relation to the genus *Nephrodium* (*Dryopteris*): "The insecurity of the diagnostic characters founded on the indusium is nowhere more prominent than here. Before all it is plain that the absence of this organ within the Aspidieae has no systematic value." We may readily assent to this from the point of view of the herbarium-systematist. In the practice of diagnosis and classification it may be useless: nor is it here possible to designate strictly under any general name those species of *Dryopteris* that are actually ex-indusiate. But from the point of view of Comparative Morphology the ex-indusiate condition has a definite significance, since it has opened the way to the recognition of the Acrostichoid development.

CHAPTER XLII

ASPLENIOID FERNS

THE Asplenieae, as grouped by Diels (*Nat. Pflanzenfam.* I, 4, p. 222), comprise an apparently compact body of fifteen genera. The diagnosis which he gives is drawn in somewhat general terms, and does not take into account the genesis of the structures there designated as "sori." Though these have a superficial similarity throughout the group thus assembled, there is now sufficient evidence to show that they have originated from two distinct sources. The one is by extension and modification of form of the individual sorus of the type of *Dryopteris*: the other is by fusion of sori into elongated coenosori, as in *Blechnum*, which may subsequently be disintegrated, though not necessarily along lines corresponding to the original fusion. There are thus two lines phyletically distinct, which are confused under the heading Asplenieae as applied by Diels.

The group is, however, divided by him into (1) Aspleniinae, and (2) Blechninae, which respectively correspond roughly with the distinction in phyletic origin. The Aspleniinae include for the most part those genera in which the sorus has been a unit throughout descent, though subject to modification in form: the Blechninae include genera in which the "sorus" is not a unit, but the whole or part of a coenosorus, soral fusion having played its part in their evolution. But the distinction as drawn by Diels is not a clean cut between the two. Certain genera, and in particular *Scolopendrium* (= *Phyllitis* Ludwig), will have to be transferred from the Aspleniinae of Diels to his Blechninae if the distinction is to be a phyletic one. The grounds for this will be stated in detail later. It will be desirable to accentuate this distinction by segregating the treatment of these phyletically distinct lines. The present Chapter will deal with those Ferns which may be designated "Asplenioid," in the sense that they share with the genus *Asplenium* the fact that the sorus is a definite unit: while a later Chapter will treat of the Blechnoid Ferns in which fusion has led to the formation of a coenosorus.

The Asplenioid Ferns thus defined include the genera *Athyrium* Roth (1799, 86 species), *Diplazium* Swartz (1801, 208 species), *Diplaziopsis* C. Chr. (1906, 1 species), *Asplenium* Linn. (1753, 429 species), *Ceterach* Adanson (1763, 4 species), and *Pleurosorus* Fée (1850, 3 species).

Fig. 669. Fig. 31, *a*, *b* = *Phyllitis Scolopendrium* (L.) Newm. *a* = very young sorus cut vertically. *b* = the same older, with veins, protecting "indusial" flaps, and the first sporangia (× 125). Figs. 32–35 = *Asplenium*, showing venation and sori (× 3). Fig. 32 = *Asplenium obtusatum* Forst. Fig. 33 = *Asplenium horridum* Kaulf. Fig. 34 = *Asplenium serra* Langsd. and Fisch. Fig. 35 = *Asplenium* (*Diplazium*) *celtidifolium* Kze. Fig. 36, *a*–*c* = *Asplenium obtusatum* Forst., sections of fertile pinnae, showing successive stages of development of the sori. (*a* and *b* × 125: *c* × 50.)

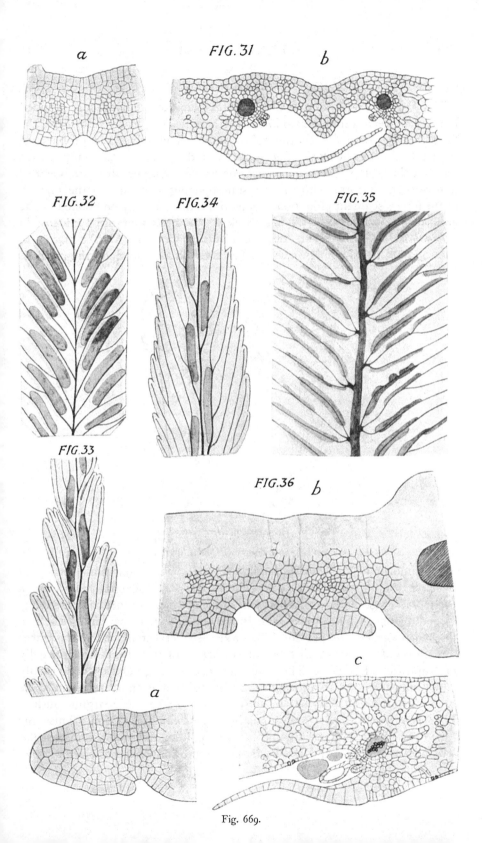

Asplenium Linn.

The typical genus around which the Asplenieae are habitually grouped is the Linnaean genus *Asplenium*. But here, as frequently elsewhere, the type that is nominally central raises rather than solves the problem of phyletic origin. The distinctive features of the Ferns of this large genus are that the sori arise as a rule singly on the fertile veins: they are linear or linear-oblong in form, with the indusium opening so as to face the mid-rib of the fertile segment (Fig. 669). Sometimes the sori may be hooked at the end, as in *Athyrium*: or double, back to back, as they are typically in *Diplazium*: or the indusium may occasionally open towards the margin of the segment instead of inwards. All of these facts bear upon the problem

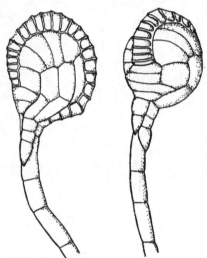

Fig. 670. Sporangia of *Asplenium Tricho-*
manes L., after C. Müller, showing the stalk
consisting of a single row of cells. (× 140.)

of their morphology. The sporangia are small, with a long stalk formed of a single row of cells (Fig. 670). The spores are bilateral, and have a perispore. These Ferns are herbaceous or firm in their texture, with creeping, ascending or upright stock. The leaves are alternate or spiral. Lattice-thickened scales are present. The axis is dictyostelic, with leaf-traces sometimes of a single meristele, but consisting of two straps which usually fuse upwards (Fig. 671). The leaves may be simple, but they are usually pinnatifid, and their branching may run to 4 or even 5 partitions. The venation is usually free, but in some there may be intramarginal fusions (*Thamnopteris* Presl), or even reticulation (*Asplenidictyum* J. Sm.). This large genus comprises 429 species, and is cosmopolitan. The habit and vascular structure readily accord with what is seen in the simpler types of *Dryopteris*,

though the sorus of *Asplenium* appears as a more difficult problem. A broad
outlook will have to be taken in its solution.

Here, as elsewhere in the Ferns, the question of soral morphology is to be
studied in relation to the widening leaf-area. The narrow highly divided
Sphenopterid type of leaf, so common among Palaeozoic vegetation, was
probably the source from which most modern Ferns originated (see Vol. I,
Fig. 93, *A* and *C*: also Fig. 75). The simpler Dryopteroid and Asplenioid
Ferns still conform closely to that early type, as do also the Hymenophyllaceae. Primarily the Ferns appear to have been mesothermic hygrophytes,
as so many of those of the present day still are. The highly-divided leaf
presented a large proportion of surface to bulk of tissue, and this would

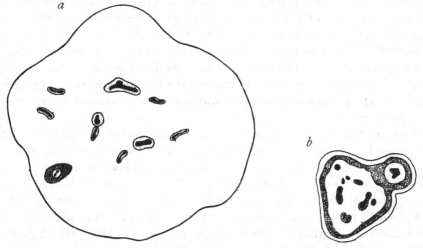

Fig. 671. *a* = transverse section of the stock of *Asplenium (Diplazium) marginatum* L., a large
and sappy species. The dictyostele presents here three meristeles, with leaf-traces opposed to
each gap, each consisting of two vascular straps (× 3). *b* = a similar section of the much smaller
Asplenium alatum H.Br. Here the dictyostele and leaf-traces are similar, except that two straps
of each leaf-trace fuse laterally soon after their departure from the leaf-gap (× 12). One root-
trace is seen in *a*, and two in *b*.

be a successful proposition under moist conditions. But exposure to the
dryer air of exposed sites would impose a more compact construction. An
increasing leaf-area, secured either by widening or by webbing of the leaf-
segments, would give not only an increased photo-synthetic area, but also
that decreased proportion of bulk to surface which exposed life demands.
Evidence of such widening and webbing, often accompanied by a thicker
texture, can be obtained by comparison of the representatives of almost any
large family of Ferns. The Asplenieae are no exception to this, and a striking
example is seen in the leaves of the huge epiphyte, *A. nidus*, where the
broad leathery leaves bear individual sori several inches in length (see
Vol. I, Fig. 52). So long as the leaf remains a general-purposes leaf, serving

both the offices of nutrition and of propagation, the sori will tend to follow the widening surface of the blade. Either they will be increased in number by some form of duplication, or they will be themselves extended. Examples of this appear in very striking form in the Marattiaceae, and they have been traced in relation to the widening leaf-area in Vol. II, Chapter XX. They illustrate the results of progressive extension of the sorus in the so-called *Archangiopteris*, *Protomarattia*, and in *Danaea*: but subdivision probably accounts for the spread of the numerous circular sori on the broadened leaf of *Christensenia*. The same alternative was presented to the Ferns now under discussion. The theory to be elaborated here is that while *Dryopteris* in its broad-leaved forms repeated or reduplicated its individual sori, in *Asplenium* the individual sorus was subject to extension along the veins of the widening area of the blade, appearing as the linear sori shown in Fig. 669. The question will be what relation these linear sori of the Asplenioid Ferns may bear to other well-known types of sorus elsewhere. The enquiry will naturally be directed towards the Dryopteroid type, partly by the similarity of vegetative structure in the two families, partly by those hooked ends and other unusual soral forms already mentioned as occurring in *Asplenium*.

If we examine a large run of herbarium specimens of the genus, certain species are found to present exceptional forms of sori in considerable numbers. A specially favourable species is found in *Diplazium lanceum* Thunb., which was figured by Hooker and Greville as *A. subsinuatum* (*Ic. Fil.* Vol. I, Pl. 27). A single leaf of this species, now in the British Museum, collected by Wallich at Napalia in 1829, yielded sori which appear to link the type of *Asplenium* unmistakably with that of *Dryopteris*. Already in the Dryopteroid genera *Didymochlaena* and *Mesochlaena*, and to some extent also in *Luerssenia*, an elongation of the sori appears, following the outwards-running veins to the margin of the widened blade. But here the receptacle remains fertile round the distal curve, and the extension appears equal on both sides. If we imagine it interrupted at the curve, the type of sorus would approach that of *Diplazium*; if only one side were developed, the sorus would rank as *Asplenium*. All of these conditions were found on a single leaf of *Diplazium lanceum* (Fig. 672, 1–6). Occasionally the general form of the sorus did not appear to differ materially from that of one of these Dryopteroids (1): but others may be unequally developed on the two sides (2, 3, 4). In others again the fertility of the distal end of the receptacle may be lost, and the sorus takes the character of *Diplazium* (5): or one side only (and here the anadromic side) of the sorus is developed, it then has the character of *Eu-Asplenium* (6). *D. lanceum* has an open venation, and the position of the sori relatively to them is the same as that in *Dryopteris*.

In all this we may see a natural development of the lop-sided or zygomorphic sorus, which was initiated from the more ancient radial type along such a series as *Gleichenia, Lophosoria, Alsophila, Hemitelia,* and *Peranema.* The zygomorphy was slight in *Peranema,* but already it was marked by those lateral pouches of the fertile receptacle, which became more elongated in the Dryopteroid Ferns (see Chapter XLI, p. 124). One or both of these find their extreme form in the actual sori respectively of *Asplenium* and of *Diplazium,* while the central part of the receptacle, which was the original fertile region, has dropped out of functional development. Meanwhile it is only the lateral lobes (both or only one) of the *indusium superum* which survive. Without the intermediate steps it would be difficult to trace in the Asplenioid indusium the correlative of a part of the Cyatheaceous *indusium inferum.*

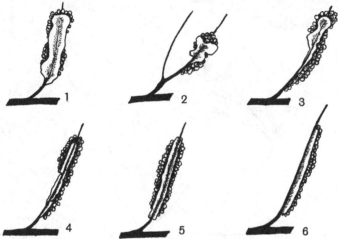

Fig. 672. 1–6=drawings from sori all borne on the same leaf of *Diplazium lanceum* Thunb., showing stages that illustrate steps between the Dryopteroid sorus (1–3) and those of *Diplazium* (4, 5), and *Eu-Asplenium* (6). (From a specimen in Kew Herbarium under name *Asplenium lanceum,* Napalia. Dr Wallich, 1829.)

ATHYRIUM Roth

Additional evidence of the correctness of these comparisons may be found in the genus *Athyrium,* which there is reason to believe is more nearly related to *Dryopteris* than is *Asplenium* itself. It comprises Ferns with creeping, but more frequently with ascending or upright stock, on which the leaves are borne spirally. Sometimes these Ferns are almost dendroid. Thin-walled scales are present, not the latticed scales of *Asplenium.* The adult stem contains a rather wide-meshed dictyostele. The petiole is traversed at its base by two broad straps of the leaf-trace, as in *Dryopteris Thelypteris* (L.) A. Gray, and *Polystichum lonchitis* (L.) Roth: these unite below the blade to form a gutter-shaped meristele. The leaves are at least once, and usually twice or thrice pinnate, and the venation is open. The sori and indusium are oblong, and are often strongly recurved, sometimes even horse-shoe-shaped, or more or less reniform, and then they hardly differ (as van Rosenburg remarks) from *Lastraea* (*Dryopteris*) (Fig. 673). In these features *Athyrium* corresponds to some of the less advanced types of *Dryopteris*, a fact that concentrates attention upon the details of its sorus.

The genus *Athyrium* was founded by Roth in 1799 (*Tent. Fl. Germ.* III, 58) to receive certain species of the older Linnaean genus *Asplenium*, which he found to be characterised not only by hooked or sometimes horse-shoe-formed

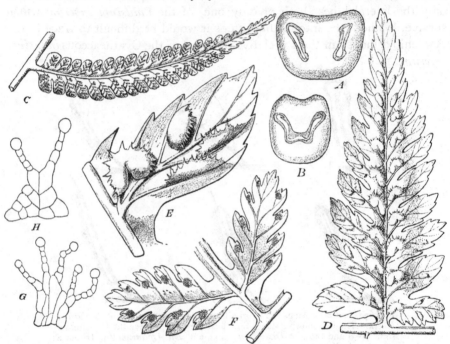

Fig. 673. *Athyrium* Roth. *A*, *B* = diagram of transverse section of the petiole of *A. filix-foemina* (L.) Roth, *A* below, *B* in upper half. *C* = *A. acrostichoides* (Sw.) Diels, primary pinna. *D*–*E* = *A. filix-foemina*. *D* = primary pinna. *E* = secondary segment with venation, sori, and indusium. *F*–*H* = *A. alpestre* (Hoppe) Nyl. *F* = two tertiary segments, with venation and sori. *G*, *H* = indusium highly magnified. (*A*, *B*, *F*–*H*, after Luerssen: *D*, *E*, after Mettenius: *C*, after Diels.) From *Natürl. Pflanzenfam.*

sori, but also by the structural fact that their receptacle has a special vascular supply. Whereas in *Asplenium* the sori are inserted on the vein direct, in *Athyrium* a small vascular strand branches off from the back of the fertile vein and passes into the receptacle, thus giving it a higher morphological individuality of its own. Mettenius notes this (*Farngattungen*, VI, p. 57, Frankfurt, 1859), but he does not on that account admit that *Athyrium* is a linking genus between *Asplenium* and *Aspidium*. He held rather that *Athyrium* is at all events nearer to *Asplenium* than to *Aspidium*. While his point may be true enough for the classification of 1859, it is to be remembered that Mettenius was writing in pre-Darwinian days. It does not now in any way detract from the interest that *Athyrium* possesses in pointing a probable genetic relation with *Aspidium,* or more nearly with *Dryopteris.* Seen in the light of more recent examination of soral structure, the extension of a special vascular supply into the receptacle is a feature of some importance for comparison: for all along the line from the Gleicheniaceae through the Cyatheaceae to the Dryopterideae a special vascular supply passes into the receptacle, and it is specially marked in the long stalk of the sorus of *Peranema.* Accordingly the presence of a special receptacular strand in *Athyrium,* which is absent in *Asplenium,* marks its nearer relation to *Dryopteris* and to *Peranema.*

Fig. 674. *Athyrium decurtatum* (Ktze.) Presl. Part of a pinna with pinnules bearing sori. Those at the base of each pinnule are sometimes Dryopteroid, sometimes they show variability leading to the Asplenioid type of the distal sori.

The basal sorus of each pinnule frequently has, like *Dryopteris,* a kidney-shaped indusium. This is seen in *Athyrium filix-foemina* (L.) Roth: but the sori borne distally on the pinnules have the one-sided Asplenioid form, with the fimbriated indusium opening towards the mid-rib. They represent the anadromic side of the horse-shoes. A species that has been specially quoted by Mettenius, as bearing basal sori of the Dryopteroid type, is *A. decurtatum* (Ktze.) Presl. This was examined in a specimen in the British Museum collected by B. Balansa, Paraguay, 1874: the variability was found as described (Fig. 674). These examples will suffice to illustrate what may be regarded as the conservatism of the basal sori. Here and there they may also be of the Diplazioid type, as Luerssen has noted (*Rab. Krypt. Fl.* III, p. 131). Such facts appear strongly to support the Dryopteroid affinity of the Asplenioid Ferns, and to indicate *Athyrium* as the nearest point of contact. That contact appears to be with those simpler types of the Dryopteroids, which share with *Athyrium* the open venation and the highly cut leaf-blade.

As in the Aspidieae so also here the indusium is sometimes liable to partial or complete abortion. This is seen in *Athyrium alpestre* (Hoppe) Rylands, where the indusium is rudimentary even in the young state, consisting of hair-like rows of cells (Fig. 673, *F, G, H*). In the mature state the sorus appears unprotected, the indusium having shrivelled (Luerssen, *l.c.* p. 143). Such reduction or abortion of the indusium may be found also in some of the larger forms of *Diplazium*, and it becomes a distinctive feature in *Ceterach* and in *Pleurosorus*.

ALLANTODIA Wall.

This problematical genus, styled *Diplaziopsis* by C. Christensen, contains a single large Eastern species. It will have to be examined in fresh and developmental material before it can be properly understood. Provisionally it would seem probable that its single species is a relatively advanced Asplenioid type, as its reticulate venation indicates. The sorus may probably be structurally like that of *Asplenium*, but the spores are liberated by rupture of the thin membranous indusium, not by inversion and shrivelling, as in *Asplenium*.

DIPLAZIUM Swartz

The distinction of this genus from *Asplenium* or *Athyrium* is not a sharp one. In *Diplazium* the veins are as a rule free, and the sori and indusia extend along both sides of some of them (Fig. 669, 35). This is exactly what might be expected to occur at times, if the Asplenioid Ferns were derived from the Dryopteroid type, along lines already suggested in the case of *Asplenium* (*Diplazium*) *lanceum*. As Diels remarks (*l.c.* p. 225), the character of the Diplazioid sorus, used by Presl to delimit the genus from *Asplenium*, has only this importance, that in *Asplenium* it appears exceptionally, or at least much more rarely than in *Diplazium*. Both may in fact be held as derivatives of the Dryopteroid type. There are about 200 species in this genus, mostly inhabitants of moist tropical countries, and often of large size. This combined with the suggested origin points to a nutritional explanation of the Diplazioid state. The easy nutrition of the moist tropics would encourage a full habit, with the development of both sides of the originally horse-shoe-shaped sorus, whereas only one side is fully developed in typical *Asplenium*, or in *Athyrium*. (Compare Fig. 672, p. 143.)

The vascular system of *Diplazium* resembles that of *Asplenium* (Fig. 671). But in a large *Diplazium* such as *D. marginatum* the two straps of the leaf-trace may take each a separate course far up into the petiole (*a*), while in a small *Asplenium* such as *A. alatum* they may fuse almost immediately on their departure, after the manner frequent in these Ferns (*b*).

In a sub-dendroid Asiatic species, *Diplazium esculentum* (Retz) Sw., a medullary system is found within the dictyostele; strands may originate independently in the pith, or as internal thickenings of the meristeles. It may be held as a concomitant of large size (Y. Ogura, Reprint from *Botanical Magazine*, Tokio, Vol. XLI, No. 483, p. 172). The detail of the sorus of *Diplazium*, compares on the one hand with that of *Asplenium*, on the other with *Dryopteris*. A section traversing the vein of *Diplazium celtidifolium* Kze. transversely shows the indusium with its arching flaps right and left, as they appear in *Dryopteris* when cut in a similar direction (Vol. I, Fig. 14): while either side of it corresponds to what is seen in *Asplenium* (Fig. 669, 36 *a–c*). Some of the Ferns of this affinity attain a very considerable size, and a high complexity of leaf-architecture. Christ quotes *D. ceratolepis* Christ as being several metres high, with dendroid habit. The leaf-segments of the larger species often appear as broad expanses in which occasional vein-fusion or even reticulation may occur as a secondary consequence. Such development is sometimes accompanied by a partial or complete abortion of the indusium,

Fig. 675. *Diplazium ceratolepis* Christ. Two ultimate segments from the very large leaf, natural size, showing vein-fusion, and elongated sori, without indusium. (After Christ.)

a condition consistent with growth in moist forest shade (Fig. 675).

CETERACH Adanson

A similar abortion of the indusium is seen as a regular feature in Adanson's genus *Ceterach*, though it is probably to be ascribed to different biological circumstances. The genus comprises small xerophytic plants with upright rhizome and simply pinnatifid leaves, their lower surface being covered by brown scales. The veins anastomose towards the margin. The sori which are of the type of *Asplenium* have the indusium partially or completely abortive (Fig. 676). Biologically it may well be that the effective protection by scale armour has made the indusium superfluous. States of partial abortion shown by the detailed drawings of Luerssen clearly demonstrate the Asplenioid origin, while the vein-fusions indicate a derivative position in that series.

PLEUROSORUS Fée

With this may probably be ranked the small genus *Pleurosorus*, which includes mountain Ferns of wide distribution, in which the indusium is completely absent (Fig. 677). The condition of these is very nearly that of *Gymnogramme*, and it is quite likely that some of the Ferns that have been ascribed to that affinity may ultimately find their natural place here. This has been broadly suggested for *Aspleniopsis* Mett. & Kuhn, in which the habit clearly supports the affinity (E. & P. I, 4, p. 272, Fig. 145, *A*). There

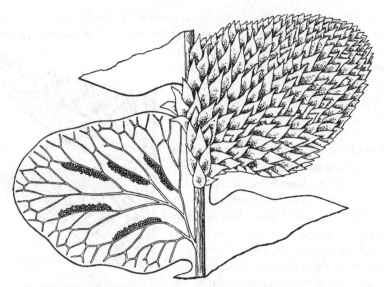

Fig. 676. *Ceterach officinarum* D.C. Two segments from the upper part of a leaf. The right-hand segment shows the scaly covering of the lower leaf-surface: the left-hand segment the nervation and sori after removal of the scales. The minute indusium is invisible under the low magnification. (× 8. After Luerssen.)

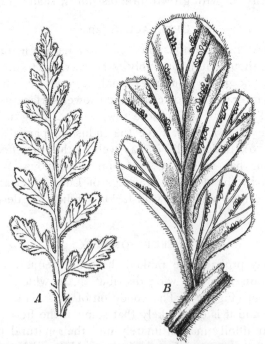

Fig. 677. *Pleurosorus Pozoi* (Lag.) Diels. *A* = a primary pinna. *B* = a secondary pinna, with venation and sori, enlarged. (After Fée, from *Natürl. Pflanzenfam.*)

is no clear indication of an Acrostichoid development from the Asplenioid Ferns, unless it be in the problematical genus *Rhipidopteris* Schott. In this genus the species described as *Elaphoglossum* (*Rhipidopteris*) *flabellatum* H. Chr. has foliage of the type of *Asplenium ruta-muraria*, while a compact development of an ex-indusiate sporophyll of the type of *Pleurosorus Pozoi* would form a ready basis for an Acrostichoid development (Christ, *Elaphoglossum*, Zurich, 1899). The spores of *Rhipidopteris* are recorded as possessing a perispore, which is a significant fact in support (Hannig, *Flora*, 1911, p. 338). It must remain for others to test this possible relationship by exact observation.

The genera *Triphlebia* Baker, *Diplora* Baker, and *Scolopendrium* L. (*Phyllitis* Ludwig), will be treated later, as being probably Lomarioid derivatives rather than of Asplenioid relationship.

COMPARISON

The genera thus retained in the Asplenieae form a naturally coherent group: there is indeed difficulty in drawing any consistent line between the genera *Asplenium*, *Athyrium*, and *Diplazium*: and these comprise by far the majority of the widely spread species. It is characteristic of them all that they maintain the individuality of the sorus. There is no evidence of soral fusion: even *Diplazium*, to which such fusion has been ascribed in explaining the apparently double sorus, is open to the much more probable interpretation of soral extension which is here suggested. All of them may be held as Dryopteroid derivatives. The general habit and the anatomy of the less specialised of these Ferns fully coincide with this view. It requires some experience to distinguish *Athyrium filix-foemina* from the common Dryopteroid Ferns in any wood without looking at the sori. The similarity of the dictyostele and of the leaf-trace between the Lady Fern and certain species of *Dryopteris* has been noted, while *Athyrium* also shares with the Shield Ferns their relatively thin chaffy scales, though *Asplenium* and *Diplazium* tend to the type with lattice-thickening, clathrate as it is called. These characters together with the open venation and the general Spheno-pterid-architecture of the leaves relates the Asplenieae and in particular *Athyrium* to the Dryopteroid Ferns.

Supported by this broad and general comparison, the soral features may be expected to show some degree of similarity. The basal sori of the pinnules in certain species of *Athyrium* have been shown to have the actual outline and construction of those of *Dryopteris*. Certain genera allied to *Dryopteris*, such as *Didymochlaena*, illustrate the power of spread of the sorus along the course of a vein. Given this, and a possible inequality in that spread on the two sides of the sorus, all that is positively required for the origin of the sorus of *Athyrium* from a Dryopteroid source is present. The steps of the inequality have been sketched by Von Goebel (*Organographie*, II, 2, 1918, pp. 1151, 1195). Abortion of the region of the receptacle nearest to the margin of the leaf, and of that part of the indusium, while one or both of the elongated sides of the receptacle retain their fertility, would give respectively the sori of *Eu-Asplenium* or of *Diplazium*. But the curious point here is that that region nearest to the leaf-margin represents the originally fertile receptacle, according to the comparative history given in the preceding pages. It is the oldest part that becomes non-functional in *Asplenium* and *Diplazium*.

The sorus of the Asplenieae may thus, on the basis of comparison, be held to be the last terms of a progressive series of changes that started from a non-indusiate radial sorus, such as is seen in *Gleichenia, Lophosoria*, or *Alsophila*. Hairs are commonly present in the last two genera at the base of the receptacle. Their place is taken in certain Cyatheoids and Woodsioids by the *indusium inferum*. A zygomorphic or lop-sided development already visible in *Peranema* becomes more pronounced in *Dryopteris* by extension of the lateral pouches of *Peranema*, and their still greater elongation in *Didymochlaena*, though the primary receptacle still retains its fertility: but as illustrated by the various soral types seen in *Diplazium lanceum*, that fertility is finally lost in *Diplazium* and *Asplenium*, and it is the product of those secondary pouches of *Peranema* which remains as the fertile receptacle of these Ferns. Finally, the indusium itself becomes partially or completely abortive in various Asplenioid Ferns, and the penultimate state appears thus as the exposed linear sorus of *Ceterach* or *Pleurosorus*. But ultimately there may be a possible Acrostichoid spread of fertility over the leaf-surface, as in *Rhipidopteris*, if that Fern has its proper place with the Asplenieae.

The essential point in this phyletic story, as witnessed by the comparisons given above, is that fusion to form coenosori plays no part in it. As the leaf widens it is the individual sorus that lengthens, so that however nearly the result may resemble the coenosori of certain species of *Nephrolepis* or those typical of *Pteris*, such similarity can only be held as homoplastic. A closer and more critical question arises, however, in the comparison of the Asplenioid with the Blechnoid Ferns. These have habitually been classed together. That there is reason to believe that those soral similarities which they show are also homoplastic, and that the elongated fertile tracts of the Blechnoids are really coenosori, will be the theme of the next two Chapters.

Turning now to the natural grouping of the Asplenioid Ferns, they may be arranged in sequence according as their soral characters appear progressively distinct from the Dryopteroid type, as it is seen in the Male Shield Fern.

ASPLENIEAE

(Excl. Blechninae: also *Triphlebia* Baker, *Diplora* Baker, *Phyllitis* Ludwig and *Camptosorus* Link.)

I. SORI INDUSIATE.

Athyrium Roth, 1799	86 species.
Asplenium Linn., 1753	429 species.
Diplazium Swartz, 1801	208 species.
Diplaziopsis C. Chr., 1906	1 species.

II. SORI EX-INDUSIATE, or showing various degrees of its abortion.

Ceterach Adanson, 1763	4 species.
Pleurosorus Fée, 1850	3 species.

BIBLIOGRAPHY FOR CHAPTER XLII

678. METTENIUS. Farngattungen, VI. Frankfurt. 1859.
679. Sir W. HOOKER. Species Filicum, III. 1860.
680. LUERSSEN. Rab. Krypt. Fl. III, p. 129. 1889. Here the literature is fully quoted up to its date.
681. CHRIST. *Elaphoglossum*. Zurich. 1899.
682. CHRIST. Farnkräuter, p. 188. 1897.
683. DIELS. Natürl. Pflanzenfam. I, 4, p. 222. 1902.
684. HANNIG. Flora, p. 338. 1911.
685. BOWER. Studies II, Ann. of Bot. XXVI, p. 269. 1912. Studies IV, Ann. of Bot. XXVIII, p. 363. 1914.
686. VON GOEBEL. Organographie, II, 2, p. 1151. 1918.

CHAPTER XLIII

ONOCLEOID FERNS

THIS Family is represented only by two genera, comprising five species: out they merit a separate treatment in virtue of the distinctive nature and comparative value of the features which they show. They have been classed by Diels in the Woodsieae, as a sub-tribe Onocleinae, parallel with the Woodsiinae; but he wisely remarks that the relationship is exceedingly doubtful. That there is some degree of affinity with these there can be no doubt, but it reflects more clearly the common source from which they both sprang, than any intimate and direct relation between them as they now stand. Both may best be held as derivatives from a common Cyatheoid ancestry. The two genera of Onocleoid Ferns are *Matteuccia* Todaro (= *Struthiopteris* Willd.), with three older species, to which C. Christensen has lately added a fourth: three are from Asia, the other is the Ostrich Fern (*M. struthiopteris* (L.) Todaro), from Europe, Asia, and America: and *Onoclea* L. with a single species from North America and Eastern Asia. They inhabit northern or mountainous stations, with which their seasonal leaf-fall is in accord. The family, though few in species, is of wide distribution: this fact suggests antiquity of origin, but there is no very early palaeontological record, though *Onoclea* has been recognised from Lower Tertiary rocks in America. (See Seward, *Mem. Geol. Surv. Scotland*, 1924, Ch. IV, p. 74.)

MATTEUCCIA Todaro

The best known species is *M. struthiopteris* (L.) Todaro, the Ostrich Fern, which has been described in detail by Luerssen (*Rab. Krypt. Fl.* III, p. 482), and by Campbell (*Mem. Bost. Soc. Nat. Hist.* IV, 1887). It is illustrated in Fig. 678, *A–F*. *M. orientalis* (Hk.) Trev. and *M. intermedia* C. Chr. are less familiar, and they will be more particularly described here, since they present features important for comparison. *M. orientalis* is a coarse-growing Fern of about the same dimensions as *M. struthiopteris*. It has an obliquely ascending or upright stock completely covered by the bases of densely tufted leaves. No runners have been observed like those which are a marked feature in the Ostrich Fern, as also in certain Cyatheoids. The leaf-bases are enlarged, and bear rough brownish outgrowths, recalling the pneumatophores of *Plagiogyria* (Vol. II, Fig. 543). The leaf-bases are covered by broad chaffy scales. The simply pinnate leaves are dimorphic: the broad pinnae of the sterile leaves are deeply pinnatifid, with open

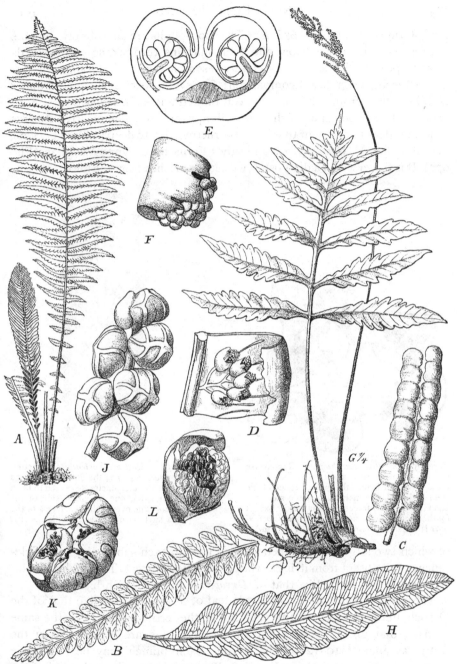

Fig. 678. *A–F = Matteuccia struthiopteris* (L.) Todaro. *A* = one fertile and one sterile leaf, showing habit. *B* = a sterile pinna. *C* = a fertile pinna. *D* = part of a fertile pinna with venation and sori. *E* = section through a fertile pinna. *F* = sorus with indusium. *G–L = Onoclea sensibilis* L. *G* = habit. *H* = sterile pinna. *J* = fertile pinna. *K* = fertile pinnule. *L* = sorus with indusium. (*A, D–F* after Luerssen: *J–L* after Bauer: *B, C, G, H* after Diels.) From Diels, *Natürl. Pflanzenfam.* I, 4.

Pecopterid venation (Fig. 678, *B*). The sporophylls are also simply pinnate, but the segments are narrow, and their margins turn strongly downwards so as to protect the sori. They closely resemble those of *Blechnum capense* (L.) Schlecht, a significant comparison: for several species now placed under *Blechnum* have from time to time been ranked under *Struthiopteris* (= *Matteuccia*). The importance of this fact will appear later. The massive stock is solid, and does not present that basket-structure, due to deep axillary pockets, described for *M. struthiopteris* and other Ferns by Gwynne-Vaughan (Fig. 679). It is traversed by a dictyostele with large meshes, from the lower margins

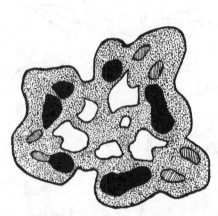

Fig. 679. *Matteuccia struthiopteris*. Transverse sections of the erect stock. The epidermal pockets are left blank, the ground tissue of the stem is dotted, the meristeles of the stem are black, and the leaf-traces cross-hatched. Except for the pockets the structure of *Metteuccia orientalis* is the same. (After Gwynne-Vaughan.)

Fig. 680. *Matteuccia orientalis*. Venation of the fertile pinna: the dots on the veins indicate the receptacles, here seen to be in straight rows, one on either side of the mid-rib. The over-arching margins of the pinna have been removed. (Enlarged.)

of which two large strap-shaped strands pass to each subtending leaf. Sclerenchyma is absent from it, in contrast to the Cyatheaceae. The whole vascular structure is strikingly like that of. *Dryopteris oreopteris* (Ehrh.) Maxon.

The general character of the fertile leaf of *M. orientalis* is like that of the Ostrich Fern, and in both the venation of the fertile pinnae is on the same plan as that of the sterile, but the branch-veins are fewer, 5–6 on the fertile as against 10–12 on the sterile. Each branch may bear a single sorus, as in *M. struthiopteris* (Fig. 678, *D*): but whereas there the sori of the lower branch-veins lie nearer to the mid-rib than those of the upper, in *M. orientalis* they are all seated at an equal distance from it, forming regular intramarginal rows (Fig. 680). This detail is matched in *Blechnum capense* (Fig. 701, 22): compare also Mettenius (*Fil. Hort. Lips.* Pl. IV, Fig. 21). Each vein is continued a short distance beyond the sorus, but it stops where the margin of the pinna curves over as a continuous protective flap. This resembles an indusium, but it is firm in texture, and is coloured brown. In addition to this protection each sorus of *M. orientalis* is covered by a shell-shaped indusium, attached on the side next to the mid-rib (Fig. 678, *L*). It is structurally like one of the ramenta or scales. A section through a sorus shows how the leaf-margin overlaps this scale, giving a very complete protection to the sori within (Fig. 678, *E*). The receptacle is convex, and a short tracheidal branch enters it. The numerous sporangia are arranged in basipetal succession, a fact already demonstrated developmentally for *Onoclea sensibilis* (Fig. 681). In fact we are here dealing with a typically gradate type of Ferns, which has both a protective leaf-margin and a basal indusium.

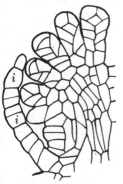

Fig. 681. *Onoclea sensibilis* L. Young sorus in vertical section, showing basipetal succession of the sporangia. *i*=indusium. (× 200.)

The development of the sporangia of the Ostrich Fern has been described in detail by Campbell (*Mem. Bost. Soc. Nat. Hist.* IV, 1887). Their orientation is not exactly maintained in the mature sorus, though it is indicated in their early development. This fact is related to the structure of the sporangial head, which is still unequally convex on its two sides, as it commonly is in gradate Ferns: but the annulus is nearly vertical, and the series of cells is definitely interrupted at the insertion of the stalk (*Phil. Trans.* 1899. Vol. 192, p. 56, Pl. 5, Fig. 91). In fact the annulus is of the type characteristic of advanced Leptosporangiate Ferns, though the sorus is like that of the Cyatheaceae. These details will be found to bo important for comparison with *M. intermedia*. The spores are large, and their number in each sporangium is between 48 and 64. The occurrence of a perispore is inconstant in the Onocleoid Ferns (Hannig, *l.c.* p. 340). It appears to be absent in *Onoclea sensibilis*: but Hannig found it present in *Matteuccia struthiopteris* and *orientalis*.

From the time of Willdenow and other early writers a relation has been recognised between the Onocleoid Ferns and *Blechnum*. There is a general similarity of habit, but an obvious objection would seem to be the presence of a true indusium in the former, and its absence in the latter. If then a non-indusiate member of the Onocleineae were found, that would materially strengthen an otherwise valid comparison instituted by the instinct of the older systematists. The new species *Matteuccia intermedia* C. Chr. (Fig. 682) proves to be non-indusiate (Christensen, *Bot. Gaz.* Vol. 56, 1913, p. 337). This Fern is intermediate in habit between *M. orientalis* and *M. struthiopteris*.

It has a massive stock, showing a wide-pithed dictyostele, and a binary leaf-trace (see Studies IV, Text-fig. 2). The axis and rachis are scaly and the leaves dimorphic, the sporophylls closely resembling those of *M. orientalis*, but they are shorter. Their details also correspond, excepting in the absence of the indusium. Early stages of development of the sorus show no sign of it (Fig. 683). The sori are superficial, of distinctly intra-marginal origin, and

Fig. 682. *Matteuccia intermedia* C. Chr. Habit of Fern as grown in Glasgow Botanic Garden, but showing only sterile leaves. (Much reduced.)

with a gradate sequence of the sporangia: moreover, sections parallel to the margin show that they are disposed in strictly regular rows, one row on either side of the mid-rib (Fig. 683, *C*): and the individual sori are in close juxtaposition. Both in structure and arrangement the sori compare closely with those of *Alsophila*, where also the sori often show a regular alignment. The sporangia are relatively complex, and compare most nearly with those of *Lophosoria* (Fig. 684). The annulus, which shows slight signs of obliquity, consists of over 50 cells, as against about 40 in *Lophosoria* (compare Vol. II, Fig. 551), and

39 for *M. struthiopteris* (Campbell, *l.c.*). The series, however, is not continuous, but is interrupted at the insertion of the stalk. Glandular hairs of the Blechnoid type are present though sparingly on the young pinnae.

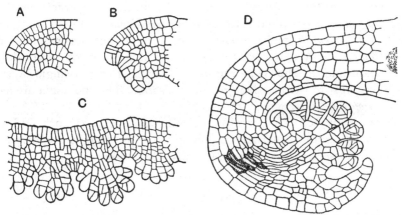

Fig. 683. Development of the sorus of *Matteuccia intermedia*. *A*, *B*, *D*=sections through the pinna-margin, showing successive stages of the superficial, gradate, non-indusiate sorus, protected only by the curved margin. *C*=a section parallel to the margin traversing a succession of the distinct non-indusiate sori. (× 125.)

There are thus good grounds, soral and anatomical, for comparison of *Matteuccia intermedia* with the simpler Blechnoid Ferns, which go considerably beyond the mere features of external habit. The relations downwards are clearly with the Cyatheoid Ferns, and especially with the ex-indusiate types such as *Alsophila* and *Lophosoria*. But on the other hand, it is now more evident than before that the relationship upwards, between *Matteuccia* and the genus *Blechnum*, so clearly recognised by the early systematists, has the support of more detailed comparison than they had instituted. The new facts appear to suggest that the Blechnoids probably sprang from some early Cyatheoid source, and the nearest indication of a connection is seen in the Fern so happily named by Christensen, *Matteuccia intermedia*.

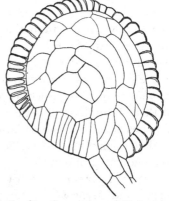

Fig. 684. Sporangium of *Matteuccia intermedia*. (× 125.)

ONOCLEA Linn.

The single species *O. sensibilis* L. is abundant in the Eastern States of America from Florida northwards. It is also found in Eastern China. It is a creeping swamp plant, with elongated rhizome showing occasional bifurcations, bearing isolated leaves which are dimorphic. The sterile leaves are

pinnate, and the pinnae sometimes pinnatifid. The pinnae are broadly
lanceolate, with a mid-rib and Pecopterid venation: but there are profuse
anastomoses of the veinlets (Fig. 678, *G–L*). The rhizome contains an
extended dictyostele, with some indications of those superficial axillary
pockets which are a marked feature in *M. struthiopteris* (Fig. 679), though
absent in *M. orientalis* and *intermedia*. This suggests that its creeping habit
is secondary. The gradate condition of the sorus has already been alluded
to: the indusium, as shown in Fig. 681, consists of only a single layer of
cells, in contrast to the massive growth in *Cyathea*. The sporangia are as in
M. struthiopteris.

The gametophyte, sexual organs, and embryology of the Onocleid Ferns
appear to follow the usual Leptosporangiate type. Details are given by
Campbell (*Mosses and Ferns*, 1918, pp. 305, etc.).

COMPARISON

It has already been suggested that the relation of the Onocleid Ferns
downwards is with the Cyatheaceae, all of them having superficial and gradate
sori. Upwards, as indicated by earlier writers, it is with the Blechnoid Ferns
which, as we shall see, have also superficial sori. In fact the Onocleid Ferns
may be regarded as a synthetic group, linking the Cyatheoids with the
Blechnoid Ferns. It has also been shown in Chapters XL, XLI, that a derivative upward line from the Cyatheaceae has led through the Woodsioid
Ferns towards the Dryopteroids; and a firm basis of fact indicates that the
Asplenioids are derivative from the latter. The question will then arise as
to the relation of these two progressive lines *inter se*: we shall enquire how
far they may be regarded as separate in their history, and therefore distinct
in their natural or phyletic classification. As an aid to the structural comparisons detailed in these Chapters a working hypothesis based on a biological
reading of those facts may be introduced here, and tested by its application
to a wide area of fact.

The comparisons already detailed in Chapter XLI indicate that the
Asplenioid Ferns were by origin delicate in texture, though certain of their
derivative types became in some degree xerophytic. This is seen in many of
the smaller species of *Asplenium*, such as *A. ruta-muraria*, as well as in the
epiphytic species, such as *A. nidus*, while xerophytism is specially marked
in the structure of *A. Ceterach*. But the simpler Dryopteroid types, from which
the Asplenioid Ferns are all believed to have been derived, had highly divided
Sphenopterid foliage, with that relatively large proportion of surface to bulk
which is suitable enough for plants that grow habitually in shade. This
condition is perpetuated in *Athyrium*, which comparison points out as the
genus probably nearest allied to *Dryopteris*: many other Asplenioid Ferns

are of like texture. The Asplenioid sorus with its delicate indusium is also in keeping with this hygrophytic habit.

The Onocleoid and Blechnoid Ferns present as a whole a strong contrast, which appears in the leathery texture of their leaves. These as a rule are not highly subdivided, and often the plants are markedly heterophyllous, while the young sori are closely enfolded in the narrow segments of the sporophylls: the lateral flaps of these are strongly recurved, and not unfrequently brown and almost woody in texture. Such features are clearly xerophytic adaptations: they are characteristic generally of the Onocleoid and frequently of the Blechnoid Ferns. Thus while the Asplenioids may be held to have sprung from types adapted to a shade-habit, the Onocleoids and Blechnoids appear to have been from the first more or less xerophytic, and stand in marked antithesis to the hygrophytic origin of the Asplenioids.

This is the hypothetical position from which the comparative study of the Onocleoid and Blechnoid Ferns may start. Both they and the Asplenioids have been very successful types, as shown by the large number of living species. They have spread into the most varied areas, even into those climatically resembling one another, and conversely differing from their hypothetical sources. Accordingly it need be no cause of surprise that they should have developed along parallel lines, and that their later derivatives should present a high degree of homoplastic resemblance. This has led to systematic confusion of the living representatives of the two stocks, which a more strict comparison may tend to resolve if the biological aspect of the problem, as here sketched, be kept clearly in view from the first.

The immediate phase of our problem deals with the Onocleoid Ferns, and more particularly with their relations downwards in the scale. The most striking differences from the Cyatheoids are: (i) the smaller size, the absence of a tall dendroid habit, and consequently their simpler anatomy: (ii) the heterophylly: and (iii) the seasonal leaf-fall of the Onocleoids. The first of these is in no way distinctive, for there are Cyatheoids of low stature, while some Alsophilas, and in particular *Lophosoria* itself, bear creeping runners of the same nature as those of *Matteuccia*, with a relatively simple vascular system. *Metaxya*, a Protocyatheoid, has actually a creeping solenostelic rhizome with an undivided leaf-trace. It is worthy of remark that the nearest parallels are found with those Cyatheoid Ferns, which are held to be the least advanced (Chapter XXXII). Heterophylly is a feature strongly impressed upon both the Onocleoid and the Blechnoid Ferns: but it does not occur in the Cyatheaceae: nor is a seasonal leaf-fall a feature with them, though there is a seasonal change of leaves. Both of these are, however, characters adaptive to season and station.

The Onocleoid Ferns may be regarded as a series of advance running parallel to that of the Woodsieae, and like them they have carried the more primitive Cyatheoid type into boreal and alpine climates. There additional protection was a necessity, which has here been achieved partly by their firm texture and by seasonal leaf-fall, partly by heterophylly, and partly by their investure with broad protective scales. The two parallel series may be held as outstanding colonists, in which the leaves more especially have been modified so as to meet the conditions of the more exacting habitats to which they have spread. They may in fact be regarded respectively as arctic and xerophytic derivatives from the more primitive Cyatheoids.

The anatomy of these Ferns accords with this view. The simple dictyostele of *M. intermedia* (Fig. 682) is only a slight advance on that of *Lophosoria*. The leaf-trace of this Fern is divided in large leaves into three straps, which unite upwards into a single meristele. In *Matteuccia*, in accordance with its smaller size, the trace divides only into two straps: but these again fuse upwards into a single meristele of essentially similar character to that of *Lophosoria* (Fig. 685). This is the type of petiolar structure described by Bertrand and Cornaille as the "Onocleoid Trace" (*Structure des Filicinées actuelles*, Lille, 1902, p. 91). Thus the vascular anatomy harmonises with the view of the Onocleoids as being derivative from an early Cyatheoid source.

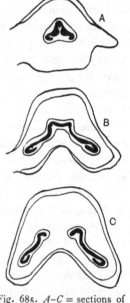

Fig. 685. *A–C* = sections of the petiole of *Matteuccia struthiopteris*: *C* is near its base, *B* from a middle position, and *A* from a higher level. (After Bertrand and Cornaille.)

But the greatest interest naturally centres round the sporophylls. The simply pinnate sporophyll with narrow enrolled pinnae is a type that is not present in the Cyatheoids. Among primitive Ferns it is seen in *Plagiogyria*. Notwithstanding its markedly oblique annulus, its dermal hairs, and as now known its mixed sorus, *Plagiogyria* was long ranked as a section of the genus *Lomaria* (Hooker, *Syn. Fil.* 1883, p. 182). The heterophyllous condition alone does not justify that actual relationship, and for reasons given in Chapter XXXI *Plagiogyria* has been provisionally placed in relation to the Osmundaceae. With this exception heterophylly of the Lomarioid type appears to be distinctive among Ferns where the sorus is a separate entity. If the venation of the sporophylls and the disposition of the sori upon them be examined, it is apparent that the Pecopterid open type is common to the Cyatheoids and Onocleoids, while in the latter the sterile and fertile leaves differ only in the sporophyll being narrower, and the vein-branchings fewer. Further, the disposition of the sori upon them follows the Cyatheoid type, where only one sorus appears on each vein, inserted laterally upon it. The distance of insertion from the mid-rib may vary, as it does in *Cyathea sinuata* (Vol. II, Fig. 557), and in *M. struthiopteris* (Fig. 678, *D*, *K*): on the other hand the sori may be in such positions as to form regular intra-marginal series. This is common in the Cyatheaceae, and particularly in species of *Alsophila* and *Hemitelia*. It is found in *Matteuccia orientalis* and *intermedia* (Fig. 680), and it becomes a constant and indeed an essential feature in *Lomaria*.

The distinctively gradate condition of both *Matteuccia* and *Onoclea* points directly towards the Cyatheaceous type. The detail of the large sporangium of *M. intermedia*, more complex than that of *Onoclea* as represented by Campbell (*l.c.* Pl. VII, Fig. 26), compares most nearly with that of *Lophosoria*, those of the other species being smaller. It is significant that the larger and more complex sporangium goes along with the absence of an indusium in *M. intermedia*, though this protection is present in the rest. These features indicate *M. intermedia* as the most primitive member of the family, while they direct comparison towards *Lophosoria* and *Alsophila* among the Cyatheoid Ferns. The inconstancy of occurrence of an indusium within a very close and natural circle of affinity raises again the question of its nature and origin. We have seen in Chapter XL how variable is the occurrence as well as the form of the indusium in the Cyatheaceae, and how Christ found, even within the genera *Alsophila* and *Cyathea*, that the absence or presence of an indusium is not a generic constant. The same now applies in *Matteuccia*, which includes both indusiate and non-indusiate species. In the case of *M. intermedia* there is no evidence of abortion to explain its absence: the structure of the young sorus, as seen in Fig. 683, is as consistent with an originally non-indusiate state as are those of *Alsophila* (Vol. II, Fig. 564), or *Lophosoria* (*Ann. of Bot.* XXVI, Pl. XXXV), or even of *Gleichenia* itself (Fig. 490). In none of these is there any reason for entertaining the idea of abortion to explain the non-indusiate state. The structural evidence indicates rather the advent of the indusium as a new structure. It may be held to have made its appearance *de novo* in *M. struthiopteris* and in *Onoclea*: but *M. intermedia* shows by its absence a primitive condition, which it shares with the series of Superficial Ferns above named. The existence of this second instance of inconstancy of an indusium in relatively primitive Ferns, related to but distinct from the Cyatheaceae, strengthens the conclusion that an indusium can be formed *de novo*, while conversely it discounts any forced comparison with the lower indusium of the marginal series as represented in the Dicksonioid Ferns.

The next Chapter will deal comparatively with the Lomarioid Ferns. It will there be apparent how important a part in the probable evolutionary sequence is taken by the Onocleoid Fern *Matteuccia intermedia*: for the view will be advanced that the Lomarioid Ferns were derived from some such source as that represented by this new non-indusiate species. Meanwhile the conclusion is inherent in the above pages that the Onocleoid Ferns are best regarded as forming a separate phyletic offshoot from some Cyatheoid source, and should accordingly be constituted a distinct Tribe, as the ONOCLEEAE. The phyletic grouping of the living Onocleoid Ferns that have been here examined should probably be as follows:

ONOCLEEAE

I. Non-indusiate: open venation.
   Matteuccia intermedia C. Chr.

II. Indusiate: open venation.
    Matteuccia struthiopteris (L.) Todaro.
    Matteuccia orientalis (Hk.) Trev.

III. Indusiate: reticulate venation.
     Onoclea sensibilis Linn.

BIBLIOGRAPHY FOR CHAPTER XLIII

687. METTENIUS. Fil. Hort. Lips. Leipzig. 1856.
688. HOOKER. Species Filicum, IV, p. 160. 1862.
689. HOOKER. Synopsis Filicum, p. 182. 1883.
690. LUERSSEN. Rab. Krypt. Flora, III, p. 480. 1889.
691. CAMPBELL. Development of the Ostrich Fern, Mem. Bost. Soc. Nat. Hist. IV, No. II. 1887.
692. CHRIST. Farnkräuter, pp. 284, 323. 1897.
693. DIELS. Nat. Pflanzenfam. I, 4, p. 164. 1902.
694. BOWER. Studies IV, Phil. Trans. B, 192, p. 55. 1899.
695. BERTRAND and CORNAILLE. La Structure des Filicinées actuelles. Lille. 1902.
696. GWYNNE-VAUGHAN. New. Phyt. IV, p. 214. 1904.
697. BOWER. Studies II, Ann. of Bot. XXVI, p. 300. 1912.
698. BOWER. Studies IV, Ann. of Bot. XXVII, p. 367. 1913.
699. CHRISTENSEN. *Matteuccia intermedia*, N.Sp. Bot. Gaz. 56, p. 337. 1913.
700. VON GOEBEL. Organographie, II, 2, p. 1148. 1918.
701. SEWARD. Hooker Lecture, Linn. Journ. p. 219. 1922.

CHAPTER XLIV

BLECHNOID FERNS

THIS Family includes nearly two hundred living species, a fact that shows its high success. They are distributed among ten genera, of which by far the largest is *Blechnum* with 138 species. The genera may be segregated into four groups according to their soral features. The genus *Blechnum* forms, together with *Sadleria*, a central point from which the rest may be held as derivative. Accordingly *Blechnum* will be examined first. The habit of the Blechnoid Ferns is very various, some are creeping, some climbing, some almost dendroid (Fig. 686). They are characterised by a stiff or leathery texture of their foliage, suggesting xerophytic adaptation. They show various degrees of that dimorphism which was a constant feature of the Onocleeae: it is most marked in the type of *Lomaria*, now held as a section of the genus *Blechnum*. There is reason to believe this to have been the original type, the segments of the sporophylls being narrow, and the margins strongly incurved, as they are in *Matteuccia*. In the rest the sterile and fertile leaves are more or less fully alike, and steps will be traced explaining how this may have come about in the course of the evolution of the tribe from the *Lomaria*-type.

These Ferns were ranked by Diels as a Sub-Tribe Blechninae, under the Tribe Asplenieae. For reasons already partly explained in Chapter XLIII and to be developed further as the treatment proceeds, they will here be held as a phylum distinct in origin from the Asplenieae, though showing signs of evolution parallel to them. Borne upon an axis of various pose the leaves are often simply pinnate, with open venation: but sometimes there is a higher pinnation: various degrees of reticulation are also seen. The common feature of them all is that linear coenosori take a parallel course, one on either side of the mid-rib, while a vascular commissure below the common receptacular line of each supplies them in a manner not unlike that seen in *Pteris*. From these Ferns, however, they are phyletically distinct (Fig. 687). Geographically the Blechnoid Ferns are very widely spread, and many are tropical. A great preponderance of their species grow in the Southern Hemisphere, and they are particularly plentiful in Polynesia.

BLECHNUM Linn. emend. Mett.

In accordance with general opinion since the time of Mettenius the genus *Blechnum* is accepted in its wider sense, as comprising three sub-genera: (i) *Lomaria*, (ii) *Eu-Blechnum*, and (iii) *Salpichlaena*. *Lomaria* will be taken first, as probably representing a prior type phyletically. It differs from *Eu-Blechnum* in the relation of the sorus to the ostensible margin of the leaf. In *Lomaria* the actual margin of the narrow fertile pinna is curved downwards as in *Matteuccia struthiopteris* (Fig. 678, *E*): this is the "indusium marginarium" of Presl's *Tentamen* (1836, p. 141). But in *Blechnum* (*l.c.* p. 101)

Fig. 686. Habit of *Blechnum tabulare* (Thunbg.) Kuhn, showing the dendroid character, with dimorphic leaves, the sporophylls being erect, the trophophylls spreading: from South Brazil. (After Wacket, from Christ.)

he speaks of an "indusium lineare, scariosum, margine libero, costam respiciente": and later (p. 102) more explicitly he says: "attamen margo frondis semper evidentissime liber est, et indusium proprium adest." Clearly Presl held the indusia in his two distinct genera of *Lomaria* and *Blechnum* to be essentially different things. But comparative study of the development shows that the indusium-like margin is constant throughout the extended genus *Blechnum,* and it will here be regarded as the "phyletic margin." This, however, comes to be apparently intra-marginal in *Eu-Blechnum,* owing to the formation of a new structure which originates along

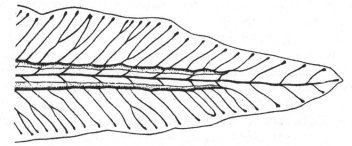

Fig. 687. Tip of a pinna of *Blechnum orientale* L., made transparent so as to show the relation of the coenosorus on either side of the mid-rib to the venation of the blade. Each coenosorus is protected by a continuous "indusium" which is phyletically the leaf-margin: the expanded leaf-surface on either side is a new formation, the "flange," by means of which the area of photo-synthesis is increased. This is a typical *Eu-Blechnum.* (× 3.)

the line of greatest curvature: this will here be called the "flange." Its origin has been traced by comparison of numerous species: but though the flange may produce a large photo-synthetic expanse in *Eu-Blechnum,* the position of the phyletic margin is not thereby altered. Its identity is

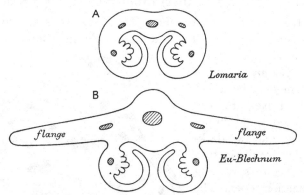

Fig. 688. *A* = diagrammatic section of the fertile pinna of the type, *Lomaria,* with incurved margins, but no flange. *B* = a similar section of the type, *Eu-Blechnum,* showing the same parts as before with the addition of flanges right and left. Compare Fig. 687.

maintained throughout, and it corresponds to the incurved margin of the fertile pinna in *Matteuccia.* The relative position of the parts is best seen in sections (Fig. 688, *A, B*), from which it appears that no membranous indusium, such as that seen in *Onoclea* or in *M. struthiopteris,* is present either in *Lomaria* or in *Eu-Blechnum.* It has been seen in the preceding Chapter that *M. intermedia* has no indusium: this species thus supplies the hypothetical starting-point for the Blechnoid series. It was necessary to state these facts at the outset so as to fix the terminology, and to aid lucid description.

Here it may fitly be noted that in no natural group of Ferns do the observations of Hannig on the occurrence of a perispore bring such inconsistent results as in the Blechnoids and Onocleoids. He found (*Flora*, Bd. 103, 1911, p. 339) that a perispore is present in *Sadleria, Doodia*, and *Brainea*, and also in *Phyllitis (Scolopendrium)*: but it is absent in *Blechnum* and *Woodwardia*, and also in *Stenochlaena (Lomariopsis)*. Such results cut across what is certainly a very natural family, as judged by many other features. It is worthy of remark, however, that the Onocleeae are similarly inconstant: for he states that a perispore is absent in *Onoclea sensibilis,* but present in *Matteuccia (Struthiopteris) orientalis* and *germanica*. Such facts need not destroy confidence in this valuable diagnostic character for general use: the facts for the Ferns at large are too consistent for that. For instance, it is very constantly present in the Aspidieae and Asplenieae, and absent in the Pteroids. But the fact that inconstancy occurs both in the Onocleoid and the Blechnoid Ferns appears to strengthen the relation between these families, a relation which underlies the whole phyletic argument of this Chapter.

BLECHNUM (§ LOMARIA) TABULARE (Thunbg.) Kuhn

In habit this is a dwarf Tree-Fern (Fig. 686). It is synonymous with *Lomaria Boryana* (Swartz) Willd., and it resembles in general characters the more familiar *B. capense* (L.) Schlecht (= *Lomaria procera* Spr.). Its stem may be 4 feet high, and the leaves are very firm and coriaceous, and simply pinnate. The narrow dark-coloured scales are distinctive. The sterile leaves are spreading and the venation open. The sporophylls are erect, the pinnae narrow, and the reflexed indusioid margin dark brown. The vascular structure of the axis is dictyostelic with a massive pith (Fig. 689). Following the succession of the leaf-traces as numbered from below, it is seen that at the base of each

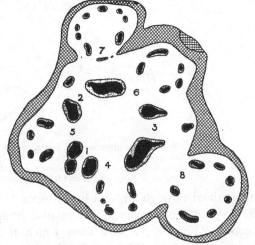

Fig. 689. Transverse section of the stock of *Blechnum tabulare* (Thunbg.) Kuhn. The numbers indicate the successive leaf-gaps, from which the leaf-traces arise as two straps, subdividing early to form a horse-shoe curve. (× 4.)

foliar gap a root-trace passes off: later the leaf-trace separates as two straps which divide upwards into a varying number of strands, arranged in a horse-shoe. This vascular system is of a type general for the genus; it is essentially the same as in *Matteuccia*, but more subdivided.

The fertile pinna of *B. tabulare* is typical of *Lomaria*, in that it has its protective margins strongly curved downwards, and there is no flange. A transverse section of a pinna of medium age appears as Fig. 690, 4, *g*. The sorus which is continuous longitudinally is seated on the vascular commissure here cut transversely. It is very perfectly protected by the marginal flap, which is here unusually thick, doubtless as a xerophytic adaptation. The succession of sporangia is gradate, a sequence rarely departed from in this species: but one exceptional instance is seen in Fig. 4, *h*. The question arises as to the relation of the flaps to the marginal segmentation. Here, as also in *B. discolor* and other *Lomarias*, its origin is marginal: but the relation to the marginal initial in any given section is sometimes indeterminate, with a bias towards the concave side. Its origin is clearly marginal in Fig. 4, *a*, *b*: but the bias towards the concave side appears in Fig. 4, *c–f*. This apparent variability may have its origin in differences of the exact plane in which the selected sections were cut relatively to the strongly circinate pinna.

BLECHNUM LANCEOLATUM (R. Br.) Sturm

The same question arises also in *B. lanceolatum* (R. Br.) Sturm, a species which shows no marked flange, though there may be a variable lateral distension of tissue. The outline and venation of the fertile pinna are shown in Fig. 690, 5. Transverse sections at such a point as (*x–y*) appear in outline as in Fig. 690, 6, *a*. The indusial flaps are here thinner than in *B. tabulare*, and the section is almost circular owing to distension of the tissues below the receptacles, as photo-synthetic tracts bearing stomata. The commissure is as before, but a vascular supply is often seen to branch off towards the upper surface, each twig of which terminates in a gland (6, *b*). The sporangia are here spread over a rather wide flat surface, and with varying indications of a basipetal succession their order of appearance graduates into a mixed state of the adult sorus. Usually the indusial flap springs obviously from the marginal segmentation (6, *d–f*): but if the section be cut near to either limit of the soral region the flap may be replaced by another apparent margin, which a reference to the basal part of Fig. 690, 5 will explain (6, *g*).

These two examples have been selected as showing the features typical of *Blechnum*, §*Lomaria*. In the form and early development of the maturing pinnae there is a close similarity in *Blechnum*, §*Lomaria* to what has been seen in *Matteuccia* (Fig. 683, *D*). In neither is there any distinct "flange." The indusial flap appears itself to represent the recurved margin, as it does in *Matteuccia*. Several other species suggest an intermediate state leading towards the development of the flange as it is seen in *Blechnum brasiliense*. For instance, *B. attenuatum* (Sw.) Mett., and *B. L'Herminieri* (Bory) Mett., detailed descriptions of which are given in Studies IV (*Ann. of Bot.* XXVIII, 1914, pp. 377–379).

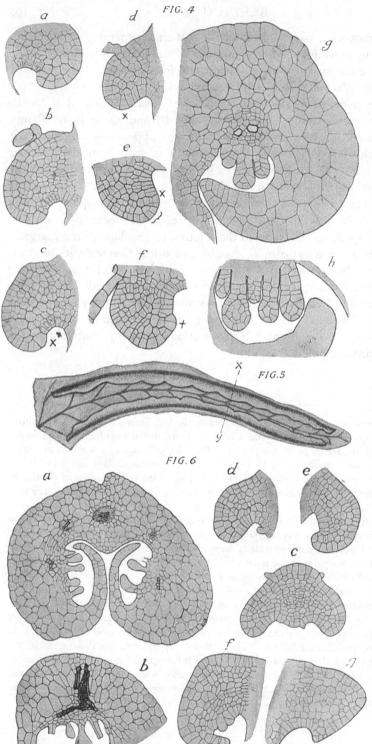

Fig. 690. Fig. 4, *a–h* illustrate the development of the indusial flap and sorus in *Blechnum tabulare. a, b* show a segmentation which indicates that the origin of the flap is marginal: in *c–f* this is less clearly seen. *g* shows a later stage with basipetal sorus: in *h* there is an occasional example of irregularity. (× 125.)

Fig. 5. A single fertile pinna of *B. lanceolatum* Sturm. *x–y* indicates the plane in which the section 6*a* might have been cut. (× 6.)

Fig. 6. *B. lanceolatum. a* shows a transverse section in plane *x–y*, Fig. 5, of a fertile pinna: no obvious flange is present. *b* = part of a similar section, showing a vascular strand passing towards the upper surface to supply a gland (compare Fig. 696). *c–g* = successive stages of development of the wings of the pinna in transverse section. (*a, b* × 50 : *c–g* × 125.)

BLECHNUM SPICANT (L.) Wither

Perhaps the best intermediate illustration is provided by the familiar Hard Fern, *Blechnum spicant*. This common species is the only representative of *Blechnum* native in North Temperate lands. Its very sclerotic upright stem contains a vascular system with binary leaf-traces, conforming to the type, though simpler than in *B. tabulare*. Its leaves are dimorphic. The general character of the fertile pinnae is seen in Fig. 691, *A*. Here there is a narrow flange on either side, so that the coenosori appear to be slightly intra-marginal. If the tip of the pinna be cleared, the sori and their vascular relations are visible as in Fig. 692, 9. Short vascular twigs pass into the flange: this is an advance on the species previously described. The sori, shown as continuous shaded tracts, stop some distance from the apex. A vascular commissure runs below each, connecting the veins : but this also stops short where the sorus ends. The relation of the commissure to the veins suggests very strongly that it is actually composite, being built up from individual extensions of the anadromic branches of the furcate veins. This was the view held by Mettenius (*Fil. Hort. Lips.* p. 60, Taf. v, Fig. 5).

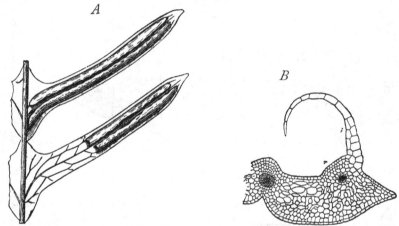

Fig. 691. *Blechnum spicant* (L.) Wither. *A*=two segments from the upper part of an immature fertile leaf: the "indusium" and sori of the lower pinna have been removed halfway up from the base to show the nervation. (× 5.) *B*=half of a transverse section of a fertile pinna, after Mettenius. *i*= "indusium." *r*=receptacle with underlying commissure cut through. The larger vascular strand is the mid-rib of the pinna. (× 30.)

The development of the fertile pinna of *Blechnum spicant* begins in the usual way, by alternate cleavages of marginal cells (Fig. 692, 10, *a*). But the marginal cells take their place quite definitely at the apex of the flange, which is here of larger size proportionately than in the previous examples (Fig. 692, 10, *b-d*). The whole flange is referable in origin to them. But soon a stronger growth appears on the abaxial face distinctly back from the margin (*b*). It consists of a broad convex weal, and the slope of it which faces the mid-rib obliquely soon shows deeper cells : these, undergoing segmentation, produce sporangia, with slight signs of a gradate sequence (*c*, *d*). On the side away from the mid-rib the "indusium" arises; it is relatively late in origin, and remote from the margin. The late appearance and intra-marginal position mark *B. spicant* as farther removed from the condition of *Matteuccia* than the species previously described. But notwithstanding this the essential identity of the several parts within so natural a genus is beyond doubt. *B. capense* (L.) Schlecht and *B. gibbum* (Lab.) Mett. give similar results (Fig. 692, 10, *a-c*). These species also show instability in their dimorphism, with widening out of the flange in fertile pinnae, so as to form a broad expanse with an extensive venation of its own.

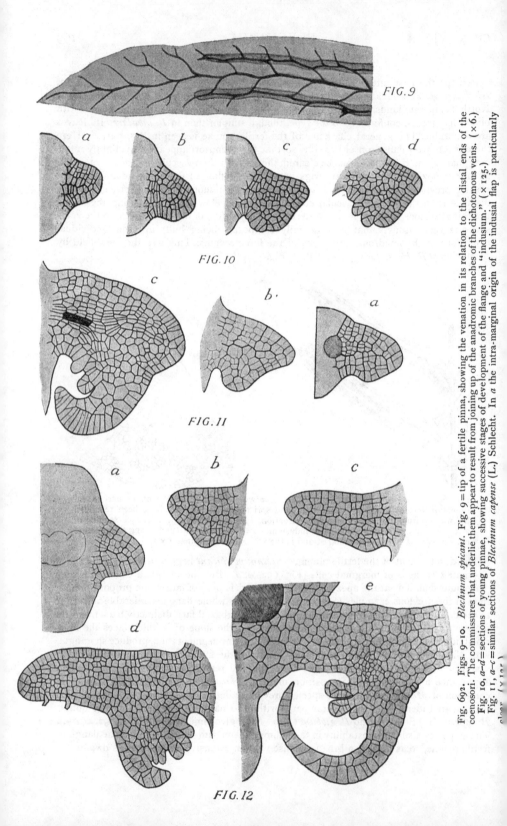

FIG. 9

FIG. 10

a   *b*   *c*   *d*

FIG. 11

c   *b'*   *a*

FIG. 12

a   *b*   *c*

d   *e*

Fig. 692. Figs. 9–10. *Blechnum spicant*. Fig. 9 = tip of a fertile pinna, showing the venation in its relation to the distal ends of the coenosori. The commissures that underlie them appear to result from joining up of the anadromic branches of the dichotomous veins. (×6.)
Fig. 10, *a–d* = sections of young pinnae, showing successive stages of development of the flange and "indusium." (×125.)
Fig. 11, *a–c* = similar sections of *Blechnum capense* (L.) Schlecht. In *a* the intra-marginal origin of the indusial flap is particularly clear. (×125.)

§EU-BLECHNUM

Such conditions as those just described lead naturally to the state characteristic of *Eu-Blechnum*, as it is seen in *B. brasiliense* Desv., or *B. occidentale* L.
(Fig. 693). These are large sub-dendroid species, with pinnate leaves, bearing
on their bases and on the stem and terminal bud dark scales, together with

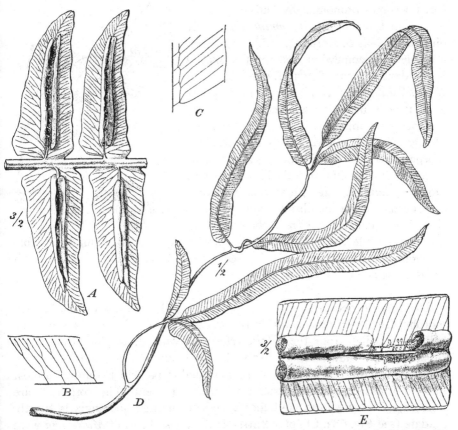

Fig. 693. *Blechnum* (L.) [§ *Eu-Blechnum*]. *A–C*=*B. occidentale* L. *A*=part of a leaf. *B*, *C*=dia-
grams of the venation, *B* of a sterile, *C* of a fertile pinna. *D–E*=*B. volubile* Kaulf. *D*=part of a
leaf, *E*=part of a pinna with venation and sori. (*B*, *C* after Mettenius, the rest after Diels.) From
Natürl. Pflanzenfam.

mucilaginous hairs (Vol. I, p. 198). Their leaves are all alike, and their
pinnae broad. The fertile pinnae bear coenosori of the same type as in
B. spicant, seated close to the mid-rib (Fig. 693, *A*). In their development
the wings of the fertile pinna arise in the same way as the sterile, by the
usual marginal segmentation (Fig. 692, 12, *a*). The sorus does not make
its appearance until the wing has attained considerable size: it originates
as a rather massive upgrowth at some distance from the actual margin,

which here takes no direct part in its formation (Fig. 692, 12, *b*, *c*). Active growth on the side remote from the mid-rib tilts it over, so that it faces centrally (*d*). It takes an indusioid form, over-arching the receptacle and the sporangia which it bears; the indusial margin, curving strongly, forms a very perfect protection for the receptacle (*e*). There is at first a gradate sequence of the sporangia (*d, e*): but later the sorus becomes pronouncedly "mixed."

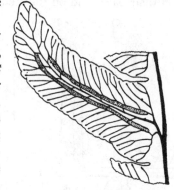

Certain types of *Eu-Blechnum* are doubly pinnate, such as *B. Fraseri* (A. Cunn.) Luerss., a species that somewhat resembles *B. spicant* in the development of its fertile segments: or *B.* (*Salpichlaena*) *volubile* Klf. (Fig. 693, *D*). The latter is an American species, which has assumed a climbing habit, somewhat like that of *Lygodium*. It shows variability in its di-morphism, while the vein-endings anastomose closely within the margin of the broad expanse. Double pinnation, but with an open venation, also occurs in the dendroid genus *Sadleria* (Fig. 694). Its coenosori are of the Eu-Blech-

Fig. 694. *Sadleria cyatheoides* Klf. A simple pinnule from the doubly pinnate leaf, showing the venation, and its rela-tion to the coenosori. (× 3.)

noid type, while the venation of the pinnule appears as a quite probable elaboration of the branched veins seen in *Matteuccia* (Fig. 678, *B*), or in *Cyathea sinuata* (Vol. II, Fig. 557).

The specific details given in these pages will have already outlined a morphological progression within the large genus *Blechnum*. This may be held as starting from the *Matteuccia*-type, with dimorphic leaves, and superficial, non-indusiate sori, disposed in two parallel rows, and covered by a simple curving downwards of the marginal flaps, which thin off almost like an indusium: but no true indusium is present. The sori here are isolated, of radiate form, and the succession of their sporangia is strictly gradate (see Fig. 683, Chapter XLIII). If the rows of sori of *Matteuccia* were linked together into continuous series the coenosori of *Blechnum* would result. In *Blechnum tabulare, lanceolatum*, and others, the fertile pinnae conform very nearly to the type of *Matteuccia*. But indications are already seen in these species that the protective flap does not always coincide with the margin as defined by segmentation: the divergence is most marked towards the apex and base of the pinna. Proceeding through the series of species described this divergence increases, both in time and place of origin of the flap, till in *Eu-Blechnum* the state is reached as seen in *B. brasiliense*. The series (more fully worked out in Studies IV than as here stated) indicates that the true margin by descent is the indusium-like flap, but that it has undergone a phyletic slide from its originally marginal position towards a position on the lower surface of the pinna. In course of this change the marginal segmentation, typical of Fern-leaves, is transferred from the true margin to that new growth, the flange, as it gradually asserts itself in the series compared. If this be the true history, then the flange, though secondary by descent, gradually assumes the prior place in the ontogeny, in accordance with its increasing importance as a photo-synthetic organ. By its formation the assimilating tissue of the fertile pinna is greatly increased, spongy parenchyma and stomata being produced upon it. The consequence of its appearance is that the nutritive disability of dimorphism is obliterated, and all the leaves, sterile or fertile, take a similar form. The biological advantage of self-nutrition of the sporophylls thus gained is too obvious to need insistence. It might of course be possible to invert the thesis, and to suggest that the Lomarioid state is derivative, by abortion of the broad lamina ; but this view would present comparative difficulties in relation to the sorus and the indusial flap, while it would conflict with physiological probability.

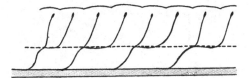

Fig. 695. Scheme of the venation of *Blechnum*, showing the mid-rib below, and forked veins arising from it. The dotted line indicates the commissure consisting chiefly of storage tracheides.

On the view thus stated the Blechnoid fusion-sorus arose from a gradate type, with isolated sori like those seen in *Matteuccia*. These were all seated on distinct veins. The formation of commissures connecting the veins, and the consequent running together of the separate sori into the Blechnoid fusion-sorus, is a step easily understood when starting from a type like *Matteuccia*, where the sori are in very close alignment in the young state (Fig. 683, *C*). The constitution of the vascular connections is partly by deflection of the anadromic branch of each forked vein, partly from a spread of storage tracheides of the receptacles. It is important to realise the distinctness of these two factors in the formation of the vascular connections (Fig. 695) (see Studies IV, p. 400). The general conclusion of this comparative study is then that the Blechnoid Ferns as a whole are an up-grade sequence: that they originated from a heterophyllous source, represented among living Ferns most nearly by *Matteuccia intermedia* C. Chr., and that the sequence culminated in the homophyllous type of *Eu-Blechnum*. The earlier steps of the sequence were characterised by adaptation to xerophytic conditions, while the latter were modified in harmony with mesophytic surroundings.

ACROSTICHOID DERIVATIVES

Certain Blechnoid Ferns have adopted an Acrostichoid character of their sori. This seems a natural sequel to the formation of coenosori; for if the production of sporangia can extend longitudinally so as to link up sori originally distinct, why not also transversely? Evidence of it comes both from the Lomarioid type, and also from that of *Eu-Blechnum*. The former is seen as a slight spread of the sporangia towards the margin in *B. attenuatum*

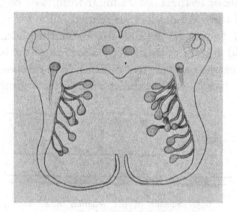

Fig. 696. Transverse section of a mature fertile pinna of *Blechnum attenuatum* (Sw.) Mett. The mid-rib is here traversed by two vascular strands, the lateral wings being so curved downwards as to give a quadrangular outline: they end in the recurved indusial flaps, while the sporangia spread over a considerable part of the inner surface. At the adaxial angles are flask-shaped glands, often present in *Blechnum*. (× 8.)

(Sw.) Mett. (Fig. 696), and still more in *B. Patersoni* (R. Br.) Mett., and in *B. penna-marina* (Poir.) Kuhn (see Studies IV, Pls. XXVI, XXVII). In none of these is the flange a marked feature. An Acrostichoid spread is more fully represented in the genus *Stenochlaena*, most of the species of which have from time to time been included in the old genus *Acrostichum* (*Syn. Fil.* p. 412). It consists, however, of nothing more than climbing Lomarioid Ferns, with the fertile pinna widened out into an Acrostichoid expanse. There is also the monotypic genus *Brainea*, which is a small Tree-Fern, undoubtedly related to *Eu-Blechnum*. Thus there is reason to believe that the transition to an Acrostichoid state has occurred repeatedly among the Blechnoid Ferns.

STENOCHLAENA J. Sm.

S. sorbifolia (L.) J. Sm. is a species widely spread throughout the tropics, and notwithstanding vicissitudes of nomenclature it may properly be placed as a Blechnoid derivative. It is a widely climbing Fern with extraordinary variability of leaf-form (Fig. 697). A Pimpinelloid form of leaf is found in the lower parts of the plant, resembling those of *B. filiforme* (A. Cunn.) Ettingsh.: but in the climbing region the sterile leaves are commonly of the usual Blechnoid type, while the fertile also resemble those of species of *Blechnum*, except that the soral region is wide and everted at maturity, exposing very numerous sporangia spread over its surface. As further illustrating the variability of leaf-form in the genus there is a vivid description by Karsten of a Moluccan species, referred by Christensen to *S. aculeata* (Bl.) Kze., where there are two types of sterile leaves in the climbing region, one of which corresponds to the sterile leaves of *S. sorbifolia*, the other is closely appressed to the surface of the support, and Karsten ascribes to them a water-collecting function (see Vol. I, Fig. 42).

Notwithstanding the elongated internodes of the climbing stem of *S. sorbifolia*, the structure accords with that usual in the Blechnoids (Fig. 698). Roots formed on the side in contact fix it to the support, while rhizoid-like hairs also assist. The periphery of the roughly polygonal section is sclerotic, interrupted here and there by lacunar tissue. The vascular system is clearly a modification of the *Blechnum* type, the meristeles being large, with very narrow leaf-gaps, opposite to which the usual root-traces are seen (Nos. 2 and 4). The leaf-traces consist usually of four to six strands forming a horse-shoe, which is thus of advanced type as compared with *Matteuccia*, *B. tabulare*, or *B. spicant*. All this accords with the position of *Stenochlaena* as a derivative type.

The youngest stages of the fertile pinnae of *Stenochlaena* have not been observed: but sufficient has been made out to give a basis for comparison with *Blechnum* (Fig. 699). The fertile pinna has either one or two vascular strands: the form of its transverse section is as in the simpler species of *Blechnum*: there is no obvious flange, but a curved wing on either side of the mid-rib, which thins off at the margin (*a*). The concave surface is covered over a very considerable area by sporangia together with glandular hairs. The sporangia are not grouped in any definite sori, nor are there any projecting receptacles (*c*). The origin of the sporangia appears to be almost simultaneous, many appearing to be of like age in a given section (*b*). But other sections show clear evidence of a mixed character, though the succession never seems to be long maintained. If a longitudinal section be cut so as to traverse one of the wings vertically to its surface, the phalanx of sporangia appears to be continuous (*d*). The veins severed transversely are

Fig. 697. *Stenochlaena sorbifolia* (L.) J. Sm. Drawings, after Christ, showing the great variability in leaf-form. *a* = end of the sterile leaf. *b* = part of a sporophyll. *c* = leaf of a young plant. *d* = type described as *Scolopendrium D'Urvillei* Bory. *e* = pinnatifid adventitious leaves. *f* = pinnate adventitious leaves. *g* = doubly pinnate adventitious leaves of Asplenioid type. *h* = trebly pinnatifid adventitious leaves. *i* = type described as *Davallia achilleaefolia* Wall. All are of the natural size. The highly branched forms of leaf may be regarded as reversions to an ancestral highly pinnate type.

wide apart, and quite distinct from one another: in fact there is no commissure linking the veins together, as is usual in *Blechnum*. The sporangia
arise from the whole surface intervening between them: the condition is
distinctly Acrostichoid. Often there is no sign of a flange in *Stenochlaena*
(Fig. 699, 17, *a–c*): occasionally, however, a vestigial outgrowth may be
found (*e, f*).

The sporangia of *Stenochlaena* are fair examples of those of Blechnoid
Ferns (Fig. 700, *A—D*). They are structurally intermediate between the type
with obliquely continuous, and that with vertical and interrupted annulus.
It will be clear that they are not far removed from the type with a continuous
oblique ring, such as is seen in *Matteuccia*, where, as here, the number of its
cells is high (Fig. 700, *E*). They are in fact just such sporangia as might have
been anticipated, following the analogy of other Ferns which have passed
from a gradate ancestry to a mixed state of the sorus. The cumulative effect of
all the evidence, foliar morphology, anatomy, soral and sporangial structure,
is to show that *Stenochlaena* is a Blechnoid type which has assumed a
climbing habit and an Acrostichoid, non-soral state.

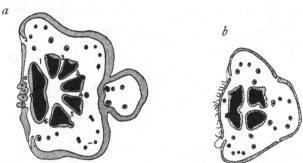

a

b

Fig. 698. *Stenochlaena sorbifolia*: transverse sections of the stock.
a and *b* show different sizes and complexities of structure.

BRAINEA J. Sm.

This monotypic genus is represented by a small Tree-Fern, *B. insignis*
(Hk.) J. Sm., from China, of Blechnoid habit. Anatomically it resembles
advanced members of the group, with a dictyostele giving off leaf-traces each
originating as two straps, which divide upwards to form some 10 or 12
strands. It bears scales and mucilaginous hairs. The leaves are all alike, but
the fertile pinnae are narrower with crinkled margins. There is no indusial
flap, and the sporangia spread over the lower surface, but stop short of the
margin. The venation is for the most part open, but with occasional fusions,
after the type of *Doodia* (Vol. I, Fig. 94, *l*). The apex of a fertile pinna
suggests the origin of the Acrostichoid state (Fig. 701, 18). The distal veins

FIG. 17

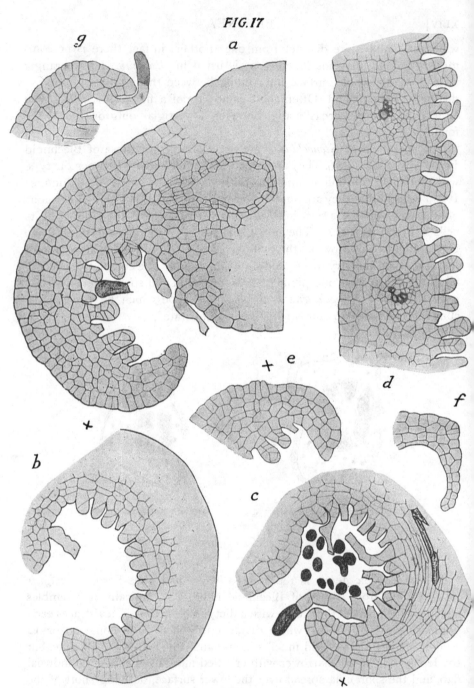

Fig. 699. *Stenochlaena sorbifolia* (L.) J. Sm. Sections transverse (*a, b, c*) and longitudinal (*d*) of the fertile pinna. They show the flattened concave receptacle, with "Acrostichoid" spread of the sporangia over a large area: the identity of the sori is completely lost. Indications of a "flange" are seen in *e, f*, but none is present in *g*. × marks the position where the flange would normally be. Glandular hairs of the Blechnoid type are present. (All × 125.)

may be sterile and open: but lower down isolated circular sori are borne on their bifurcated branches: farther down again the veins are connected by arched commissures, which merge into a connected soral line on either side of the mid-rib. Farther back still from the apex, these soral lines may extend outwards from the commissures to form broad Acrostichoid tracts. A section of such a young pinna will appear as in Fig. 702, *B*: comparison with a like section of *Woodwardia* (*W*) shows how close the similarity is, except for the absence of the indusial flap.

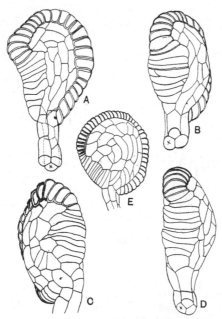

Fig. 700. *A—D* = sporangia of *Stenochlaena sorbi-folia*. *E* = a sporangium of *Matteuccia intermedia* for comparison (all × 60). The mark (×) on the sporangial stalk indicates the "peripheral" side of the sporangium. As in *Dryopteris* the other two rows of the three-rowed stalk pass upward to the "central" face. Compare Fig. 657.

The development of the fertile pinna, as seen in sections of *Brainea*, shows the usual marginal segmentation (Fig. 701, 19, *a*). Soon cells project on the lower surface, forming a rounded growth on which sporangia appear, with indications of a gradate sequence (*b*): but there is no sign of an indusial flap. Thus the sorus is not restricted peripherally; it spreads towards the margin, following especially the outward course of the veins, and with the ages of the sporangia intermixed (*c*). Hence it may be concluded that *Brainea* is descended from a gradate ancestry with a sorus restricted as in *Eu-Blechnum* or *Woodwardia*: and that it has progressed to a mixed condition of the sorus, with Acrostichoid spread towards the margin of the pinna. It is clearly

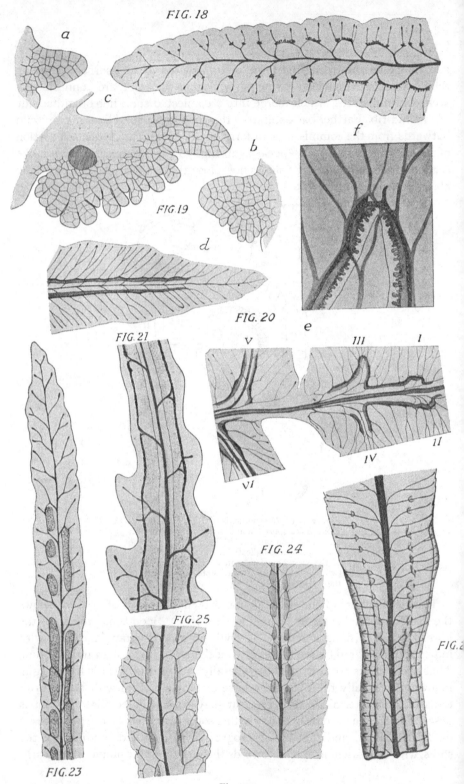

FIG. 18

a

c

b

f

FIG. 19

d

FIG. 20

e

FIG. 21

V

III

I

II

IV

VI

FIG. 24

FIG. 25

FIG. 23

FIG. 2

Fig. 701.

Blechnoid, and ultimately Matteuccioid in its origin: but the progression to the Acrostichoid state has probably been reached along a line quite distinct from that of *Stenochlaena*: for in the latter it is the "flange" that is absent, while in *Brainea* the indusial flap is abortive.

Fig. 702. *B*=transverse section of the mid-rib and sori of a pinna of *Brainea insignis* (Hk.) J. Sm., where the sori are exposed, there being no indusial flaps. *W*=a similar section of *Woodwardia radicans* (L.) Sm., in which indusial flaps are present. (×35.)

DISINTEGRATION OF THE COENOSORUS

Reasons have been advanced above for holding the linear sori of *Blechnum*, § *Lomaria* to be fusion-sori derived from some such heterophyllous type as *Matteuccia intermedia* (p. 173), and that the homophyllous type of *Eu-Blechnum* resulted from further modification of pinna-structure in harmony with mesophytic surroundings (p. 165). The widening of the leaf-expanse that characterises the latter has frequently been accompanied by disintegration of the linear-fusion-sorus, and displacement of the resulting parts; thus giving the types of *Woodwardia* and *Doodia*. A group of smaller genera centering round *Phyllitis* (*Scolopendrium*) may also be traced to a like source. It has been shown in Studies IV that these Ferns have the Matteuccioid-Blechnoid type of vascular construction: in particular this appears in the stock of *B. punctulatum* var. *Krebsii*, which itself gives the key to these developments (Fig. 703). Sometimes the disintegration may appear to be such as to resolve the fusion-sorus into the original sori, each seated on its own vein, as in the flattened fertile pinnae of *B. capense* (L.) Schlecht (Fig. 701, 22): but in most cases the disintegration results in irregular fragments which present no strict relation to the primary veins (Fig. 701, 23), while in others again the fragments are related rather to the commissures than to the primary veins (Fig. 701, 24, 25).

Fig. 701. Top right-hand drawing (Fig. 18) shows the distal end of a young pinna of *Brainea insignis* (L.) J. Sm., with venation, and the relation of the non-indusiate sori to it (×10). Fig. 19, *a–c*, sections of pinnae, showing the initial stages of sori resembling those of *Blechnum*, but without indusial flap (×125). Fig. 20. *Blechnum fraxineum* Willd. *d*=apical region of pinna with venation, showing its relation to the coenosori. *e*=lower region showing partial pinnae I-VI. *f* shows detail of pinna III, under a higher power (*d, e* ×2: *f* ×15). Fig. 21. Portion of a pinna of *Blechnum Fraseri* Luerss., showing the relation of the venation to the commissure (×15). Fig. 22. Part of a pinna of *Blechnum capense* (L.) Schlecht, showing the transition from the sterile to the fertile state, with disintegration of the coenosorus (×2). Fig. 23. Apical region of a pinna of *Blechnum spicant* (L.) Wither, showing venation, and partial disintegration of coenosorus (×3). Fig. 24. Part of a pinna of *Blechnum cartilagineum* Sw., showing disintegration of the coenosorus (×3). Fig. 25. *Woodwardia areolata* (L.) Moore, part of a pinna transitional between sterile and fertile, showing vein-fusions, and the indusial flaps here much narrower than in the sterile type (×2).

The vascular connections of the coenosorus arise from two sources. The first is the primary venation of the Matteuccioid type, where each separate sorus springs from the original vein with its long conducting tracheides, extended as shorter storage tracheides into the receptacle. The second consists chiefly of these shorter, almost brick-shaped tracheides characteristic of the receptacle. They link up the veins by commissural loops, and may sometimes be seen to take a course of their own independently of the longer conducting tracheides (Fig. 701, 20*f*, 21). The relation of the two vascular constituents is suggested by the diagram Fig. 695. Taking the instance of soral disintegration seen in *B. spicant* (Fig. 701, 23), in which species it is not an infrequent condition, each soral fragment is covered by its own indusial flap, which is a portion of the phyletic leaf-margin. Frequently a vascular process projecting on the anadromic side represents a part of the derivative commissure. These details, combined with the varying length of the fragments, show that the disintegration of the coenosorus is not a mere resolution into the original sori, but a breaking up of it into arbitrary parts.

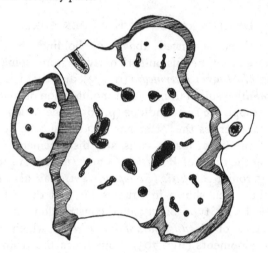

Fig. 703. Transverse section of the stock of *Blechnum punctulatum* Sw. var. *Krebsii* Kze., showing a structure normal for *Blechnum*. (× 3.) Compare Fig. 689.

WOODWARDIA AND DOODIA

Such disintegration of the coenosorus as appears sporadically in many species of *Blechnum* is stereotyped in *Woodwardia* and *Doodia*, becoming in them generic characters. These genera may be held as derivatives from the *Blechnum*-type, distinguished by a regular partition of the coenosorus into isolated parts, each of which corresponds as a rule to a single loop of the commissure (Fig. 704). Both genera consist of terrestrial Ferns, usually with an upright paleaceous axis; but occasionally it is creeping, as in *W. areolata* (L.) Moore. The foliage is usually stiff and cartilaginous, especially in *Doodia*: but in *Woodwardia*, in accordance with its shady habitat, the leaves are more herbaceous. The two genera are distinguished by the sori of *Woodwardia* being sunk in the mesophyll, while those of *Doodia* are superficial. The primary veins are linked together by the commissures that bear the sori: the veins passing out from these may be free, as in *D. media* (Fig. 704, *D*), or reticulate with wide meshes, as in *W. radicans* (Fig. 704, *A*) and *areolata*. The leaves are usually homophyllous, but sometimes dimorphic, as in *W. areolata*.

The hypothesis that these Ferns represent a Blechnoid type in which the coenosorus has been disintegrated is supported primarily on general grounds of habit, and of the arrangement and structure of the sori. Secondly by the

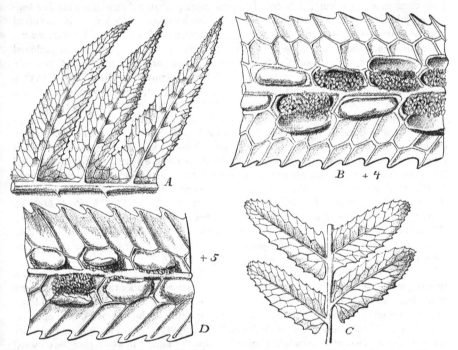

Fig. 704. *A, B = Woodwardia radicans* (L.) Sw. *A* = secondary pinnules. *B* = part of one pinnule with venation and sori. *C, D = Doodia media* R. Br. *C* = primary pinna. *D* = part of a pinna with venation and sori. (After Diels, from *Natürl. Pflanzenfam.*)

fact that they show the Blechnoid anatomy, with leaf-traces springing from the sides of the leaf gaps, as two strands, one or both of which soon abstrict smaller strands, the number of these varying roughly according to the size of the petiole (Studies IV, p. 403, Figs. 15, 16). The hypothesis is also supported by the details of development of the fertile pinnae, which correspond closely to those seen in *Eu-Blechnum*. The lateral pseudo-wing of the fertile pinna arises by segmentation directly from the marginal series, as in *B. brasiliense* (Fig. 692, 12, *a–e*). It attains considerable size before any sign of the sorus appears: and when it does the similarity between *W. radicans* and *B. brasiliense* is seen to be extremely close (compare Studies IV, Pl. XXV, Fig. 12, and Pl. XXX, Fig. 27).

A further line of comparison may be drawn from the apical region of the sporophylls in *Doodia* and *Woodwardia*. In the former the "sori," or rather the fragments of the coeno-sorus, are frequently disposed not in one but in two rows on each side of the mid-rib, the one nearer, the other farther removed from it. Where the sori are placed alternately it would be possible to regard them as fragments of a single Blechnoid coenosorus, each alternate one being removed farther from the mid-rib. But this simple suggestion will not explain all the facts; for, as Mettenius showed (*Fil. Hort. Lips.* Pl. VI, Fig. 5), the sori of the outer row may be superposed on those of the inner, while in *D. dives* Kze. an incom-plete third row may occur (*l.c.* p. 66). It seems more probable that with a widening leaf-surface additional sori may be initiated *de novo*. Such duplication has probably occurred also elsewhere on a broadening leaf-surface, particularly in *Dryopteris* and *Polypodium*.

On the other hand, evidence bearing upon the hypothetical disruption can be gathered from the apical region of many fertile leaves and pinnae : and particularly in *D. caudata*, where the tip is prolonged and entire. Here, while the coenosori are disrupted below, they become continuous for long distances upwards: thus they appear to reconstruct in the simpler distal region the probable source from which the more complicated state originated. This is seen also in *Woodwardia*, and with remarkable clearness in a species from Hong-Kong, described by Sir W. Hooker as *W. Harlandii* (*Sp. Fil.* Vol. III, p. 71). He tells how the oblong sori become confluent distally as a continuous chain close to the mid-rib, and frequently "sending out nearly opposite pairs of sori even where there is no distinct costule." This condition appears in the photograph from a specimen in the Kew herbarium (Fig. 705). It links readily with Mettenius' drawing of the bipinnate *W. virginica*, though the latter displays a more advanced state of disruption than *W. Harlandii* (*Fil. Hort. Lips.* VI, Figs. 1, 2). The comparison of these species with the simply pinnate leaf of *W. blechnoides* (Mettenius, *l.c.* Figs. 3, 4) at once suggests that the state seen in *W. Harlandii* is con-nected with the extent of the leaf-expanse. Similar details confirming that relation are shown for *Blechnum fraxineum* Willd., in Fig. 701, 20 *d–f*.

The last-named species has leathery leaves bearing few pinnae : its normal *Eu-Blechnum* type of pinna is seen in (*d*). Such pinnae appear successively smaller towards the distal end of the leaf (*e*) : a few imperfectly formed pinnae are seen at the base of the terminal lobe : the last of them are represented only by outward archings of the coenosori, each opposite to the departure of a lateral vein from the mid-rib (*e*, I–VI). It may be a question whether these represent nascent or decadent pinnae, perhaps the latter : in which case the outward archings seen in *B. fraxineum* would be the expression of an ancestral tendency to pinnation imperfectly carried out at the distal leaf-tip, and the relation of the veins to them would be held as defining their pinna-character. But other developments exist where archings of the coenosorus arise between the veins, and in no definite relation to them. These are seen particularly in *Blechnum punctulatum* var. *Krebsii*, as they have also been noted in *W. Harlandii* by Sir W. Hooker. The latter have no obvious relation to pinnae. Both types of arching of the coenosori appear in leaves of broad expanse, and it is probably correct to see in the archings of both types a mode of extension of the coenosorus in rela-tion to an expanded rather than merely a condensed leaf-area.

The irregularities of the coenosorus described in the preceding paragraphs demonstrate it as an entity, which is subject to modifications where a wide leaf-expanse is present. The examples cited taken from three distinct genera suggest that in the Blechnoid affinity a disintegration of the fusion sorus is liable to follow on the broadening of the fertile blade, combined with a condensation of its branching. The recognition of this forms a fitting prelude to the study of those cognate developments which culminate in the soral condition characteristic of the genus *Phyllitis* (*Scolopendrium*).

Fig. 705. *Woodwardia Harlandii* Hook. from Hong-Kong. Nat. size. The continuous Blechnoid coenosorus is extended into pairs of lateral sori facing one another. Disruption of these would give a condition closely resembling *Woodwardia virginica*. The type specimen represented by Hooker (*Fil. Exot.* Pl. VII) is more disrupted than this, showing the condition to be variable. (From a photograph supplied from the Royal Gardens, Kew.)

BLECHNUM PUNCTULATUM Sw. var. KREBSII Kunze

The normal plant *B. punctulatum* Sw., native in South Africa and Java, has the ordinary characters of a Blechnoid Fern of the Section *Lomaria*. It is strongly dimorphic, with narrow fertile pinnae: but these occasionally show interruptions of the fusion-sori: these accompany a widening of the pinnae that points to a state intermediate between §*Lomaria* and §*Eu-Blechnum*. This feature seen occasionally in the normal species is specially developed in a variety with broader fertile leaves; discovered near Grahamstown, Natal, by Krebs, it was described and figured by Kunze (*Farnkräuter*, 1847, p. 176, Taf. LXXIV). He gave it the name of *Scolopendrium Krebsii* Kunze. Mettenius (*Fil. Hort. Lips.* 1856, p. 67, Taf. V, Fig. 7) adopted this name. But the plant has since been ranked as a variety of *B. punctulatum*

Swartz. Illustrations of the pinnae of the Natal Fern appear in Vol. I, Figs. 228, 229. In all its general features, including the vascular anatomy (Fig. 703), the Natal variety is typically Blechnoid. Its peculiarity lies in the variability in width of the fertile pinnae, and with this goes a variability of the soral characters, ranging from narrow pinnae with typical Lomarioid coenosori, to broad pinnae with soral characters which resemble *Scolopendrium*. Hooker (*Species Filicum*, III, p. 30) has described his own hand-lens observations on these, in a passage that finds its natural illustration in the drawings shown in Fig. 706. The authors quoted above were all pre-Darwinian writers, to whom the varietal similarity of a Lomarioid Fern to *Scolopendrium* would appear as a striking phenomenon, rather than as an evolutionary sign. It seems strange, however, that with the facts already demonstrated nearly a century ago, no nearer approach should now be shown in systematic works between *Blechnum* and *Phyllitis* than at that early date. The former genus is still ranked with the Blechninae, and the latter with the Aspleniinae.

The slightest deviations from the normal fusion-sorus seen in *B. punctulatum* var. *Krebsii* appear upon pinnae of greater than the normal width: they consist in an outward arching of it between the veins connecting the commissure with the mid-rib (Fig. 706, 28, *a*). Sometimes the curvature is very slight: but where it is pronounced it is commonly associated with a partition of the coenosorus into short lengths and very irregular (Fig. 706, 28, *b*). The usual point of interruption is towards the anadromic end of the arch. Several intermediate states are seen in Fig. 706, 28, *c*: in one of them the vascular commissure is still complete, but the indusial flap is interrupted: in another the flap is still continuous, but the commissure is interrupted, with an isolated tracheide lying in the gap: in a third both flap and commissure are interrupted (28, *e*). In such cases, while the vein-endings run out towards the margin, a process of the storage-xylem of the commissure underlies the detached ends. More advanced states of disruption appear in Fig. 706, 28, *c, d*, and here the detached portions of the coenosorus extend out towards the margin of the widened pinna. Consequently, in extreme cases they appear in pairs, with the indusial flaps facing one another, but still without any vein between them, as in *Phyllitis* (Fig. 706, 30).

Fig. 706. Fig. 28. *Blechnum punctulatum* Sw. var. *Krebsii* Kunze. Portions of pinnae showing various states of disintegration of the coenosorus. *a* = a condition very near to that normal for *Blechnum*, especially near the apex, but below the coenosorus is strongly arched outwards. *b* = a rather more advanced state of disintegration. *c, d* = still more advanced arching and disintegration, so as to resemble *Phyllitis*: at (×) in these drawings supernumerary sori are present (*a–d* × 2). *e* = outward arching and disruption shown in greater detail. (× 10.)

Fig. 29. *f–h* = sections of young fertile pinna for comparison with the normal type of *Blechnum* (10–12, Fig. 692) and with *Phyllitis* (Fig. 669, 31). (× 125.)

Fig. 30. *Phyllitis Scolopendrium* (L.) Newm., portion of the lateral flap of a leaf, with mid-rib, showing the venation and sori: slightly enlarged.

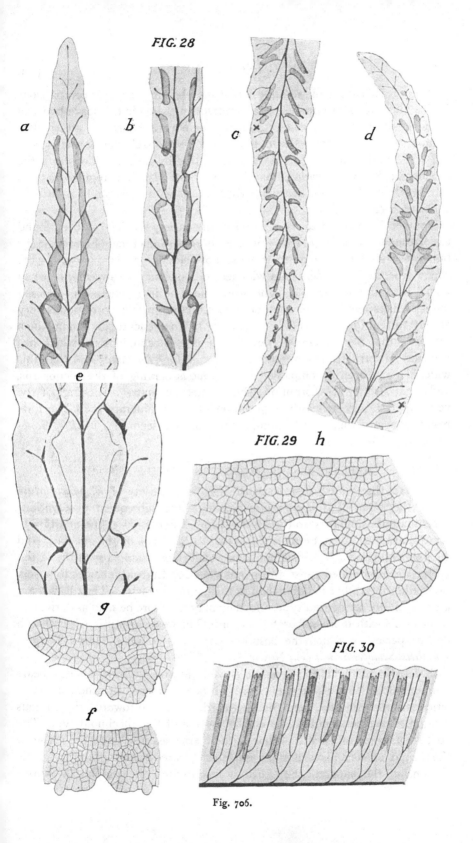

FIG. 28

FIG. 29

FIG. 30

Fig. 706.

Other details are liable to appear that may be important for comparison: for instance secondary branch-veins are frequent, lying in the space between the forks of the primary veins, and supernumerary sori may be attached to them as at (×), Fig. 706, 28, *c*, *d*. These sori are probably new formations in the sense that they are not directly derivative by disintegration from the coenosorus. Such innovations may be found established as regular features in *Phyllitis*, and they occur frequently in Ferns with sporophylls of widely continuous expanse.

The development of the coenosorus of the variety *Krebsii* has been found to correspond to that of *B. spicant* where a normal pinna is cut (Fig. 706, 29, *g*): but if sections be cut so as to traverse both sides of a narrow loop of the coenosorus, such as occurs on a widened fertile pinna, the appearance when young is as in Fig. 706, 29, *f*, and when older as in *g*, where the opposed indusial flaps overlap one another. The structure in fact corresponds very closely to that in *Phyllitis* (Fig. 669, 31, *a*, *b*). The general conclusion which follows is that *B. punctulatum* var. *Krebsii* is a Blechnoid Fern which has widened its fertile pinna beyond that normal for § *Lomaria*: and that this widening has been accompanied by a sinuous curving of the coenosorus, and often also by its disruption, while sometimes extra branchings of the veins appear, together with a formation of supernumerary sori. The final result is a structure closely comparable to that seen in *Phyllitis Scolopendrium*.

PHYLLITIS Ludwig (= SCOLOPENDRIUM Adanson)

The fact that Kunze first described the Natal variety of *B. punctulatum* under the name of *Scolopendrium Kunzii*, while subsequent observations accentuate its similarity to the Hart's Tongue Ferns, raises the question of their relation to other Ferns, in particular to *Blechnum* on the one hand and to *Asplenium* on the other. They have usually been ranked with the latter. In the most extended sense the genus *Phyllitis* Ludwig includes four sections, differing rather in habit than in their soral characters. Their leaves are all of a highly condensed type, and they may therefore be held as derivative from Ferns with more elaborate leaf-form. The genus comprises less than a dozen species, of which the best known is the common Hart's Tongue, *Phyllitis Scolopendrium* (L.) Newm.

The habit of the Hart's Tongue is well known: its upright stock bears homophyllous leaves, almost entire and broadly winged, with the characteristic sori extending far from the well-marked mid-rib towards the sinuous margin. The stock contains a vascular system of the Blechnoid type. The root-steles come off each from one of the meristeles just below the leaf-gap. As the gap opens it gives off right and left the paired strands of the leaf-trace. Passing up the leaf-stalk the pair may fuse to form the complex X-shaped structure characteristic of the genus, a feature also found in *Asplenium*. Lastly, the young parts are densely covered with ramenta, which bear at their apices large mucilage glands similar to those of *Blechnum*. These characters give general support to the relationship of *Phyllitis* to *Blechnum*, while at the same time the petiolar structure appears to reflect towards *Asplenium*.

The relation of the sori to the venation of the Hart's Tongue and to the sinuosities of the margin is shown in Fig. 706, 30. The primary veins arising from the mid-rib bifurcate twice, or sometimes more. On the outermost branches of each group, and facing outwards from the centre of the branch-system of each primary vein, are the elongated sori. As the sori connected with the successive primary veins face one another, the consequence is the arrangement typical of *Phyllitis*. The slight indentations of the sinuous margin correspond to these pairs of sori, and the convexities to the regions lying between the forks of the primary veins: these relations in the Hart's Tongue coincide with those seen in the variety *Krebsii* (Fig. 706, 28, a–e). In both they may be held as evidence of a suppressed pinnation still present in a widening blade. Support of the correctness of this view is found in many monstrous forms of the Hart's Tongue, and especially in those designated *laciniata*. In them the position of the laciniae is very constantly inter-soral, and they themselves appear to represent suppressed pinnae.

The development of the double sorus of *Phyllitis* follows the same lines as in the variety *Krebsii*. A depression of the lower surface of the young sporophyll first appears, at the margins of which the indusial flaps arise as upgrowths with the usual segmentation: below them right and left are procambial strands (Fig. 669, 31, a). The region between them is clear of vascular tissue, and its surface rises into a ridge (Fig. 669, 31, b). The indusial flaps soon overlap, covering the receptacles which lie directly over the vascular strands; these produce sporangia at first with indications of a basipetal succession, but this soon merges into a mixed character.

These details of similarity between *Phyllitis* and the *Krebsii* variety of *Blechnum punctulatum* indicate that there is here something more than homoplastic likeness. They appear rather to prove a near relationship between *Phyllitis* and *Blechnum*, among forms which have had in common a widening of the leaf-area with suppressed pinnation: and this has led to complications of the coenosoral structure. The change has worked out in arching curvatures, combined with disruption; the whole being carried out in leaves which have been undergoing condensation from narrow highly pinnate branching towards a broad, and finally an entire expanse.

Other Ferns, ranked with the common Hart's Tongue under the genus *Phyllitis*, or closely related to it, illustrate various modifications of its simple plan. For instance, *P. hemionitis* (Lag.) O. Ktze., though still retaining the open venation and the sori arranged as in *P. Scolopendrium*, bears auricles at the base of the shorter and wider blade (see Christ, *Farnkr.* Fig. 667). It is but a step from this to the state of the sporophyll of the North American *Camptosorus rhizophyllus* (L.) Link, or the Chinese *C. sibiricus* Rupr., with their partially reticulate venation (Fig. 707, *A*), but towards the margin and the caudate tip of the blade the veins are free. In the broader reticulate region the sori appear irregularly disposed, though still a reminiscence of the *Woodwardia*-type is seen in the soral fragments borne on the veins parallel to the mid-rib. Towards the caudate tip the arrangement is simplified, passing in *C. sibiricus* (Fig. 707, *B*) first into an arrangement reminiscent of *B.*

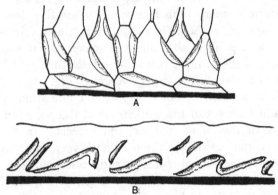

Fig. 707. Irregular soral tracts of *Camptosorus*. *A* = *C. rhizophyllus* (after Mettenius), showing reticulate venation, bearing irregular soral tracts, often associated in opposite pairs as in *Phyllitis.* *B* = *C. sibiricus* Rupr. Drawing from the narrowed distal end of a blade, from China (Cowdray, 1920), Kew Herbarium. Here there is a distinct approach to the condition seen in var. *Krebsii.*

punctulatum v. *Krebsii*, and finally it may merge into a simple Blechnoid coenosorus. The most highly modified type of all is that of *P.* (*Schaffneria*) *nigripes* (Fée) O. Ktze., where there is no mid-rib in the spathulate blade. The venation is reticulate, especially near to the margin, and the disposition of the soral tracts is irregular (Fig. 708). The broad scales still bear the glandular tip as in *Blechnum.* All of these appear as results of further condensation of leaf-form from that seen in the common Hart's Tongue. But yet they show in their details traces of the Blechnoid origin, with *B. punctulatum* v. *Krebsii* as the index of that source. The reticulation may be held as secondary, indicating such types as derivative: and with it habitually comes a less regular disposition of the coenosoral fragments. These may frequently owe their origin to innovation rather than to any direct disintegration of the Blechnoid coenosorus. Nevertheless, the simpler and especially the distal parts of the sporophylls, where the venation is open, point clearly to the Blechnoid source for them all.

Lastly, there remain certain small genera which have been ranked with the Asplenieae (Diels, E. and P. I, 4, p. 222): but Copeland has more recently pointed out their relation to *Phyllitis* (*Philipp. Journ.* 1913, Vol. VIII, p. 151). They are *Triphlebia* Baker and *Diplora* Baker. Already these had been placed by Christ under the generic heading *Scolopendrium* Sm. (*Farnkr.* p. 212). After examination of the specimens in Kew and the British Museum, I assent to the statement of Copeland that these Ferns are not unstable in frond-form alone: even the characters used in founding the genera are not invariable on single plants. His conclusion is that the proper name of the Ferns

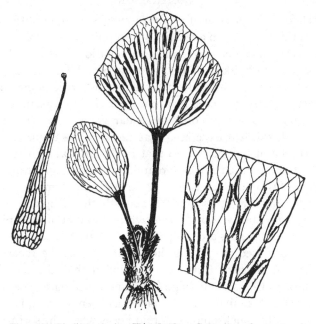

Fig. 708. *Phyllitis nigripes* (Fée) O. Ktze. Centrally a plant, natural size, showing habit. To the right part of the blade with reticulate venation, and irregular partial coenosori. Left a scale with distal glandular cell. (After Fée, from Christ.)

styled *Diplora* and *Triphlebia* is probably *Phyllitis Durvillei* (Bory) O. Ktze. (*l.c.* p. 152). At that we may leave them for further investigation, for they do not appear to throw any important light on the questions in hand.

On the ground of their linear sori a number of rather heterogeneous genera have been associated under the heading of the Taenitideae, and their relation has been suggested in many quarters with the Blechnoid Ferns. Professor Von Goebel has recently examined them in detail, and considered their probable connections critically (*Buit. Ann.* Vol. XXXVI, 1926, p. 107). The result has been to increase the doubts already entertained as to the relationship of any of them to the Blechnoid series, while some, such as the epiphytic *Heteropteris* and *Hymenolepis*, may now find a definite place elsewhere (see p. 222). The remainder, including *Taenitis* itself, may be provisionally regarded as *"genera incertae sedis."* For the present purpose it is best to confine the discussion to those genera of which the relations to the Blechnoid Ferns are generally accepted: but it seems appropriate to remark that semi-xerophytic types, such as the Blechnoid Ferns are, might be expected to adopt an epiphytic habit, and to undergo such consequent modifications as would make the recognition of their affinity difficult.

COMPARISON

The Blechnoid Ferns, cleared as above suggested of doubtful allies, form a coherent series based upon a relatively xerophytic character of the vegetative system, with dimorphic leaves and open venation. The dermal appendages are broad scales and hairs, with a large distal glandular cell. Two coenosori run one on either side of the mid-rib of the narrow fertile pinna: and they are closely invested by the recurved margins. These broad features of the section *Lomaria*, including those also of the vascular system, are all in accord with what is seen in *Matteuccia*, excepting that *Lomaria* has coenosori in place of separate sori. But it has been seen that in *M. intermedia* the indusium is absent, while the separate circular sori are disposed in regular intra-marginal rows. A lateral fusion of such sori, together with the formation of a vascular commissure linking the veins together beneath the receptacle, is all that is wanting to convert the type of *M. intermedia* into that of a simple *Lomaria*. In both the sori themselves are basipetal, as they are in *Alsophila*, where also they are often disposed in parallel lines, but without the recurving of the pinna-margin. From such facts it is concluded that the simple type of *Lomaria* originated by soral linkage from a non-indusiate type such as that of *M. intermedia*. A similar soral linkage has been seen to have occurred repeatedly among the Dicksonioid-Pteroid Ferns, where the sori are marginal: but here they are superficial in origin, as they are in the whole Cyatheoid series and their derivatives. The conclusion seems then clear that the structural advance to soral linkage has been homoplastic in the marginal and the superficial series; resulting in the one case in Pteroid, in the other in Lomarioid types of Ferns. Both have been successful innovations, as measured by number of species.

The type of *Lomaria* was also successful as shown by the wide geographical spread of the numerous species. But it suffered from the disability that nutritive material elaborated by the trophophylls had to be transferred from the point of production, through the leaf-base and axis, to the sporophyll. An obvious amendment would be to secure to the attenuated sporophyll the power of self-nutrition sacrificed to the protection of the naked sori. This relaxation of a xerophytic feature would naturally follow on life under less exacting conditions. The "flange" with its broadening expanse of photo-synthetic tissue secured this. Within the genus *Blechnum* all degrees of its origin may be seen, from the state of the highly coriaceous types, such as *B. tabulare* and *discolor*, where the convex upper surface of the fertile pinna shows a smooth curve and no flange; through such states as are seen in *B. attenuatum* and *L'Herminieri*, where slight outgrowths appear at the point of strongest convexity, or *B. spicant* where the outgrowth is larger; to the full type of *Eu-Blechnum* as seen in *B. brasiliense* and *occidentale*. Here the "flange" has the appearance of a normal blade, with epidermis, mesophyll, and a system of open venation extending outwards from the commissure, but sometimes linked towards the pseudo-margins. The true margins, still strongly recurved over the coenosori, retain their identity throughout the whole series of species that illustrate this new development. The flange, however highly organised, is bounded by a pseudo-margin, and is itself a pseudo-lamina as regards its phyletic origin. By its formation the sporophyll assumes an outline similar to that of a foliage leaf, and *Eu-Blechnum* appears as though homophyllous. This origin of an innovation at the point of greatest curvature of a reflexed margin is not without parallel among Ferns, for instance in *Pellaea intramarginalis* (Klf.) J. Sm. (Hooker, *Second Century of Ferns*, Plate LXXII); while signs of it are seen also in *Cassebeera*, and in *Cheilanthes lendigera*. The point of greatest curvature will naturally be that of least resistance by pressure of the superficial tissues.[1]

A very interesting feature morphologically is the fact that the true margin, after taking the character of an indusium, is not only displaced but is actually delayed in time of its appearance, as the flange increases in proportion to it. If comparison be made between the figures illustrating *B. tabulare*, *B. spicant*, *B. capense*, and *B. brasiliense*, it is seen that by gradual steps the marginal segmentation is diverted from the "true margin" to the "flange." The former is also delayed in its appearance, so that the marginal segmentation is continued directly into the flange, upon which the morphological margin appears as a secondary development. There has in fact been a "phyletic slide" of the true margin to a superficial position, involving changes both in time and place of its origin. Such a progression is not without its parallels elsewhere among Ferns and particularly in the Pteroids; while it finds its biological justification in the fact that nutrition must needs precede the formation of propagative organs.

Some species may be doubly pinnate, or they may show intermediate states between simple and double pinnation. This is seen in *Blechnum Fraseri* and *diversifolia*: it also appears in the leaves of *Stenochlaena*

[1] A significant homoplastic parallel to this origin by enation of an elaborate photosynthetic organ, with stomata but naturally without any vascular tissue, is seen in the sporogonium of the Moss *Splachnum luteum*, described by Vaizey (*Ann. of Bot.* Vol. v, Plate II, p. 1. 1890).

sorbifolia (L.) J. Sm. (Christ, *Farnkr.* p. 40, Fig. 96), and it is a regular feature of the dendroid genus *Sadleria*. A sinuous or dentate margin elsewhere may be held to represent a further suppressed branching in simply pinnate species. Such facts are intelligible enough according to the hypothetical origin of the Blechnoids from Onocleoid Ferns, and finally from some Cyatheaceous source: for in these families varying degrees of pinnation are exemplified, down to a simple entire blade, as in *Cyathea sinuata* (Vol. II, Fig. 557).

An Acrostichoid development may be traced by easy steps from the Blechnoid coenosorus. It involves in simple cases merely a widening and flattening of the receptacle, while the indusioid margin still covers it, as it is seen to do in *B. attenuatum* (Fig. 696), and in *B. penna-marina.* It is but a step from these to *Stenochlaena* (Fig. 699), in which the extended receptacle is concave, actually following the concavity of the curved wing of the pinna. But at ripeness the curve is everted, giving so clearly an Acrostichoid appearance that the genus was included by Sir W. Hooker under the old comprehensive heading of *Acrostichum.* In all of these, which have the narrow Lomarioid pinnae, the old marginal flap is maintained. But in *Brainea* the start was made from a full Eu-Blechnoid state, with a well-developed flange. Here, however, the indusial flap is absent, and developmental study shows that it is not even initiated (Figs. 701, 702). The distal region of the pinnae of *Brainea* may often reveal the coenosoral origin, with its narrow spread of sporangial development: but lower down the sporangia extend outwards from the commissure so as to cover a considerable area of the flange, or even the whole of it. It thus appears that an Acrostichoid state has been initiated more than once, and with differences of detail in the Blechnoid Ferns. Further, it may be remarked how in *Brainea* the progression has been the converse of that in *Acrostichum praestantissimum* (Chap. XXXVIII, Figs. 621, 622). The spread of the fertile area in the latter is from the margin inwards, a fact naturally related to its Pteroid origin with marginal coenosorus; but in *Brainea* it extends from within outwards, starting from the superficial coenosorus seated near to the mid-rib.

A comparative study of the leaves of Ferns at large leads to the recognition of a progressive integration of the blade. Originally derived by branching of narrow parts, these became either webbed or widened: or more commonly both changes may have worked coincidently. The venation, naturally open in the first instance, was liable to undergo anastomosis, forming first a coarse network, and finally in many families a finer reticulation. The final result of such progressions often resulted in types having an entire blade, and usually with a reticulate venation. An attractive morphological problem relating to the origin of these will be to decide what part lateral extension and what part webbing has taken in producing the broad blade as it is.

Towards that end each example will need to be subjected to a separate comparative analysis, a line of study that has been advanced of late by Von Goebel (*Botanische Abhandlungen*, Heft 1, Jena, 1922). Sir William Hooker, in constructing his scheme for systematic use rather than embodying any such evolutionary ideas as those here contemplated, placed the simplest-leaved types first, and so they are still disposed in the *Synopsis Filicum*. But phyletically it seems probable that they should stand last, as the later and it may often be the final result of phyletic advance. This is the general view which will apply in most families of Ferns, and it finds some degree of illustration in the Blechnoids, particularly in those of the family in which soral disintegration is a feature. For with the formation of broad leaf-surfaces modification of the sorus is apt to follow, particularly where such surfaces result chiefly from widening or extension of parts already present. The sori may be expected to follow the expansion of the part that bears them. These general reflections form a proper introduction to the study of that soral disintegration which plays so prominent a part in the morphology of the more advanced Blechnoid Ferns.

The general hypothesis entertained for these Ferns is that they originated under xerophytic conditions, demanding reduction of the proportion of surface to bulk. This led to a compact and coriaceous type of leaf Commonly they are simply pinnate: but comparison has indicated a more definite origin from Onocleoid and Cyatheoid forms with circular superficial sori borne upon leaves often showing a high degree of pinnation. Evidences of this are still clearly to be seen in the leaves of the Onocleeae, and even of the Blechnoids themselves. It has been seen how the narrow and strongly recurved fertile pinnae of the *Lomaria*-type have freed themselves from a position so extreme as to be unpractical, by the innovation of the flange, giving the Eu-Blechnoid sporophyll. This has resulted not from webbing but from widening of the pinna, which thus takes the lead in the formation of a broad surface. But apical growth may still elongate the pinna indefinitely. Thus the soral tracts will be subject to extension either longitudinally or transversely, or both. The forms which they take may be examined with advantage from this point of view.

If the growth of the pinnae in length outstrip the capacity of soral development, disintegration of the linear coenosorus of *Blechnum* would result: the effect of this is seen not unfrequently as an abnormality, and it is illustrated in *Blechnum spicant* (Fig. 701, 23). It becomes a generic character for *Woodwardia* and *Doodia* (Fig. 704). A wider comparative interest attaches, however, to the changes in outline and continuity of the coenosorus in *B. punctulatum* var. *Krebsii*. The normal species belongs to the section *Lomaria*, and has narrow pinnae, "often not more than one-eighth of an inch broad": but the breadth is variable. In the variety *Krebsii* the pinnae are much broader than in the normal species, but of about the same length: the chief interest lies in the effect of the increasing breadth upon the soral tracts. The effect near to the narrow apex of the pinna may be slight; and the parallel coenosori may be nearly normal. But passing downwards, as the greater width is reached the coenosorus is seen to be thrown into arches, more or less strongly curved, with or without disintegration. With the more extreme width discontinuity is the rule, interruptions appearing at the summit and base of each arch (Fig. 706, 28). The result of this is a series of isolated and paired soral tracts facing one another: in fact the sorus characteristic of the Hart's Tongue (*Phyllitis scolopendrium*) (compare Fig. 706, 30). The fertile pinnae of the var. *Krebsii* have an open venation which also corresponds to that of the Hart's Tongue, though this has a much broader and entire blade. It is reasonable to suggest that so peculiar and at the same time so similar a structure in the two Ferns has been attained along similar lines of increasing breadth: and that it occurred in both genera in a part which, like the pinna of *Blechnum*, has probably been derived ultimately by condensation from a more elaborately branched ancestral leaf.

Before this elucidation can be accepted it will be wise to consider other species than the common Hart's Tongue, and in particular those with acuminate leaves. The apex of the leaf in *Phyllitis scolopendrium* is blunt, but those of *Camptosorus rhizophyllus* and *sibiricus* are elongated into a prolonged tip. If we examine this, its structure may often be found to approximate very nearly to that of a Lomarioid pinna: for there is at the distal end a simple coenosorus on either side of the mid-rib. It has been shown how as the leaf widens downwards an outward arching with interruptions of the coenosorus appears, and the details are very like those seen in the var. *Krebsii* (compare Fig. 707, *A*, and Fig. 706). But as the leaf widens still farther, the rather complex reticulate venation is seen to bear short soral tracts, which at first sight appear irregular (Fig. 707, *B*). Examination shows, however, that the vascular loops, which run closely parallel to the mid-rib, bear tracts similar in position to those of *Woodwardia*: while the rest are frequently paired as in *P. scolopendrium*. Lastly, in another species, viz. *Phyllitis nigripes*, the apical growth is arrested early, and the broadly spathulate lamina is again reticulate: but many of the soral tracts are still paired as before (Fig. 708). The conclusion thus appears justified that *Phyllitis*, even in its most condensed and derivative leaf-forms, is a natural genus sprung from a Blechnoid source, and that *B. punctulatum* var. *Krebsii* gives a true key to its origin. In *Phyllitis*, if our comparisons be correct, the "indusium" is not by origin a true indusium, but a part of the original leaf-margin reflexed.

In face of these comparative conclusions we shall ask, What is the relation of *Phyllitis* to *Asplenium*, with which genus it has usually been ranked, or even included? This question must be decided not by the comparison of extreme types of either, but by the comparative study of their probable origins. The preceding paragraphs indicate that in the history of the soral tracts of *Phyllitis* there was in the first instance the formation of a Blechnoid coenosorus with an underlying vascular commissure, which is itself a secondary development from the original venation: and that *each soral tract is an isolated fragment of that coenosorus seated upon a portion of that commissure, and covered by a length of the original leaf-margin.* In origin it is not a single sorus at all, but a highly derivative body. On the other hand, it has been shown in Chapter XLII that *the Asplenioid sorus is referable in origin to a single sorus of the Dryopteroid type, by extension of the fertile tract along the course of the vein that bears it.* The clue to this is seen in *Diplazium lanceum* (Fig. 672), in which the entire leaf is "attenuated gradually upwards and downwards", while the sori are "irregular and linear" (*Syn. Fil.* p. 229). This is a type which gives the same opportunity for studying the effect of increasing width of the leaf-expanse as that afforded by *B. punctulatum* var. *Krebsii.* Here, however, it appears that definite sori extend along the regular veins of the widening leaf, sometimes on one side sometimes on the other, sometimes on both sides of it, and usually (but not always) with obliteration of the median, that is, the marginal region of the sorus. Thus are produced the types of *Eu-Asplenium* and of *Diplazium. Here the indusium is throughout the homologue of a part of that seen in Dryopteris, in fact it is of the nature of a true indusium.*

It thus appears that however similar the soral tracts may seem to be in *Asplenium* and *Phyllitis*, their origin has been quite distinct. The former is *a highly specialised single sorus* extended along the line of a normal vein, and covered by a true indusium: the latter is *a tract segregated from a coenosorus*, extended along a portion of a secondary vascular commissure, and covered by a segment of an original leaf-margin. The two types provide one of the most remarkable instances of homoplastic development. As the leaves of the Ferns in question become widened and their form condensed —that is, as they depart farther from the original highly branched types— these quite different structures become more and more alike; and it is this which has led to the systematic confusion of two phyletically distinct series. The only really homogenetic factors in these soral tracts are the sporangia, and the receptacle that bears them.

It will probably be objected that the well-known upward fusion of the paired vascular strands of the petiole, and the formation of those peculiar X-shaped conducting tracts seen in *Asplenium* and in *Phyllitis* are real indications of affinity (see Vol. I, p. 166, Fig. 157. Also Luerssen, *Rab. Krypt. Fl.* III, pp. 120, 150). But these anatomical features may fairly be regarded rather as consequences of parallel condensation of leaf-structure from a more diffuse and highly branched ancestry, than as signs of real affinity. Most varied examples of such condensation and vascular fusion are to be found illustrated by Bertrand and Cornaille (*Struct. d. Filic. actuelles*, Lille, 1902, Figs. 51, 54, 59, 65, 78), while the same is now seen also in *Diellia* (Fig. 596). Gwynne-Vaughan has also noted the caliper-wise fusion of the paired vascular straps in the petiole of *Ceropteris calomelanos* (L.) Und. = *Gymnogramme chrysophyllum* Klf. (MS. research notes). A like state appears also in the thin petiole of *Adiantum* (Luerssen, *Rab. Krypt. Fl.* III, p. 82). Such evidences of condensation may be held as cognate to that seen in the petioles of climbing Ferns systematically quite distinct from one another, such as *Gleichenia*, *Lygodium* and *Odontosoria* (Vol. I, p. 171, Fig. 165). All of these may be held as resulting from homoplastic contraction of the conducting system in the elongated leaf-stalk. Thus notwithstanding this point of anatomical similarity the Asplenioids and Blechnoids may still be held as consequences of convergent evolution rather than as nearly akin. The ancestry of the one appears to have been relatively hygrophytic, with elaboration of the individual sorus (Dryopteroids); of the other xerophytic with a tendency to soral fusion to form coenosori (Onocleoids and Blechnoids). Nevertheless the ultimate source from which these all took their origin was probably that now represented by the Cyatheoid Ferns: in these, as in all the Ferns here discussed, the sori are superficial in origin, as they are also in their adult position.

The Blechnoid Ferns may be grouped under four heads, according to the characters disclosed in this chapter. They are disposed as follows in a roughly phyletic sequence.

I. Coenosori continuous, running parallel to the mid-rib of the fertile pinna: with or without a photosynthetic "flange."

    (1) *Blechnum* Linn., 1753 ...   .   138 species.

    (2) *Sadleria* Kaulfuss, 1824   ...   .   4 species.

II. Coenosori showing an Acrostichoid spread over the lower surface of the pinna or of the flange.

    (3) *Stenochlaena* J. Smith, 1841 ...   .   11 species.

    (4) *Brainea* J. Smith, 1856   ...   .   1 species.

III. Coenosori secondarily interrupted, the parts forming two or more rows parallel to the mid-rib.

    (5) *Woodwardia* Smith, 1793   ...   .   7 species.

    (6) *Doodia* R. Brown, 1810   ...   .   6 species.

IV. Coenosori secondarily interrupted: the isolated parts facing one another as a rule, but sometimes less regularly placed: venation open or reticulate. (This includes *Blechnum punctulatum* Sw. var. *Krebsii* Kze.)

(7) *Phyllitis* (*Scolopendrium*) Ludwig, 1757 ... 9 species.

(8) *Camptosorus* Link, 1833 ... ... 2 species.

N.B. The inclusion of *Triphlebia* Baker and of *Diplora* Baker under *Phyllitis*, as suggested by Copeland, is provisionally accepted.

BIBLIOGRAPHY FOR CHAPTER XLIV

702. PRESL. Tentamen, p. 141. 1836.
703. KUNZE. Die Farnkräuter, p. 176, Taf. LXXIV. 1847.
704. METTENIUS. Fil. Hort. Lips, p. 60, etc. 1856.
705. HOOKER. Species Filicum. III, p. 1. 1860.
706. HOOKER. Garden Ferns, Plates XV, LXI. 1862.
707. BURCK. Indusium der Varens. Haarlem. 1874.
708. LUERSSEN. Rab. Krypt. Flora, III, p. 111. 1889.
709. PRANTL. System der Farne, p. 16. 1892.
710. KARSTEN. Buit. Ann. XII, p. 143. 1895.
711. CHRIST. Farnkräuter, pp. 40, 183. 1897.
712. DIELS. Natürl. Pflanzenfam. I, 4, p. 245. 1902.
713. BERTRAND and CORNAILLE. Struct. d. Filic. actuelles. Lille. 1902.
714. CHANDLER. Ann. of Bot. XIX, p. 373. 1905.
715. COPELAND. Phil. Journ. VIII, p. 151. 1913.
716. BOWER. Studies IV, Ann. of Bot. XXVIII, p. 363. 1914.
717. Frau E. SCHUMANN. Flora, Bd. 108, p. 201. 1915.
717 bis. SIM. Ferns of South Africa. 2nd. Ed. 1915. p. 181. Plates 78-81.
718. LEONARD. Taenatidinæ. Proc. Roy. Dub. Soc. XV, p. 255. 1918.
719. VON GOEBEL. Gesetzmässigkeit im Blattaufbau. Bot. Abhandl. Heft I. Jena. 1922.
720. VON GOEBEL. Morph. u. biol. Studien. Buit. Ann. Bd. XXXVI, p. 107. 1926.

CHAPTER XLV

DIPTEROID FERNS

THOUGH *Matonia* and *Dipteris* have long been recognised as probable survivals from an earlier age, they were till recent years regarded as isolated and perhaps derelict types. Little attempt had been made to connect them with Ferns of the present day. Hitherto this may appear true for *Matonia*: its nearest relatives are probably found amongst the Dipteroids and Gleicheniaceae. But discoveries among the Mesozoic fossils, together with a better knowledge of certain other living Ferns which share in more or less degree with *Dipteris* its general features, have combined to show with high probability that the Dipteroid type survives in a considerable section of the modern Leptosporangiate Ferns. In particular it is rapidly becoming clear that many of those ranked in the comprehensive but phyletically confused genus *"Polypodium"* are really Dipteroid derivatives, which never in the course of their descent were gradate, nor possessed an indusium protecting their naked superficial sori.

The first step in building up such conclusions will be to search for the nearest connecting forms, and to work upwards from them. An aid to their recognition is the existence of a rich synonymy, applied it may be to some type peculiar in form, and difficult of classification. Such a type is found in *Cheiropleuria bicuspis* (Bl.) Presl, which after vicissitudes of terminology is now recognised as the sole representative of a substantive genus *Cheiropleuria* Presl, 1849 (see Studies V, *Ann. of Bot.* XXIX (1915), p. 495).

CHEIROPLEURIA BICUSPIS (Bl.) Presl

This Fern is widely distributed in the Malayan region, where it is commonly associated with *Dipteris conjugata*, a fact that is probably more than a mere coincidence. It was figured by Hooker (*Lond. Journ. Bot.* Vol. V, p. 193, Pls. 7, 8), and his drawing has been widely quoted. The sterile leaf there depicted is bicusped, but comparison of specimens shows that it is variable in its outline, as will be gathered from the photographs shown in Fig. 709, *a–d*. Not uncommonly the sterile blade is quite unbranched.

The axis is elongated, with internodes of varying length. It is densely clothed with silky yellow hairs, and bears many brown roots. The plant hovers between a terrestrial and an epiphytic habit; as Van Rosenburgh says, it is "creeping or subscandent" (*Malayan Ferns*, p. 732, 1909). Dichotomous branching of the axis has not been observed, but lateral buds are frequent, arising from the abaxial face at the base of many leaves, though not

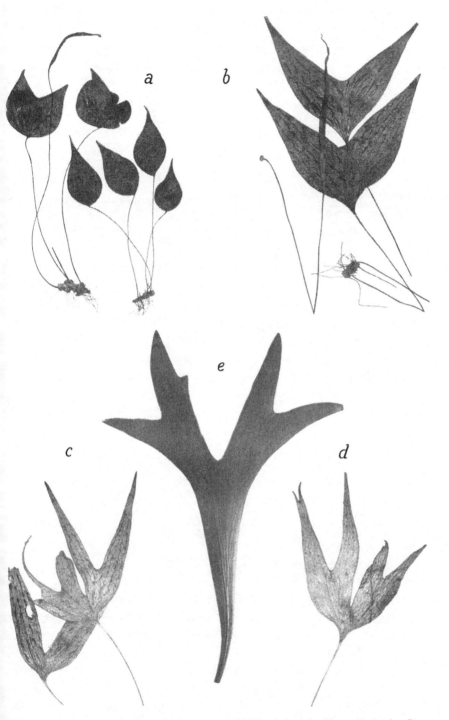

Fig. 709. *a—d*. Specimens of *Cheiropleuria bicuspis* (Bl.) Presl, from the Lingga Mountains, Borneo, illustrating the general habit. Reduced to ¼. *a* shows a number of juvenile leaves, with only one cusp: one with two cusps, and one erect and simple sporophyll. *b* = plant with two sterile leaves of the two-cusped type, as shown in Hooker's figure, *Journal of Botany*, 1846: also one fertile leaf. *c* = two irregularly lobed leaves. *d* = leaf irregularly furcate. *e* = an erect leaf of *Platycerium Hillii* Moore, reduced to ⅓, for comparison with Figs. *c*, *d*.

of all (Fig. 709 a): in this it compares with *Lophosoria* and *Metaxya*, and also with *Platycerium* (Hofmeister, *Higher Crypt.* p. 252): but it differs from *Matonia* and *Dipteris*. The leaves are strongly dimorphic, the fertile being the taller, and very narrow (Fig. 709, a, b). The sterile leaves have a leathery blade borne on a wiry petiole. The blade may be entire, or two-lobed as the specific name implies, or many-lobed. The most complex are found on fully matured

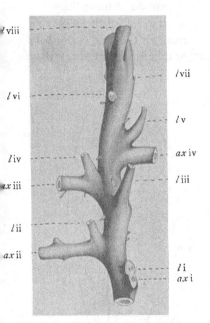

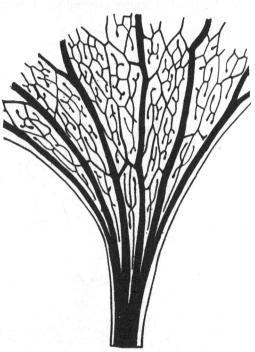

Fig. 709 a. Drawing by Dr J. McL. Thompson of a rhizome of *Cheiropleuria bicuspis* (Bl.) Presl with the superficial hairs removed so as to expose the leaf-bases, which are numbered *l.* i—*l.* viii, and the lateral axes which spring from the bases of some of them, numbered *ax.* i–*ax.* iv. The leaves iii, vi, vii, viii have no associated axes. The leaf-arrangement is alternate, and the climbing shoot is seen from the side facing away from the support. (× 2.)

Fig. 710. Trace of the vascular system at the base of the lamina of a large sterile leaf of *Cheiropleuria*, showing the pedate relation of the main veins, after the manner of *Matonia*. The smaller veins show the "*venatio anaxeti*". (× 4.)

plants, their form having obvious relation to dichotomy. Comparison with *Dipteris conjugata* shows unity of type, which the venation confirms (Vol. II, Fig. 568). The main veins dichotomise with distinct pedate sequence, after the manner of *Matonia* (Fig. 710). From them branches arise which anastomose freely, the branchlets terminating blindly within the meshes, after the type of "*venatio anaxeti*" so well represented in *Dipteris*. This is already initiated in the juvenile leaves, and it finds its like in the juvenile leaves of *Platycerium*

(Fig. 711, *A*, *B*). The narrower fertile leaf is unbranched: its lower surface is covered by a dense mass of sporangia, which will call for more special description later.

The hairs of *Cheiropleuria* conform more nearly to those of *Matonia* than of *Dipteris*, in both of which they are simple, and indurated at the distal end, while the basal cells retain the power of growth and division. Here each hair is long and unbranched, and is composed of some 20 or more thin-walled cells: in fact it is of a more primitive type than in either of these genera.

The vascular anatomy is of outstanding interest, for the axis is protostelic (Fig. 712, *A*), the stele resembling in structure and in form that of *Gleichenia*.

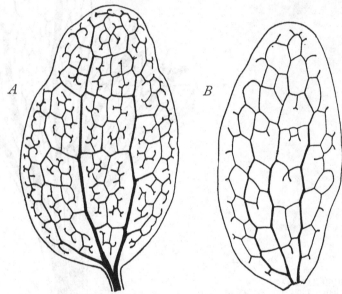

Fig. 711. *A* = a juvenile leaf of *Cheiropleuria* showing the venation, for comparison with that of *Dipteris conjugata*, Vol. II, Fig. 570. (× 3.) *B* = a juvenile leaf of *Platycerium Veitchii*, for comparison with these. (× 3.)

Obliquely from the upper side of it the leaf-traces pass off alternately right and left, their origin being as follows. First, a group of protoxylem-tracheides appears, some 3 or 4 layers within the outer limit of the metaxylem, while its contour swells into a rounded hump. Parenchyma-cells then aggregate internally to this, and a loop of xylem is thus formed, with the protoxylem lying centrally within it (Fig. 712, *A*). This, together with some of the internally-lying tracheides, moves outwards, a lateral constriction appearing on each side, till the leaf-trace is shut off from the stele by the intruding phloem and sheaths (Fig. 712, *B*). On separation it consists of an oval tract of tissue, enclosing within a ring of metaxylem a parenchymatous island, with the protoxylem (which has meanwhile divided into two strands) at its periphery. The ring very soon opens out by an adaxial fission, the margins withdrawing till the whole leaf-trace takes the form of a crescent, as in *Matonia* and *Dipteris*. But in *Cheiropleuria* the trace is narrower, and the protoxylems only two. Later the trace itself divides into two equal strands, and so passes out into the petiole (Fig. 713).

A   *B*

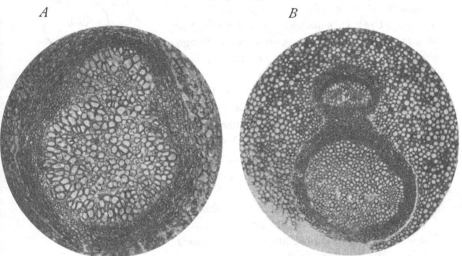

Fig. 712. *A* = a transverse section ot a protostele ot *Cheiropleuria* showing a leaf-trace being given off from it. (× 45.) *B* = a similar section, with the leaf-trace completely separated. It has two internal protoxylem groups, with the metaxylem surrounding them completely—a condition which holds only for a short distance. (× 30.)

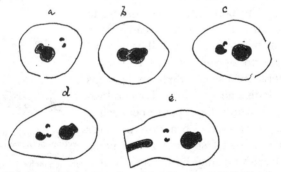

Fig. 713. Transverse section of the rhizome of *Cheiropleuria*, to show the relations of the leaf-trace and of the lateral bud to the stele of the axis. *a* shows the relation of two normal leaf-traces, where no bud is formed. *b—e* show successive sections from below upwards, in the case of a leaf-trace where a lateral bud is borne on the abaxial side of it. (× 3.)

It is most unusual, in a Fern of relatively robust habit, and of so advanced a type as *Cheiropleuria* proves to be, that the protostelic state should be continued from the sporeling to the adult state (as in *Gleichenia*), without solenostelic or other expansion: also that the origin of the leaf-trace should be so similar to that in *Platyzoma* (Vol. II, Fig. 483), for both of these features are characteristic of the Simplices. But the most striking peculiarity lies in the initial structure of the leaf-trace itself, which is most nearly matched by certain early fossils. The newly separated trace of *Cheiropleuria* passes through a Clepsydroid phase, corresponding to that seen in *Thamnopteris* (Vol. II, Fig. 432, 3 to 6). It is not clear what significance these facts may bear: such comparisons, though remarkable, should not be unduly stressed phyletically.

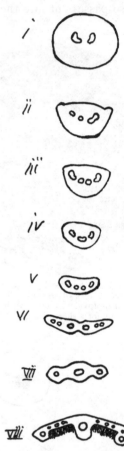

The vascular supply to the lateral bud arises in near relation to the binary leaf-trace of its adjoining leaf (Fig. 713). The nature of this branching has been considered in Vol. I, pp. 70–78. The facts are not clearly distinctive: but it is reasonable to hold that the branching arose from dichotomy with unequal development of the shanks, and with a close relation of the leaf to the base of the arrested shank. This would bring the branching of *Cheiropleuria* into line with that in *Matonia* or *Dipteris* on the one hand, and with *Lophosoria* and *Metaxya* on the other.

The leaf-trace undergoes further segregation in its course up the petiole, and widens out in the sterile blade into the venation already noted (Fig. 710). In the narrower sporophyll its upward course is suggested by the sections vii, viii, in Fig. 714. There are usually three main veins in the fertile blade, one median and one following each margin, while the expanse between is served by smaller veins, and bears on the lower surface a dense mass of sporangia and paraphyses. This condition may be compared as a whole with the fertile lamina of *Dipteris quinquefurcata* or *Lobbiana* (Vol. II, Figs. 571, 572). If we imagine the branched sporophyll of either

Fig. 714. i—vi = successive transverse sections of the petiole of a sterile leaf of *Cheiropleuria*:—i = at the base; ii = about 3 inches up; iii = 4 inches; iv = 6 inches; v = 8 inches: vi = 9 inches from the base. Figs. vii, viii are from a fertile leaf, at levels corresponding to v, vi of the sterile leaf. (× 8.)

of these species represented by a single segment and the sori spread over the whole surface, the result would be the unbranched sporophyll of *Cheiropleuria*. There are, however, specimens in the Kew Herbarium, with two equal main veins, or three apparently equal. This is in accord with the inconstancy of form and main venation of the sterile blade, while it readily finds its parallel among the living species of *Dipteris*, or the fossils attributed to the family.

In the sporophylls of *Dipteris Lobbiana* the sori form a single row on either side of the mid-rib and, though they are variable in extent, they do not spread beyond the individual mesh from the centre of which each arises (Fig. 571, Vol. II). In *D. quinquefurcata* they are not restricted to a single row, but still each is limited to a single mesh (Fig. 572, Vol. II). It is here that *Cheiropleuria* shows an advance on them: for as the sori extend over the whole leaf-surface they spread beyond the limits of the single mesh. A good index of what has happened is found in the veins, which are extended into masses of storage tracheides of the receptacle: these lie in a plane nearer to the lower surface than the substantive venation. The receptacular tracheides are thus able to

Fig. 715. *a* = part of a sporophyll of *Cheiropleuria*, seen as a transparency showing the vascular tissue. The steady lines are the venation; the irregular patches are the storage-xylem of the soral receptacles. At the points (×) these have crossed the veins at a lower plane than that in which the veins lie. (× 8.) *b* = part of a sporophyll of *Platycerium angolense*, showing the origin of the receptacular xylem from the ends of the blind veins. The receptacles are more elongated than in *Cheiropleuria*, and pass to a lower level in the mesophyll, there extending frequently across the course of the original venation, which is here represented by thin steady lines. (× 4.)

pass, in a lower plane, across the course of the latter, and so the sorus may extend beyond the limits of the individual mesh (see the points marked (×) in Fig. 715). Thus is initiated that "diplodesmic" state described in Vol. I, p.233. The receptacular system is not developed largely in *Cheiropleuria*, but it becomes a marked feature in *Platycerium* and in *Leptochilus tricuspis*. It may be held as a concomitant of the Acrostichoid development of the sori seen in many Dipteroid derivatives. In a transverse section of the fertile area of *Cheiropleuria* the two vascular systems can be readily distinguished (Fig. 716).

The fertile area is not defined as distinct sori; it extends uniformly over the lower surface. The numerous sporangia are of "mixed" origin, and they are associated with simple hairs. The individual sporangium is larger and longer-stalked than in *Dipteris*: the stalk is four-rowed as in that genus, and the annulus forms a continuous oblique ring: but its induration is not continued past the stalk, while the lateral stomium is distinctly below the equator of the sporangial head. The number of annular cells is large, 26 to 36 have been counted (Fig. 717). The cell-cleavage of the young sporangium is, as in *Dipteris*, by two rows of segments, giving the four-rowed stalk, a feature that is shared with many Cyatheoids (see Vol. II, pp. 291, 302). The spore-output has been shown by H. H. Thomas to be typically 128: actual counts were 124, 123, 108. This is an important link with the fossil Dipteroids, coinciding as it does with certain archaic features in the sporangium (Vol. II, p. 321).

There is probably no living Fern which is so clearly a synthetic type as *Cheiropleuria*. Its characters present a singular combination of the primitive and the advanced. The former include the protostelic axis, with a branching susceptible of interpretation in terms of dichotomy: the presence of simple

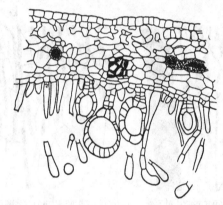

Fig. 716. Transverse section of part of a fertile blade of *Cheiropleuria*. (× 50.) It shows the diplodesmic state, the storage tracheides of the receptacle being nearer to the lower surface than the strands of the venation, which lie right and left of it.

hairs and the absence of dermal scales: the peculiarly primitive leaf-trace: the prevalent dichotomy of the main veins of the blade: the absence of any indusium protecting the superficial coenosori: the large size, two-rowed segmentation, and oblique annulus of the sporangium, and the relatively high spore-output. These features collectively confirm the Matonioid-Dipteroid affinity of *Cheiropleuria*, while the vascular structure strongly suggests the further link with the living Gleichenias, and a still further reference to certain Mesozoic fossils on the ground of the relatively high spore-output. Moreover, all of these Ferns have superficial sori, a feature which has been constant for them at least from Mesozoic times.

Characters of advance appear in the webbed leaf-blade, with few and irregular branchings, or none at all: the highly reticulate venation; and the mixed character of the Acrostichoid coenosorus, of which the extended system of receptacular xylem gives rise to the "diplodesmic" structure. These features serve to link *Cheiropleuria* with a number of more modern types of Ferns, some of which will now be described and compared. The first of these will be the remarkable Stag's Horn Ferns, of the genus *Platycerium*.

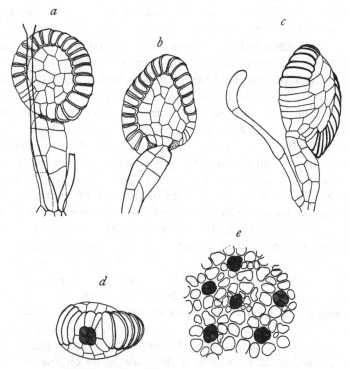

Fig. 717. Sporangia of *Cheiropleuria*. *a—d* show various aspects of the sporangium, with its 4-rowed stalk, and relatively large capsule and continuous oblique ring. *e* shows part of a sorus cut transversely to the sporangial stalks. These are shaded, while the hairs are left clear. (× 80.)

PLATYCERIUM Desvaux, 1827.

This genus owes its systematic permanence to the very distinctive vegetative characters which it presents. The earlier systematists, though linking it with *Acrostichum,* appear to have regarded *Platycerium* as an isolated and bizarre phenomenon. Even in later times Diels has remarked that "the genus stands quite isolated among the Polypodiaceae of the present day": and he specifically states that *Cheiropleuria* has no near relation to it (*Nat. Pflanzenfam.* I, 4, p. 339). More recently Von Straszewski concludes that *Platycerium* does not belong to the Acrosticheae, nor does he hold it akin to *Dipteris* or *Cheiropleuria* (*Flora,* 1915, Bd. 108, p. 304). But Christ places

Platycerium and *Cheiropleuria* in close relation, with a pointed reference to *Dipteris* (*Farnkr.* p. 128). Such divergent expressions of opinion will serve to stimulate interest in the examination of these remarkable plants.

Platycerium is represented by some 14 species, which are all tropical epiphytes: the genus extends to both eastern and western hemispheres, but chiefly in the Malayan area. The habit, with its nest-leaves appressed to the tree trunk on which the plant grows, and the erect or pendant "foliage-leaves" which habitually bear the sori, is familiar from text-books, and from the specimens commonly grown in greenhouses: so that the specific details need not be described here (compare Vol. I, Fig. 54). It may suffice to say that revision of the large series of specimens in the herbarium at Kew confirms very strongly its Dipteroid character both in form and venation of the leaves, and notably in their bifurcate branching. How near the similarity of the erect leaves may be to those of *Cheiropleuria* will be seen from comparison of the photographs represented on Fig. 709. Von Straszewski has traced the development from the sporeling: he finds the first leaves to be upright, and of simple form, with a single vein: the third leaf is round or kidney-shaped, with reticulate venation (compare Fig. 711, *B*, above), and it is furnished with the usual stellate hairs. The creeping stem bears the later leaves in bi-seriate order upon its upper side. They are differentiated at once as nest-leaves and erect or foliage-leaves: but the two types do not follow in any constant sequence. In close relation to the erect leaves, lateral buds arise from the stem, and their vascular supply has been found to spring from it. These relations appear to resemble those of *Cheiropleuria* and *Lophosoria*. The soral patches which usually appear on the underside of the erect or pendant foliage-leaves, are of irregular outline, and sometimes they are borne on special leaf-lobes (*P. coronarium*). These fertile areas may attain very large size: on a specimen in Kew of *P. grande* (Cunn.) J. Sm., from Singapore, it is more than a foot across. But fertility is not restricted to the "foliage-leaves": Poisson has recorded the production of sporangia on the nest-leaf of a *Platycerium* grown in Paris (Von Straszewski, *l.c.* p. 300). This fact suggests that the two types of leaf have differentiated from a common prototype, which may have been not quite like either of them. Such a prototype probably resembled the homophyllous leaves of the Dipteroids. I see no sufficient reason for regarding the nest-type of leaf as the more primitive.

In Ferns so highly specialised as *Platycerium*, and of a rather gross habit, it is natural to anticipate that there will be a disintegrated vascular system: and that is what examination shows. Even in the sporeling we need not look for structure of a strict type: nevertheless, the sporeling, after a short protostelic stage, passes to a condition with a solenostele interrupted only by the departure of the leaf-traces (Von Straszewski, *l.c.*

Figs. 16, 17). But this orderly sequence is soon departed from, and the adult stems show a high degree of vascular disintegration with numerous perforations. The complexity of the result is related to size. In *P. alcicorne* (= *P. bifurcatum* (Cav.) C. Chr.), one of the less robust species, the axis is traversed by a simple dictyostele, but highly perforated. The leaf-trace arises from this as a group of small strands rather irregularly disposed (Fig. 718, *A*). But in *P. aethiopicum* (= *P. stemaria* (Beauv.) Desv.), one of the most robust species, there is a complex medullary system, in addition to the outer circle of strands which represents the primary solenostele: this medullary system contributes its quotum to the outgoing leaf-trace (Fig. 718, *B*). As Miss Allison remarks, "it would be quite consistent with

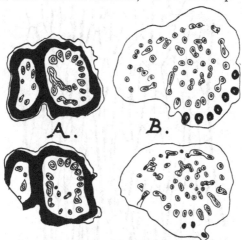

Fig. 718. *A* = sections of the rhizome of *Platycerium alcicorne* showing the relation of two leaf-traces to the ring of meristeles of the axis. *B* = similar sections from *Platycerium aethiopicum* showing the greater structural complexity, with numerous medullary strands enlarged. (After Miss Allison, *New Phytologist*, 1913, Vol. XII, p. 311.)

the structural facts if we were to consider *Platycerium* with its complicated dictyostele as the dictyostelic type of a series of which *Dipteris* and *Matonia* are the solenostelic types." The inner cycles of the polycyclic axis of *Matonia* would then be regarded as providing the prototype of the medullary system seen in the larger *P. aethiopicum*. The leaf-trace of *Platycerium* consists of many strands, some ten in *P. Hillii*: but already in *Cheiropleuria* there are two, and *Dipteris Lobbiana* has four (see Vol. II, Fig. 574, p. 316). It thus appears that the vascular structure of *Platycerium* may be held as cognate with that of the Dipteroids, allowance being made for the larger size and higher specialisation.

As the numerous strands of the leaf-trace pass up the broadening leafstalk of *Platycerium* they spread out like a fan, with many anastomoses,

210   DIPTEROID FERNS   [CH.

finally settling down into a network of *venatio anaxeti,* which as such
supplies the vegetative expanse (see Studies V, Text-fig. 10). But in the
fertile or soral patches an additional vascular supply arises in the form of
branches which, running parallel to the lower surface, supply the sori. This
has long been known, and has been described by Mettenius (*Fil. Hort. Lips.*
1856, p. 26) and by Hofmeister (*Higher Cryptogamia,* 1862, p. 252), as well
as by others. Examination of the fertile patches shows, with varying distinctness in different species, that the sporangia are restricted to more or
less definite parallel lines. This is seen most clearly where the area is less
prolific, and the leaf has the appearance of being only half fertile. There the

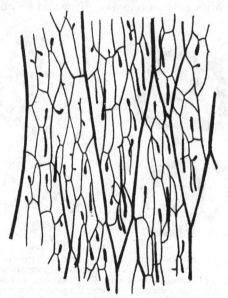

Fig. 719. Part of a lamina of *Platycerium aethiopicum* which is only half fertile, i.e. with the sori
isolated and small. The diplodesmic state is less
pronounced than in Fig. 715, *b.* (× 3.)

linear sori may appear isolated, and relatively short: and these give the
clue to the real constitution of the fertile areas of *Platycerium.* But it is
obvious even in the fully developed fertile patches of *P. aethiopicum.*

If such a semi-fertile area be cleared and stained, the course of both
vascular systems can be followed (Fig. 719). The substantive system is
of the *Dipteris*-type, but the free twigs are elongated in the form of receptacular tracheides, which bore downwards so that their distal ends
approach the lower surface, along which they extend. Often each remains
restricted to its own areola: but in not a few cases the receptacular strand
extends past the vein limiting its own areola, thus crossing in a lower
plane to the next. This is what has been seen occasionally in *Cheiropleuria*

(Fig. 715, *a*): but it is a much more pronounced feature here, even in the partially fertile patch of *Platycerium*. In a fully fertile patch the course of these receptacular strands is much longer; and they may even branch, forming thus an elaborate second system of venation in a plane closely within the lower surface: so that the patch is "diplodesmic" (Fig. 715, *b*). The appearance presented in transverse section is shown in Fig. 720, where the strands of the two systems are seen to lie at quite distinct levels, while the sporangia are attached close to the superficial receptacular strand. The similarity to what has been seen in *Cheiropleuria* cannot be missed, but with the difference that here the sori, though elongated, maintain their identity: in *Cheiropleuria* the sporangia are spread over the whole fertile area, after the manner usual in Acrostichoid Ferns, and soral identity is lost.

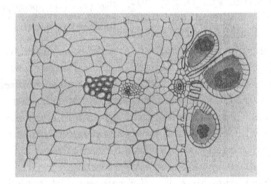

Fig. 720. Transverse section of a fertile leaf of *Platycerium willinkii* Moore, showing vascular strands belonging to the two diplodesmic systems: the strand nearer to the upper surface belongs to the main system of the lamina, that nearer the lower surface is a receptacular strand. (× 66.)

Hofmeister (*Higher Cryptogamia*, p. 252) described a diplodesmic structure in the humus-leaves of *P. alcicorne*. It is a very remarkable thing that this condition of the humus-leaves should be repeated in the sporophylls, but only in their fertile regions (see "Studies on Spore-producing Members," *Phil. Trans.* Vol. 192, 1899, p. 86). It suggests that possibly all the types of leaf in *Platycerium* sprang from a "general purposes" type.

The dermal appendages are hairs, which are branched in a stellate manner. Their origin by steps of increasing complexity, from unicellular glandular hairs to the multicellular stellate state, has been traced by Von Straszewski in the sporeling, where they are already present on the first leaves. Scales are not described for any of the species, even upon the rhizome. In this *Platycerium* accords with *Cheiropleuria*, *Dipteris*, and *Matonia*. The stellate hairs are associated in particular with the sporangia, as paraphyses. Sometimes they are very long and form an almost woolly

felt, as in *P. biforme* Bl. [= *P. coronarium* (Koenig) Desv.], where it covers
the discoid, cake-like fertile lobes, protecting the young sporangia. The
origin of the sporangia is almost simulta-
neous, but yet slight differences in time
are indicated by their relative size, as
seen in Fig. 720. The adult sporangium
is shown in Fig. 721. Here the stalk
consists of only three rows of cells, as
against the four rows in *Cheiropleuria*.
This is, however, not an essential difference, though important for comparison
elsewhere. The sporangia of *Platyzoma*
show both types (see Vol. II, Fig. 492).
Moreover there is an inherent probability
that the change should follow in any line

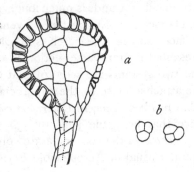

Fig. 721. *a*=a sporangium of *Platycerium
aethiopicum* as seen from the side. *b*=trans-
verse sections of its stalk. (× 80.)

of descent where reduction of the sporangium is involved. While the spore-
output in *Cheiropleuria* is typically 128, that in *Platycerium* is only 64. But
the size and spore-output are not absolutely determining causes of the change
of structure: in *Dipteris*, with smaller sporangia and an output of 64 or less,
the four-rowed stalk is retained. The sporangium of *Platycerium* is pear-shaped,
and the annulus, which consists of more numerous cells than in *Cheiropleuria*,
shows only slight obliquity, and is almost interrupted at the insertion of the
stalk, though not actually so. These facts collectively indicate for *Platycerium*
a more specialised state than in *Cheiropleuria*, but they indicate a type
cognate with it. There is no perispore in *Platycerium* (Hannig, *l.c.*).

The sporangia of *Platycerium alcicorne* show considerable regularity of
orientation, so that the plane of the annulus cuts the direction of the
underlying vein at right angles. They originate for the most part simultaneously, and in two rows, as though each sorus were an extension of a
simple radiate-uniseriate sorus. This may very probably have been the
original source.

The germination in *Platycerium* has been observed by Bauke (*Bot. Zeit.*
Bd. XXXVI) and by Von Straszewski (*l.c.* p. 272). The young prothalli are
at first filamentous and often branch early, but the filaments develop
unequally. A flattened expanse is soon formed laterally, or more than one:
these bear pluricellular glandular hairs on the lower side, as in the
Cyatheaceae and in *Diacalpe*. The prothalli are sometimes dioecious, but not
always. The fact that the lid-cell of the antheridium is here divided, as it is
in *Woodsia*, suggests, as do also the glandular hairs, a relation to other
relatively primitive Ferns, and in particular to the Cyatheoids. The hairs
are similar to those first borne on the sporeling, while in the latter an early
apospory was seen to follow on mechanical damage to the young leaf. Such

facts accord with the relatively primitive origin ascribed to *Platycerium*, but they do not bring any very distinctive material for further comparison.

CHRISTOPTERIS TRICUSPIS Christ

This fine Fern was first described, from Sikkim specimens, in Hooker's *Species Filicum*, Vol. V, p. 272, Pl. CCCIV, as *Acrostichum (Gymnopteris) tricuspe* Hooker: it stands in Christ's *Farnkräuter*, p. 49, and in *Natürl. Pflanzenfam.* 4, I, p. 199, as *Gymnopteris tricuspis* (Hook.) Bedd.: but it finds its best place in Copeland's genus *Christopteris* (*Philipp. Journ. Botany*, Vol. XII, 6, 1917). There has never been any doubt as to its identity, nor as to its close relation with *Cheiropleuria*. Describing it in 1864 Sir William Hooker writes: "This very fine and new species, with not a little of the habit and venulation of *A. bicuspe*, differs remarkably in being trilobed or tripartite, and it has always a solitary central costa to each lobe. One of my specimens has the three segments only partially contracted and fertile."

The photographic Figure 722 shows the habit of this upstanding, ground-growing species, which may attain some three feet in height. The petioles rising from the creeping rhizome are doubtless elongated, as in *Dipteris* and *Cheiropleuria*, in relation to the herbage amongst which it grows. At its base the leaf-stalk enlarges into a mammillary swelling, which persists with a terminal scar marking a smooth abscission. There is occasional branching of the fleshy axis, but it is independent of the leaf-bases, as it is also in *Dipteris*. The rhizome is covered by brown scales, which fall away with age.

The leaves are strongly dimorphic, though sometimes only half fertile, the soriferous part being then contracted, as in *Hymenolepis* (*Syn. Fil.* p. 422). The ternate form of the blade with a large terminal lobe appears at first sight far removed from the dichotomy prevalent in Dipteroid Ferns: occasionally more than three lobes may appear (see Frontispiece). But a reference to Vol. I, Chapter V, and particularly to Figs. 77 and 80, will show how nearly related in origin a ternate, and even a more richly lobed leaf, may be to equal dichotomy. In this, as in its dermal scales and other features to be described later, *C. tricuspis* is in advance of *Cheiropleuria*: nevertheless, the more detailed examination of to-day will be found to confirm the original forecast of Sir W. Hooker.

The rather fleshy axis, which is about ¼ inch in diameter, is traversed by a cylindrical, highly perforated dictyostele, which presents in transverse section a ring of some 12 small meristeles. When a leaf-trace departs, some five of these arch outwards and pass into the petiole, the two marginal being the largest, with the usual hooked xylems. After this the ring again closes. The whole construction is such as would naturally follow if a solenostele, giving off a leaf-trace such as that seen in *Dipteris* or *Metaxya*, were profusely perforated—which is the actual fact for *C. tricuspis*. The parenchyma of the rhizome is crowded with nests of black sclerenchyma: but these are not continuous strands, and they extend only a short way up

a

b

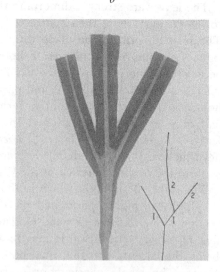

Fig. 722. *a* = a whole plant of *Christopteris tricuspis* Christ, one-sixth of its natural size. *b* = basal region of the fertile blade, with scheme of its branching. Reduced to ½.

the petiole. The same is found in *Neocheiropteris*. The course of the vascular
strands up the petiole to the blade is indicated by Fig. 723: it presents a
close analogy to what is seen in *Platycerium* (compare Studies V, Text-
fig. 10), and may be regarded as essentially similar to that of *Cheiropleuria*,
though complicated by more elaborate perforation (compare Fig. 714). In
the sections i–vi the gap between the two larger marginal strands remains
constantly open, its position being marked (×). Complications arise at the

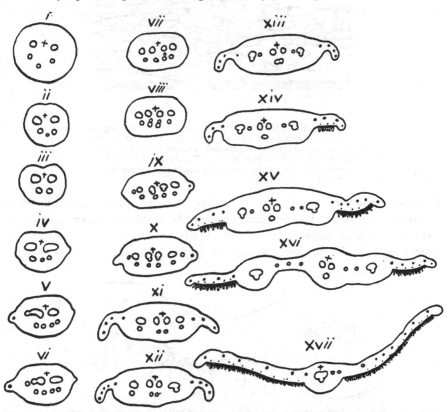

Fig. 723. *Christopteris tricuspis*: a series of sections illustrating the vascular system of the sporophyll,
the adaxial system uppermost. (×4.) The mark (×) indicates the "heel" of the horseshoe, that is,
the adaxial side of the petiole.

point of branching of the blade, and Fig. 723, vii–xvi, illustrate the course
upwards into a sporophyll. The gap (×) is maintained till the marginal
strands finally fuse in the mid-rib of the middle lobe (xvii). This last section
represents the structure of a single fertile lobe, with its enlarged mid-rib:
each lateral flap has its primary venation, while the lower surface on either
side is covered by the Acrostichoid sorus (Fig. 724). It thus appears that
the course of the strands is a modification of the primitive horseshoe,
complicated by segregation, and by various fusions: and that these are

comparable with the simpler state of *Cheiropleuria*, and, on the other hand, with a still more complex state seen in *Platycerium*.

After passing into the sterile blade the *venatio anaxeti* is like that in the Ferns last named. But more special interest is found in the fertile blade, which is again diplodesmic. The rather fleshy expanse is traversed towards

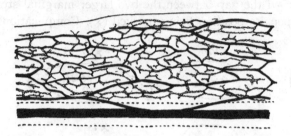

Fig. 724. *Christopteris tricuspis*: portion of the soral region of the sporophyll seen as a transparency. The heavier continuous lines represent the normal venation of the leaf which lies nearer to the upper surface. The lighter broken lines represent the receptacular system extended in a plane nearer to the lower surface. (× 5.)

its upper surface by a reticulum of stronger strands, corresponding to those of the sterile blade. But in addition there is a second system connected with it, which in sections is seen to ramify in a plane below, spreading immediately under the soral surface (Fig. 725). The relation of these two

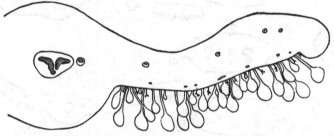

Fig. 725. Part of a transverse section of a sporophyll of *Christopteris tricuspis*, showing the diplodesmic structure and the branched hairs associated with the sporangia. (× 16.)

systems in plan is seen in Fig. 724. The primary network connected directly with the mid-rib is represented in heavier lines. The subsoral system arises from the intra-areolar twigs of this by branches which run obliquely to the lower level. Sometimes such a twig branches little, and the whole falls within a single areola: but frequently they branch freely, even fusing with strands originating elsewhere and crossing from one areola to another. Thus a continuous subsoral network, more elaborate than that of *Platycerium*, is produced. It is represented in Fig. 724 by lighter and broken lines. Thus *Cheiropleuria, Platycerium,* and *Christopteris tricuspis* mark three progressive steps in complexity of the subsoral system. A number of other Acrostichoid Ferns belonging to *Chrysodium, Leptochilus,* and *Elaphoglossum* have been examined (Studies VI, p. 11). The absence in them of a diplodesmic state shows that this is not inherent in the Acrostichoid spread of the sorus, while it brings into greater relief those Ferns where it is present.

The constituents of the sorus of *C. tricuspis* are sporangia and paraphyses: the latter are septate and branched, but they are relatively few, and are barely one-third the height of the mature sporangium. The chief constituents are thus the sporangia, and their crowded heads only are seen from without. The various ages are intermixed, and the sporangia show no grouping that suggests any primitive soral relation (Fig. 725). Again this appears as a feature of advance over *Platycerium,* though the hairs are here less prominent.

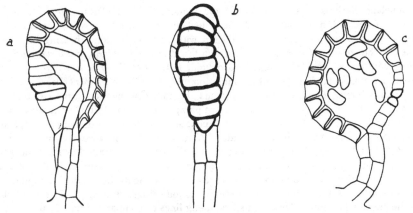

Fig. 726. Sporangia of *Christopteris tricuspis.* *a*=seen obliquely from the stomial side. *b*=from a reverse direction. *c*=a sporangium in longitudinal section. (× 125.)

The sporangium itself is of an ordinary Leptosporangiate type with a three-rowed stalk longer than the capsule. Its segmentation is three-sided: the annulus consists of about 21 cells, of which about 13 are indurated: it is not definitely interrupted at the insertion of the stalk, but the stomium is highly organised (Fig. 726). The spore-output is typically 64, and, as in *Dipteris, Cheiropleuria* and *Platycerium,* there is no perispore (Schumann, *l.c.* p. 250; Hannig, *l.c.* p. 339).

The sum of the characters thus described confirms in detail the comparison of *C. tricuspis* with *Cheiropleuria.* It strongly upholds the natural relation of both with *Platycerium,* notwithstanding the difference of habit. But the divergent mode of life of these Ferns must be kept in mind. In *Cheiropleuria* we see a Dipteroid hovering between a terrestrial and an epiphytic habit: in *Platycerium* the habit is definitely epiphytic: but in *C. tricuspis* it is as definitely terrestrial. The two last-named are both more advanced types than *Cheiropleuria*. This comes out in *C. tricuspis* in the presence of scales upon its fleshy rhizome, in its highly disintegrated stele and leaf-trace, its ternate leaf-form, its more highly organised diplodesmic sporophyll, its smaller but more specialised sporangia and lower spore-output. The features that link it more especially to *Cheiropleuria* are the underlying general scheme of construction, the diplodesmic sporophyll, and the absence of a perispore. These also connect *Christopteris* with *Platycerium*, notwithstanding its peculiarities, which are no doubt related to habit. As illustrating the effect of this the sporophyll of *Platycerium* may be compared as regards outline with that of the epiphytic *Ophioglossum palmatum* (Vol. II, Fig. 343). No one would suggest any near phyletic relation between these bizarre epiphytes : but such similarity as they show gives one of the most striking examples of homoplastic adaptation to like conditions.

Copeland (Perkins, *Fragmenta*, 1905, p. 188) founded the genus *Christopteris* on *C. sagitta* (Christ) Copeland, a species already described by Christ as *Polypodium sagitta* (*Bull. Herb. Boiss.* p. 199, 1898). This is a Fern with creeping rhizome bearing bristle-like paleae. The sterile leaf is long-stalked with a broad triangular blade, and *venatio anaxeti*. The fertile leaves are narrow and Acrostichoid. Later (*Philipp. Journ. Botany*, Vol. XII, No. 6, 1917) Copeland suggested the inclusion of *Leptochilus tricuspis* (Hk.) C. Chr. and *L. varians* (Mett.) Fournier in his new genus ; he finds confirmation in the fact that they also are diplodesmic, and that they all have paraphyses, usually simple and small, as in *C. tricuspis* : also that their paleae, though peltate at the base, are drawn out at the apex into long bristles. Later there were added *C. cantoniensis* Christ from Canton, *C. Copelandi* Christ from the Philippines, and *C. Eberhardtii* Christ from Annam. The genus thus constituted has a rather wide distribution, though clearly centred in the Malayan region. One species is Himalayan, others are from Siam, South China, the Philippines, and from New Caledonia. This Copeland regards as indicating antiquity : and he refers them to the ancient Matonioid-Dipteroid stock, along lines of comparison based upon those of my Studies VI, Jan. 1917.

It may be that as detailed knowledge increases other Ferns of "Acrostichoid" or "Polypodioid" character may ultimately find their place with those already in *Christopteris*. They may either be new species discovered in that prolific Eastern region, or others already assigned, as *Leptochilus varians* has been, to some other affinity. It is interesting meanwhile to note that Frau Schumann (*l.c.* p. 250) has already placed *L. varians* in close relation with *L. tricuspis*, both being without perispore.

NEOCHEIROPTERIS Christ

The Fern named *Neocheiropteris palmatopedata* (Bak.) Christ is a native of Yunnan. It was originally referred to *Polypodium* by Baker, but was given generic rank by Christ. It has a rather thick fleshy hypogean rhizome, bearing ovate scales: the leaves are solitary, with a long petiole bearing a pedatifid blade, deeply lobed with the apparently middle lobe erect, as in *Matonia*, while the lateral lobes are patent and pedate (Fig. 727). The well-marked mid-ribs of the successive lobes, right and left, form a katadromic helicoid system. The whole of this characteristic blade is in fact constructed on the same plan as that of *M. pectinata*. But the venation is reticulate, of

Fig. 727. *Neocheiropteris palmatopedata* (Bak.) Christ. Habit-figure of plant from South China (⅓ natural size). After Christ.

the *anaxeti* type, as in *Dipteris*. The Fern is homophyllous, and bears large naked sori disposed in rows, one on either side of the main veins: in fact, as in the simple type of *Dipteris Lobbiana*.

The dermal scales repay attention. They are peltate, and are borne on emergences, so that they project slightly from the surface. The cells forming the single layer have thickened dark-brown walls, except near to the attachment, and at the extreme margin: they are, in fact, of the clathrate type, and resemble those of *Christopteris tricuspis* (Fig. 728, *a*, *b*): but in this Fern there are marginal glands. Christ describes the margin in *Neocheiropteris* as ciliate: but old scales do not show this, possibly owing to shrivelling with age; for occasional appearances suggest collapsed glands. On the other hand, in *C. tricuspis* they are not constant. Such glands are present in *Platycerium*, and in some Polypodioid Ferns (Fig. 728, *c*, *d*).

Anatomically this Fern is in advance either of *Matonia* or *Dipteris*. The rhizome contains a highly perforated and disintegrated stele, represented in transverse section by a circle of about a dozen small meristeles: of these about five constitute the highly disintegrated leaf-trace, which passes off

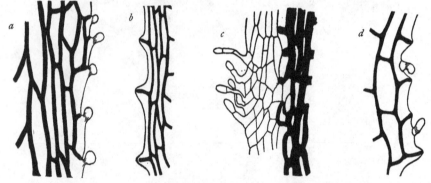

Fig. 728. Margins of scales of various Ferns, showing structure and marginal hairs. *a = Christopteris tricuspis.* *b = Neocheiropteris.* *c = Platycerium* sp. *d = Goniophlebium* sp. (× 85.)

obliquely upwards. In fact the vascular system closely resembles that of *Christopteris*, or of *Phymatodes*. There are also numerous sclerotic nests, especially towards the upper side of the rhizome.

The relation of the sori to the venation of the blade presents points of interest. The basal region of a leaf is shown in Fig. 729. The sori lie close on either side of the main veins, but only on the acroscopic side of those forming the helicoid curve: they vary greatly in size, and are specially elongated at the base of the blade. They have never been seen seated on a vein-ending, but always on a continuous vein, or on a plexus of veins: in fact, their relation is variable. Frequently a large vascular loop underlies a large sorus, but a small one may be seated on a single vein which still continues its course (Fig. 730). These facts point to a spread of the sorus over an enlarging area. It will be seen later that the vascular elaborations beneath the sorus find their parallel in some related "Polypodioid" types (p. 225).

The sporangia are very numerous, with indications of a "mixed" origin. They have three-rowed stalks, about twice or thrice the length of the capsule, and there is an interrupted annulus with from 14 to 17 indurated cells. The spores are bilateral, without perispore, and there appear to be 48 to 64 in each sporangium.

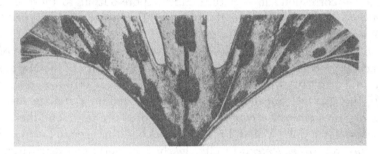

Fig. 729. Base of the fertile lamina of *Neocheiropteris* showing the branching and sori. (× 2.)

In view of the facts now before us relating to *Neocheiropteris* there cannot be any doubt as to its general phyletic relations. It shares with *Matonia* and *Dipteris* so many features that we naturally look first to them. The sum of characters does not, it is true, point directly to either genus, though a position

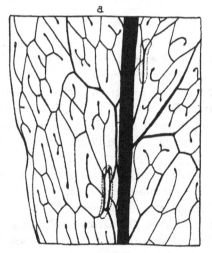
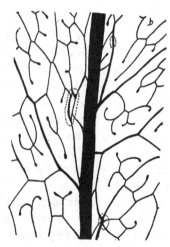

Fig. 730. *a, b* = portions of the fertile pinna of *Neocheiropteris* showing the venation, and the outlines of the large sori, with underlying vascular supply. (× 4.)

nearer to *Dipteris* is indicated by the venation and by the nature of the sorus. *Neocheiropteris*, however, bears signs of advance on either of them in the dermal scales in place of bristle-like hairs: in the highly disintegrated vascular system, and in the extended area of the "mixed" sorus. Indications of the latter state have, however, been seen in *Dipteris conjugata* (Vol. II, p. 317). In *Cheiropleuria* and in *Christopteris*, on the other hand, the soral area is greatly extended, with a "mixed" character fully developed. But while *Cheiropleuria* is singularly archaic in its anatomy, as well as in its sporangial structure and spore-output, the sporangia of *Christopteris* and of *Platycerium* correspond in structure and in spore-output to those of *Neocheiropteris*, all having adopted the ordinary Leptosporangiate type. They also all agree in the absence of a perispore.

One of the most distinctive features in these Ferns is the elaboration of the receptacular vascular tract, which seems to be a peculiarity of the Matonioid-Dipteroid Ferns. In *Matonia sarmentosa* the sorus is terminal on a vein (Vol. II, Fig. 497, *a*): but in *M. pectinata* it is seated on a stellate fusion of strands (*b*). In *Dipteris Lobbiana* though the sorus is a large one there is no special vascular supply (Vol. II, Fig. 571). The condition seen in *Cheiropleuria*, *Platycerium* and *Christopteris* results clearly from soral extension from the Dipteroid type, and it is accompanied by extensions of the vascular tissue of the receptacle, resulting in the diplodesmic state seen in their sporophylls. It seems probable that the soral supply in *Neocheiropteris* may have arisen similarly, but from a Matonioid source, with extension of the vein-junction to form the vascular ring. Whether this be so or not, it will be found to provide an interesting point for comparison with certain Polypodioid Ferns.

HYMENOLEPIS SPICATA (L. fil.) Presl

Few Ferns have suffered under so varied a synonymy as this species. It has lately been examined in detail by Von Goebel (*Ann. Jard. Bot. Buit.* XXXVI (1926), p. 108), with results which appear to fix its natural position more clearly than before: so that the history of its systematic changes need not be followed in detail here. It may suffice to say that it will now take its place as a specialised Dipteroid-derivative on the one hand; while, as Von Goebel concludes, it has much in common with the Polypodioid Ferns comprised under *Pleopeltis*.

This plant is one of the commonest epiphytes in the Malayan region. It was described by Linnaeus as *Acrostichum spicatum*, and was well figured by Hooker under Presl's name of *Hymenolepis spicata* (*Exotic Ferns*, Pl. LXXVII). From its creeping rhizome spring in two rows the narrow, unbranched, lanceolate and leathery leaves, with marked mid-rib, and *venatio anaxeti*. When fertile the leaf is contracted at its distal end, and its margins reflexed to cover the two broad linear coenosori (Fig. 731, 34, 36, 38). From the thickened receptacle arise the numerous sporangia, associated with long-stalked peltate scales (40, 41). The rhizome is described by von Goebel as containing an attenuated vascular system, associated as in so many Dipteroid Ferns with dark masses of sclerenchyma. The venation of the sterile

and fertile regions of the blade is shown in Fig. 731, 35, 36. The protection
of the sporangia is very complete, partly by the folded lateral wings, partly

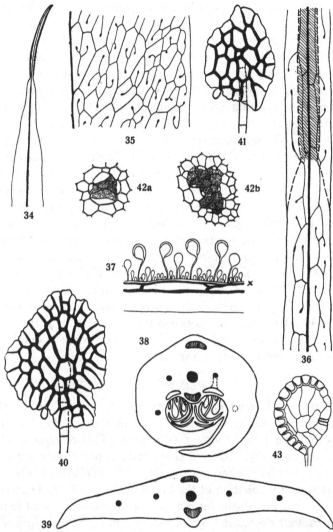

Fig. 731. 34 = *Hymenolepis spicata*, end of a leaf slightly reduced. 35 = ditto,
nervation of sterile leaf, slightly enlarged. 36 = nervation of the distal reduced
part of the leaf: the position of the coenosorus is shaded. 37 = longitudinal
section through a fertile region: x = the vascular strands branching in the fertile region: the scales are not represented. 38 = transverse section of a young
sporophyll. The sporangia are covered by scales and in-curved leaf-margins: the
vascular strands, corresponding to those of the sterile leaf, are black, those
peculiar to the fertile region are only outlined. The sclerenchyma is shaded.
39 = transverse section below the fertile region. 40, 41 = scales from the sorus,
with mucilage-papillae. 42, *a*, *b* = sections of sclerotic nests in the rhizome.
43 = sporangium. (From Von Goebel.)

by the overlapping peltate scales (38). Mucilage-glands are found on the margins of the scales (40, 41): these may be compared with those shown in various Ferns of Dipteroid affinity (Fig. 728).

The vascular system contracts below the fertile region of the sporophyll: but branches, non-existent in the sterile blade, enter the receptacle in a plane below that of the primary system, and pursue a course below the coenosori (36, 37, 38). To those who have followed the origin of the diplodesmic state in *Cheiropleuria, Platycerium* and *Christopteris* it will be apparent that this is a diplodesmic state, cognate with their subsoral developments, but it is carried out in a leaf that is more contracted. The sporangia are of an advanced Leptosporangiate type, with a three-rowed stalk.

Professor Von Goebel discusses the relation of *Hymenolepis spicata* to *Pleopeltis* rather than to the Dipteroids. By arguments that appear convincing he assigns to it a near relation to that Section of Polypodioid Ferns. But a broader interest lies in the fact that it may now be accepted as a synthetic type, linking the Dipteroid and Matonioid Ferns with a Section of the comprehensive genus "*Polypodium*." The Dipteroid affinity of *Hymenolepis* is no new suggestion: the genus was placed by Christ next to *Cheiropleuria* in his *Farnkräuter* (p. 129). But Diels included it in the Taenitidinae (*l.c.* p. 305), a group which Von Goebel regards as unnatural and therefore he detached it (*l.c.* p. 148). To this a ready assent may be given. Alternatively, the new facts support a near relation of *Hymenolepis* to *Christopteris*, the differences between them being such as their difference of habitat might be expected to induce: for the former is epiphytic while the latter is a ground-growing Fern.

PLEOPELTIS

In my Studies VI (*Ann. of Bot.* 1917, p. 18), writing on the Dipteroid derivatives, the probability was pointed out that many Polypodioid Ferns would find their phyletic grouping with them. This opinion was based on form, venation, anatomy, and soral characters: in particular comparison was then made between *Neocheiropteris* and *Polypodium (Phlebodium) decumanum* Willd. as regards the lobation of the broad blade, so clearly Dipteroid in its character. Such comparisons form a natural introduction for the new facts relating to *Pleopeltis* disclosed by Von Goebel in 1926 (*Ann. Jard. Bot. Buit.* Vol. XXXVI, p. 107). But *Pleopeltis* gives a closer comparison with the Dipteroids than does *Phlebodium*, both on the ground of venation and of geographical area: for it is richly represented in the Malayan region, while *Phlebodium* is absent.

The Section *Pleopeltis* of "*Polypodium*," as described by Van Rosenburgh (*Malayan Ferns*, 1909, pp. 623–675), includes over 100 species in the Malayan region. The genus was founded by Humboldt and Bonplond for the American species *Pl. angustata*, the name being based upon the peltate scales that protect its sori. The venation of the genus is anastomosing, having irregular areolae, mostly with included veinlets spreading in various directions: the sori are round or oblong. These features are broadly comparable with what is seen in *Neocheiropteris*, and with the Dipteroids generally. They are mostly wide-creeping, leathery-leaved Ferns, of epiphytic habit, and some species are myrmekophilous, as in *P. lecanopteris* Mett., though this is often maintained as a substantive genus (E. and P. I, 4, p. 326)[1]. The species more particularly treated by Von Goebel were *Pl. phymatodes* L. and *Pl. nigrescens* Bl., and the chief comparative interest centres round the sori. Suffice it to say in general that in the long, tongue-like leaf-form, advanced stelar disintegration, the prevalent sclereid-nests, and the general venation, *Pleopeltis* compares with the Dipteroid Ferns. The sori are of relatively large area, and circular or oval in outline. The numerous sporangia, of mixed ages,

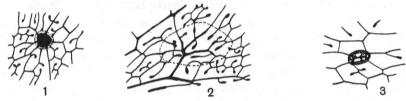

Fig. 732. *Pleopeltis*, after Von Goebel. 1 = *P. nigrescens*, the plate of tracheides underlying the sorus is shaded, the outline of the sorus dotted. 2 = *P. phymatodes*, a "dictyosorus" in surface view, its limit indicated by a dotted line. 3 = *P. schraderi*, the tracheidal plexus beneath the dictyosorus is dotted.

are interspersed with hairs, as they are also in the sori of *Christopteris*. But here the hairs may assume the form of flattened scales, a development which culminates in the peltate type seen in *Pl. angustata*, or in *Pl. macrosphaera* (Von Goebel, *l.c.* Fig. 44). Glandular cells may appear at the apex or margin of these, comparable to those seen on the clathrate scales of other related Ferns (Fig. 728).

Interesting features for comparison are found in the vascular supply which underlies the sori. In *Pleopeltis nigrescens* the receptacle is wide and flat, and a broad vascular plate underlies it, resulting from an enlarged junction of converging primary veins of the blade. This state might readily follow from widening of a punctiform type of sorus (Fig. 732, 1). The structure seen in *Pl. phymatodes* is more complex. The sorus is again circular or oval, and of large area. Branches of the primary venation here also form a junction below the receptacle, but the intra-areolar veinlets also take part, enlarging their ends, which terminate blindly, with or without branching. The receptacular area thus covers a considerable plexus of veins, sporangia being borne on the leaf-surface between the veins, and extending even

---

[1] A full description of its habit and structure has been given by Yapp (*Ann. of Bot.* XVI, p. 185. See also Vol. I, Fig. 50).

beyond them (Fig. 732, 2). Sections show that here the whole vascular development lies in one plane: the condition is essentially the same as has been seen in *Neocheiropteris*, though the details are slightly different (compare Fig. 730). A still more interesting structure for comparison with the diplodesmic state of *Cheiropleuria* or *Christopteris* is seen in *Pl. schraderi*: for here the vascular development is not restricted to a single plane. The primary venation of the leaf extends in a plane above, and parallel to the secondary network, which supplies the receptacle (Fig. 732, 3). It is in fact diplodesmic, after the manner which has been seen initiated in *Cheiropleuria*, and amplified to an elaborate double system in *Platycerium* and *Christopteris*.

These observations on species of *Pleopeltis* indicate states of increased elaboration of sori, probably in the first instance punctiform. Von Goebel raises the question of soral fusion being sometimes involved, and this cannot be excluded. But a lateral spread has clearly been a factor, leading even towards an Acrostichoid state, wherever the insertion of sporangia extends beyond the area of vascular supply. Thus in these Ferns, with sori unrestricted by indusial protections, there has been extension of the soral area, together with increased elaboration of the receptacular conducting system. This has progressed in *Pleopeltis* along lines generally comparable with those seen in Dipteroid Ferns, but with some independence of detail. When this is put in relation to the similarity that exists in other respects, such as form, venation, and habitat, the probability appears strong that *Pleopeltis* is also a derivative from the Matonioid-Dipteroid phylum. How much further such comparisons may extend must be left for the present open. Possibly *Phlebodium* may also prove to be of Dipteroid affinity, and the suggestion may also be extended to *Niphobolus*. Certain observations of Giesenhagen point in that direction (Die Gattung *Niphobolus*, Jena, 1901, p. 61, Fig. 11, *B*). Though such questions will require much critical and detailed research before they can be advanced beyond the region of suggestion, it seems not improbable that a considerable part of the collective genus "*Polypodium*" will ultimately have to be allocated to a Matonioid-Dipteroid origin.

It has been seen how in *Dipteris* and *Cheiropleuria* there is a relatively thick-stalked sporangium: it has four cell-rows, while in *Platycerium* and *Christopteris* there are only three rows. The sporangia of "*Polypodium*," produced in large numbers in the crowded sori, illustrate a further step of simplification. This is seen in *Phlebodium aureum* (Fig. 733). Compare also Vol. I, Fig. 240. Here the sporangium is initiated as a narrow unicellular papilla, in which the first oblique segmentation may often extend below the level of the epidermal wall, or it may impinge on the lateral wall distinctly above that level. In the former case the stalk at its base may be found to consist of more than one cell: but in the latter the lower part of the stalk would be a simple cell-row, though the upper might expand to the three rows, bearing distally the usual type of sporangial head. Thus comparatively the sporangium of *Phlebodium* would provide the most advanced type of the series, the steps of reduction in complexity of the stalk following roughly the course of advance in other features disclosed in the preceding pages.

A parallel series culminates similarly in the one-rowed stalk of *Asplenium* (compare Fig. 670).

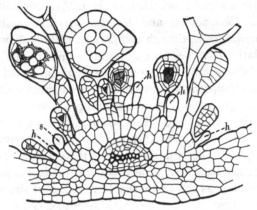

Fig. 733. Young sorus of *Polypodium* (*Phlebodium*)
aureum L. Cut vertically.

COMPARISON

The external features upon which the comparison of the Ferns here attributed to a Matonioid-Dipteroid affinity has been based have been taken up as the description proceeded. Progress has also been remarked from dermal hairs to scales, from protostely (so curiously retained in *Cheiropleuria*), through solenostely, to a highly perforated state in *Platycerium*: this is seen particularly also in *Neocheiropteris* (see Studies III, *Ann. of Bot.* 1913, p. 474, Text-fig. *B*). Nevertheless the venation of the expanded blade shows a high degree of uniformity from *Dipteris* onwards. But it is the comparison in respect of the sorus, and of the sub-soral vascular supply and of the sporangium, that gives the most striking results; these may now be summarised for the whole series.

If we compare the constitution of the superficial sorus, as it is seen in *Gleichenia, Matonia, Lophosoria, Dipteris, Cheiropleuria, Christopteris, Platycerium, Hymenolepis,* and again in *Neocheiropteris* and related forms included under "*Polypodium*"—such Ferns, akin in other characters, illustrate progressive changes in its elaboration. Though they may not themselves constitute a simple phyletic series, yet these Ferns may be held as being so far related that they indicate probable steps of transition from a simple radial sorus, of a type common among the Simplices, to a type characteristic of the highest Leptosporangiate Ferns. They all belong to the Superficial Series, and in all of them (excepting the peculiar case of *Matonia*) indusial protection is absent, while in none of them is there a gradate sequence of the sporangia. This general position may be presented in converse by saying that all the steps in advance which their sori show have been carried out without any suggestion of a marginal origin, and without reference to those indusial flaps which so often characterise marginal sori. Ultimately an Acrostichoid state is reached in *Cheiropleuria*, *Christopteris* and *Hymenolepis*, comparable with that of *Acrostichum aureum* itself, though comparison clearly shows that the steps leading to the latter have been along quite distinct phyletic lines. There is good reason to believe that the progression here indicated has pursued its own course in these Superficial Ferns or their derivatives, quite independently of any similar progressions that may have been executed among Marginal Ferns or their derivatives. For the Gleichenioid type existed certainly in early Mesozoic time, and it probably had a Palaeozoic origin in some such type as *Oligocarpia*. In fact, the sequence of these Superficial Ferns contemplated here stretches back to a time when the Leptosporangiate type itself was yet in its infancy, as shown by the more complex sporangia.

The superficial sorus had in the early Mesozoic Gleicheniaceae already settled to a definite type, foreshadowed by the Palaeozoic *Oligocarpia* (Vol. II, Fig. 493). This was repeated in *Laccopteris* and *Matonidium*, and it is represented to-day (though with a distal indusium) in *Matonia*. These were all Simplices, and the sorus was radiate. But we see in certain living Gleichenias how the centre of the radiate sorus has been filled by additional sporangia, even to the point of a mechanical dead-lock (Vol. II, Figs. 486–489). For further advance in effective fertility some structural change became necessary, such as enlargement of the circumscribed receptacle, lengthening of the sporangial stalk, diminution of size of the sporangia, or some revised method of dehiscence from the primitive median slit (Vol. II, p. 210). Some combination of these would be still more effective than any one change alone. *Matonia*, with its simple radiate sorus, adopted an ill-defined lateral slit. In *Dipteris*, with its crowded sorus, the rupture is more definitely lateral, though the annulus is still a complete ring (Vol. II, Figs. 571, 576). The like is seen in *Lophosoria* (Vol. II, Fig. 551). But while *Gleichenia*, *Matonia*, and *Lophosoria* have all retained a circumscribed central receptacle, in some species of *Dipteris* this has been widened to accommodate more sporangia (*D. Lobbiana*), or the sori themselves have been multiplied (*D. conjugata*, Vol. II, Fig. 573). Yet another feature of advance is seen in *D. conjugata*, for here the simultaneity of origin of the sporangia characteristic of the Simplices is given up, and a slight succession of age introduced (Vol. II, Fig. 575). A still greater advance in soral elaboration, involving both the innovations last mentioned, is seen in *Cheiropleuria*. Here the individual sori, as indicated by their sub-receptacular conducting tracts, are greatly extended (Fig. 715 *a*). The limits of the individual sori are no longer recognisable superficially, though they may be traced by their separate vascular supply. This extension of the receptacular tracts parallel to the leaf-surface may be held as an alternative to the elongation of the receptacle vertically to the leaf-surface or margin, as in the Gradatae. The result is in either case increased accommodation for sporangia. The course and the measure of the extension are best followed in *Platycerium* (Figs. 715 *b*, 719), and reach a climax in such sporophylls as those of *P. stemaria*. In both of these genera the identity of the sorus is still structurally maintained: but in *Christopteris tricuspis*, and apparently also in *Hymenolepis spicata*, the individual vascular systems of the sori may fuse among themselves, thus constituting a sub-soral network, or connected tract (Fig. 724). *Christopteris* may be held as a most complete example of the *diplodesmic* sporophyll, presenting in its fullest development the Acrostichoid state.

With this soral spread there goes commonly, but not always, a "mixed" condition of the sporangia. An exception is seen in *Platycerium*, in which the sporangia arise as a rule simultaneously (*P. alcicorne*), with regular orientation, in rows on either side of the underlying receptacular vein. This gives the impression of the Platycerioid sorus as an extreme extension of the radiate-uniseriate type: in fact a long drawn out representative of that in *Matonia* or *Gleichenia*. But this retention of the simple character is exceptional: in *Cheiropleuria* or *Christopteris* and others the "mixed" character of the sorus is attained early, along lines already indicated in *Dipteris conjugata*, and it is very fully developed. In fact the Dipteroid Ferns, together· with the Gymnogrammoids, are the most conspicuous examples of the direct transition from a simple to a mixed state of the sorus; in them it arises without the intervention of that gradate condition which appears in the Hymenophyllaceae, the Dicksonioids and the Cyatheoids. Its alternative, viz. the increasing soral spread, together with the development of underlying vascular tracts, may be held as the leading soral feature of the Dipteroid Ferns. An extension in area, and as Von Goebel has shown of the vascular supply also, characterises the *Pleopeltis* types of *Polypodium*. These may be held as pursuing a like method, though they carried it out in a less efficient manner.

Along with such progressions among the Dipteroid Ferns go changes in the sporangium and spore-output. The primitive sporangia of *Gleichenia* are few in each sorus: each is short-stalked, with a large head, oblique annulus, median dehiscence, and a large spore-output. *Matonia* has a non-specialised lateral dehiscence, and a small spore-output of very large spores. The Mesozoic Dipteroids had also few sporangia in their sori, with large heads, short stalks, oblique annulus, and a large spore-output: typically 512–256 for *Dictyophyllum exile* (Halle), or 128 for *D. rugosum* (H. H. Thomas). It was then a point of peculiar interest to find in *Cheiropleuria* that the typical number is 128 (Thomas); this is a fact so exceptional among living Leptosporangiate Ferns that it points a definite connection with the Mesozoic Dipteroids. The swing to a lateral dehiscence seems to have occurred in Mesozoic time, and it is characteristic of all modern Dipteroids, while the organisation of the stomium becomes more precise in the more advanced of them. The sporangial stalk underwent changes from the short massive stalk of *Gleichenia* (*a*), and that almost equally complex in *Matonia* (*f*), to the four-rowed stalk of *Dipteris* and *Cheiropleuria* (*k*), and onwards to the three-rowed in *Platycerium* and *Christopteris* (*l*), and a simpler structure still is seen in the Polypodioid derivatives (see Vol. I, Fig. 243, to which the letters refer). This reduction is not directly proportional to the spore-output, though it follows roughly the diminishing number of spores in each sporangium. It appears to be more closely related to the proportion of size of the sporangial head to the length of stalk. The change from a four-rowed to a three-rowed stalk is linked with the change of the initial segmentation of the sporangium: it has been discussed at length in Chapter XXXIII (Vol. II, p. 303). Among the Dipteroids the transition lies between *Dipteris* or *Cheiropleuria* on the one hand, and *Platycerium* and *Christopteris* on the other: and a parallel may be drawn with a like change on passing from the Cyatheoids to the Dryopteroids. The spores themselves are without perispore throughout the series (Hannig, *l.c.* p. 339).

The comparison of the facts drawn from the living Dipteroid Ferns as described above, when it is combined with the rapidly growing body of facts relating to the Mesozoic Dipteroids, will leave little doubt that in them all we see a natural phyletic sequence, leading from primitive Simplices to advanced Mixtae. The series in fact pursued a morphological course of its own, and preserved its identity from the Triassic Period to the present day. It is this which gives a peculiar value to its study: for in this series are illustrated a large body of progressive steps, both in form and structure of the vegetative organs, and also in the characters of the sori and sporangia. Not only do these follow along lines parallel in the main among themselves, but they are also parallel with similar advances in other phyla of Ferns, with which these have no near relationship. This remark applies to the general morphology of the shoot: to the progression from hairs to broad scales: to the growing disintegration of the vascular tracts: to the elaboration of the sorus from the simple uniseriate type: to the indications of change from the longitudinal to the transverse dehiscence: to the diminution of size and spore-output of the sporangium, and to the change from its simultaneous to its mixed origin. All of these progressions stand in this very natural series of Ferns upon their own footing, and a close observer would deduce them from observations on the Dipteroids themselves, even if no other Ferns were in existence. There is reason to believe that the Dipteroids have come down through the ages following their own evolutionary sequence as a clear line. That the progressions which they show are so nearly parallel with those seen in other phyla, though these appear to have been distinct from them since Mesozoic time, affords one of the most impressive aspects of their study. And yet in this the Dipteroids do not stand alone. They are, however, among the most striking witnesses to the far-reaching effect of homoplastic development. If any one feature were called for that should stamp the independent character of this phylum, it would be wise to point out the amplification of the sorus in relation to the sub-soral vascular system; for this appears to be a method of advance peculiarly its own.

The palaeontological and geographical history of the Matonioid-Dipteroid phylum of Ferns has been discussed with special knowledge by Seward in his Hooker Lecture (*Linn. Journ.* 1922, p. 227). Notwithstanding their present localisation in the Malayan region these Ferns appear to have been cosmopolitan in their spread in former days (compare Map *B*, Vol. II, p. 329). Their geological record clearly establishes the fact that certain genera were formerly inhabitants of many parts of the world in which they are now unknown, and their present restricted distribution is best interpreted as evidence of declining vigour, or as an expression of inability to hold their own in competition with more recent products of evolution. The problem of the original home of the Matonioid-Dipteroid stock is not easy of solution. When the fossil forms first appear among the records of the rocks, certain genera had already reached a vigorous stage of development in Europe and North America: by the Rhaetic period they were thoroughly established in the Tonkin region, also in Germany, Greenland, and Scania. This wide distribution at an early time will aid the interpretation of some of the suggested derivatives of the present day. It would appear possible that while these are represented in the Malayan region by such genera as *Cheiropleuria, Christopteris, Hymenolepis* and *Neocheiropteris,* and ultimately by *Pleopeltis* with its eastern preponderance, *Platycerium* had had a wider spread from early times. On the other hand, the Polypodioid stock *Phlebodium* may represent Ferns of similar origin which progressed in the American region, along lines parallel to but distinct from their preponderantly eastern relatives of the genus *Pleopeltis.*

Here, however, we approach problems which lie beyond the scope of the present work. The pursuit of them into phyletic and systematic detail must be left in the hands of those who come after, armed with special knowledge and provided, perhaps, with fresh criteria of comparison.

It will be clear from the descriptions given above that the living Dipteroid derivatives cannot be placed naturally in any linear sequence. Each genus appears to have evolved along its own line of advance from a common source, which was probably not unlike certain Mesozoic genera, having much in common with the living genus *Dipteris*. The living genera may be arbitrarily seriated as follows:

DIPTEROID DERIVATIVES

I. *Cheiropleuria* Presl, 1849   ...   .   1 species.
II. *Platycerium* Desvaux, 1827   ...   .   14 species.
III. *Christopteris* Copeland, 1905   ...   .   6 (?) species.
IV. *Hymenolepis* Kaulfuss, 1824   ...   .   4 species.
V. *Neocheiropteris* Christ, 1905   ...   .   1 species.
VI. *Pleopeltis* Humb. & Bonpl., 1810   ...   ...   90 species.
The latter include those from Malaya alone. Van Rosenburgh.

This list may very probably be added to, as the result of further investigation: and already it seems permissible to suggest that *Phlebodium* may be held as a Dipteroid derivative.

BIBLIOGRAPHY FOR CHAPTER XLV

721. METTENIUS. Fil. Hort. Lips. p. 26. 1856.
722. HOFMEISTER. Higher Cryptogamia, p. 252. 1862.
723. HOOKER. Species Filicum, v, p. 272, Pl. CCCIV. 1864.
724. CHRIST. Bull. Herb. Boiss. VI, p. 199. 1898.
725. BOWER. Studies on Spore-producing Members, IV, Phil. Trans. 192, p. 86. 1899.
726. SEWARD & DALE. Dipteridinæ. Phil. Trans. Vol. 194, p. 487. 1901.
727. GIESENHAGEN. *Niphobolus*, Fischer. Jena. 1901.
728. YAPP. *Lecanopteris*, Ann. of Bot. XVI, p. 185. 1902.
729. DIELS. Engler and Prantl, I, 4, p. 199. 1902.
730. COPELAND. J. Perkins' Frag. Fl. Phil. p. 188. 1905.
731. VAN ROSENBURGH. Malayan Ferns. Batavia. 1909.
732. CHRIST. Geogr. d. Farne, Pl. 87. 1910.
733. SEWARD. Fossil Plants, II, p. 380, where the Palaeontological literature is fully quoted to date.
734. BOWER. Studies III, V, VI, Ann. of Bot. 1913–1917.
735. VON STRASZEWSKI. *Platycerium*, Flora, Bd. 108, p. 304. 1915.
736. COPELAND. Philipp. Journ. Botany, XII. Nov. 1917.
737. HALLE. Arch. Bot. K. Svensk. Vetenkaps. Akad. Hand. Bd. 17. 1921.
738. H. H. THOMAS. Proc. Camb. Phil. Soc. p. 109. 1922.
739. SEWARD. Hooker Lecture, Linn. Journ. XLVI, p. 219. 1922.
740. VON GOEBEL. Ann. Jard. Bot. Buit. XXXVI, p. 185. 1926.

CHAPTER XLVI

METAXYOID FERNS

THE great bulk of living Ferns may be attached naturally by comparison to certain major evolutionary sequences, and may be traced through successive phases of transition which many allied forms have alike shared. Nevertheless we should be prepared to find also other minor lines of less continuous descent, comprising fewer genera and species. The Metaxyoid Ferns, so called because in their characters they appear related to the genus *Metaxya*, appear to be such a minor line. This monotypic genus has been

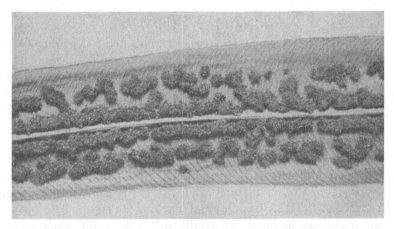

Fig. 734. Part of a pinna of *Metaxya rostrata* Pr., showing the relation of the sori to the veins. More than one sorus may be borne on a single vein. (× 2.)

described in Vol. II, Chapter XXXII as one of the Proto-Cyatheaceae, related on the one hand to the Gleicheniaceae, and on the other to the Cyatheaceae, with which family it has in fact been ranked by some of the best authorities. On the other hand, it has certain features reminiscent of *Matonia* and *Dipteris*, being in point of fact one of the superficial Simplices: but standing somewhat apart from the rest, it provides a stock open to extension and improvement as the evolution of Ferns pursued its course. The leading features of *Metaxya* are the creeping solenostelic rhizome, which bears hairs not scales, and simply pinnate leaves, with a bud attached on the abaxial side of the leaf-base. The broad pinnae have an open pinnate venation, with usually one naked and simple sorus seated on each vein (Fig. 734). But sometimes as many as four sori are borne on a single vein, as seen in a specimen of *Metaxya* in the British Museum. The sporangia have a four-rowed stalk, and a slightly oblique annulus. Such characters ensure for it an archaic position.

The old genera *Syngramme* and *Elaphoglossum* appear to be related to *Metaxya* by similar primitive features: but in the latter the circumscribed sorus is replaced by an Acrostichoid state, such as has been found to be derivative in several other phyletic lines. In particular a parallel may be traced in this respect with the Dipteroid Ferns described in the preceding Chapter. But the venation marks off the Metaxyoid Ferns from these: for here it is simply pinnate, either without any vein-fusions as in *Metaxya* itself, or, where fusions occur and areolae are formed, there are no free vein-endings within them. This is in fact the most obvious character which defines the Metaxyoid series from the Dipteroid Ferns.

SYNGRAMME J. Smith (1845)

This genus includes about 16 species of rare and local Ferns from the Malayan region. They have mostly a short and creeping rhizome, bearing bristle-like hairs (*S. vittaeformis*); but sometimes the bases of the hairs may be widened into small scales bearing a terminal bristle (*S. quinata* (Hk.) Carr). The leaves are stalked and often coriaceous: in most species the blade is undivided, but *S. quinata* bears three to five long pinnae, one of which is terminal. These pinnae are lanceolate, about $1\frac{1}{2}$ inches wide, with pinnate venation, but showing irregular marginal fusions: except for this last, and their entire margin, these pinnae resemble those of *Metaxya* (Hooker, *Sp. Fil.* Vol. v, Plate CCXCVII). It is the same with the simple-leaved *S. borneensis* (Hk.) J. Sm., except that here the marginal fusions are inconstant, the veins being free towards the base of the blade, where numerous free endings occur (Fig. 735, *a, b*). In other species the fusions may be more numerous, so as to constitute a marginal reticulum, as in *S. alsinifolia* (Pr.) J. Sm. (Fig. 735, *c, d*): but it is irregular, and may show free marginal twigs mixed with closed loops. Such facts suggest a relatively recent progression from an open Pecopterid venation, and it is noteworthy that there are no intra-areolar vascular twigs with free endings, as in the Dipteroid Ferns.

Observations on herbarium material show that, as in *Metaxya*, the rhizome s solenostelic in *S. borneensis* and *alsinifolia*: but three separate strands have been seen in the basal region of the petiole, which thus shows a structural advance on the condition of *Metaxya*. The sori follow the veins, but stop short of the leaf-margin. If a string of sori extended along one vein of *Metaxya* were linked together, the result would be what is seen in *Syngramme*. In both genera the sporangia, which are accompanied by paraphyses, have a slightly oblique annulus interrupted at the stalk, and the spores are tetrahedral. But in *Syngramme* the stalk is three-rowed, while in *Metaxya*

it is four-rowed. The sum of these characters, and particularly the habit of
S. *quinata*, reveals a general likeness, allowance being made for *Syngramme*
being advanced in certain details. The comparison must be held as suggestive rather than finally convincing: it may serve a useful end, leading
to more exact investigation later. It must not, however, be forgotten that
the geographical distribution of the present day raises difficulty: for *Metaxya*
is from tropical America, while *Syngramme* is essentially Malayan.

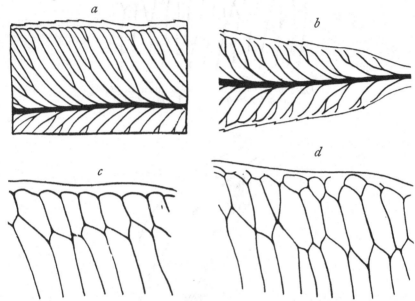

Fig. 735. *a*=venation of part of a leaf of *Syngramme borneensis* (Hk.) J. Sm., showing marginal
fusions. *b*=basal region of a similar leaf showing the fusions omitted. *c*=venation of *Syngramme
alsinifolia* (Pr.) J. Sm., from a herbarium specimen from Singapore. *d*=ditto, but showing more
complex reticulation. (× 4.)

ELAPHOGLOSSUM Schott (1834)

Unlike the more strictly localised *Syngramme* this large genus is widely
spread throughout the tropics, both eastern and western, but particularly
in the latter. It is represented in the Malayan region, but not so richly as
elsewhere. It comprises some 280 species. The simple leaves have a well-
marked mid-rib and simple or forked venation, sometimes anastomosing,
but more frequently looped near to the margin (Fig. 736). The leaves
are more or less dimorphic, with pronounced Acrostichoid soral surface
(Fig. 737). In his monograph of the genus Christ plainly states his opinion
that *Elaphoglossum* is far removed from other Acrosticheae. The only
Ferns which he recognises as resembling them are those of the genus
Syngramme, which correspond to it in habit and venation. The chief differ-

ence consists in the elongated sori of *Syngramme* being restricted to the veins. If in Ferns of the type of *Syngramme* the sporangia were spread over the whole surface, then such types would rank as *Elaphoglossum*.

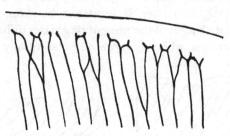

Fig. 736. Venation of the leaf-margin in *Elapho-glossum latifolium* (Sw.) J. Sm. (× 4.)

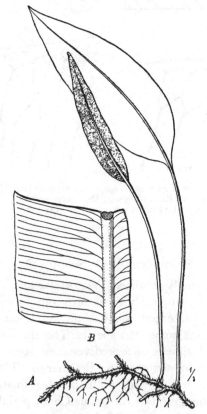

Fig. 737. *Elaphoglossum conforme* (Sw.) Schott. *A* = habit. *B* = part of a leaf with venation. (After Diels, from *Natürl. Pflan-zenfam.*)

The rhizome of *E. latifolium* (Sw.) J. Smith contains a stele nearly approaching the solenostelic state (Fig. 738, *a*), but with leaf-gaps sometimes overlapping (*b*). The leaf-trace is much subdivided (*c*), and it is so from the first, being given off as distinct strands (*a, b*). Their arrangement suggests comparison with a fully solenostelic type, such as *Metaxya*, but broken up by perforations of the meristele at the leaf-base. As in *Metaxya* and *Lophosoria* a bud may be formed on the abaxial side of the leaf-base in *E. latifolium*, its vascular supply coming from the stele below the leaf-trace itself (*b*, Fig. 738, *b*).

The systematic arrangement of this large genus has been based on the venation (Diels, *l.c.* p. 257), the steps of elaboration being similar to those seen in *Syngramme*, and both being referable to a simple open venation as in *Metaxya*: the latter is actually present in *E. heliconiaefolia* (Christ, *l.c.* p. 133, Fig. 78). But *Elaphoglossum* shows advance on the other genera in its dermal scales, which are often very elaborate structures.

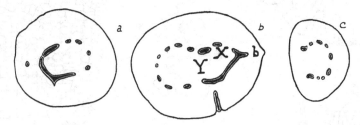

Fig. 738. *Elaphoglossum latifolium* (Sw.) J. Sm. *a*=rhizome in transverse section, showing only one leaf-gap. *b*=ditto, showing two leaf-gaps, *X, Y*, while from the larger meristele the vascular supply is passing off to a bud (*b*). *c*=transverse section of the base of a petiole. (×4.)

Such advances run parallel with the fully Acrostichoid soral state, the whole lower surface of the young fertile blade being covered with a uniform coating of sporangia, together with a few hairs. They arise from a deep epithelium (Fig. 739, *a*): single cells grow out and enlarge (*b, c*), with the usual three-sided segmentation and consequent three-rowed stalk: sometimes it appears only two-rowed at the base (*d*). The distribution of the sporangia appears to be quite independent of the veins (*e*): later on by interpolation of younger sporangia a "mixed" condition of the soral area is attained. The spores have two flattened sides and a prickly surface, as against the tetrahedral smooth spores of *Syngramme* and *Metaxya*.

The conclusion suggested by the sum of these facts is that the three genera form a series of which *Metaxya* is the most primitive, *Syngramme* taking an intermediate place, while *Elaphoglossum* is the most advanced. They illustrate progress in the substitution of scales for hairs: in the subdivision of the leaf-trace: in the increasing fusion of the veins of the blade, while the latter is itself simplified in outline: in the extension and fusion of the plural sori on the veins of *Metaxya* into the elongated sori of *Syngramme*, and in the spread of those sori over the region between the veins, as in *Elaphoglossum*: and finally in the introduction of the mixed sorus. All these advances are such as appear in other distinct phyla, such as the Dipteroid Ferns. But as they occur here in Ferns which do not show, as the latter do, the *venatio anaxeti*, the three genera may best be held as a separate phyletic progression of Metaxyoid derivatives. Further enquiry is needed to confirm or correct this view, while possibly other derivatives may be added later to the series. In particular, the relations to the Gymnogrammoid Ferns will have to be carefully considered.

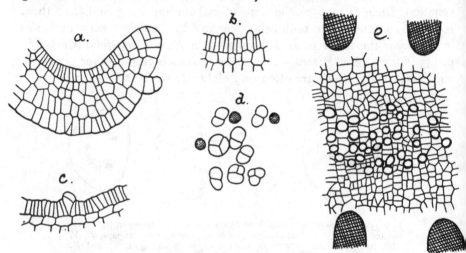

Fig. 739. *Elaphoglossum latifolium* (Sw.) J. Sm. *a*=vertical section of young lamina. *b*, *c*, show the origin of young sporangia superficially. *d*=transverse sections of sporangial stalks and hairs. *e*=tangential section of a young soral area, showing the sporangia originating between the veins. (×75.)

METAXYOID FERNS

I. Sori following the course of the veins.
    Syngramme J. Smith, 1845   ...   .   16 species.

II. Sori Acrostichoid.
    Elaphoglossum Schott, 1834: emend Christ, 1899   280 species.

BIBLIOGRAPHY FOR CHAPTER XLVI

741. J. SMITH. Lond. Journ. Bot. Vol. 4. 1845.
742. HOOKER. Species Filicum, v, p. 152, Plate CCXCVII. 1864.
743. CHRIST. Monographie des Genus *Elaphoglossum*. Zurich. 1899.
744. DIELS. Natürl. Pflanzenfam. I, 4, pp. 256, 331. 1902.
745. SCHUMANN, Frau EVA. Die Acrosticheen, Flora, Bd. 108, p. 201. 1915.
746. BOWER. Studies VI, Ann. of Bot. XXXI, p. 1. 1917.

CHAPTER XLVII

VITTARIOID FERNS

THERE is no habit that produces a more marked effect upon the structure of plants, and on Ferns in particular, than the epiphytic, often closely related as it is to a rupicolous habit. The genera that group themselves round *Vittaria* show this in high degree, and the effect extends to both of their generations. But since other Ferns are liable to similar modifications under like conditions, it is natural that there should have been difficulty in distinguishing the Vittarieae systematically from these homoplastic rivals. Few groups of Ferns have suffered more than they under this difficulty: for their creeping habit, simple anatomy, leathery unbranched leaf, of linear or ovoid form, frequently with sori that give the minimum of distinctive features, throws back diagnosis upon a balance of apparently minor characters. It is this which has led to the checkered systematic history of the Vittarieae (compare Diels, *Natürl. Pflanzenfam.* Fig. 157). The vicissitudes of their classificatory history need not be traversed here, for they are fully recorded by Von Goebel (*Flora*, 1896, p. 67), and by Williams (*Trans. Roy. Soc. Edin.* 1927, p. 173). It must suffice to say that it was the former who suggested grouping together under the heading of the Vittarieae the five genera, *Vittaria* Smith, *Monogramme* Schk., *Antrophyum* Kaulf., *Hecistopteris* J. Smith, and *Anetium* (Kunze) Splitgerber, which are now included. He excluded *Pleurogramme* Bl., and other superficially like, but actually homoplastic Ferns. It may now be held that those five genera form a homogeneous group, which is characterised by a creeping rhizomatous habit, with alternate leathery leaves of simple outline: by the presence of spicular cells in the epidermis, and an absence of sclerenchyma: by clathrate scales: a relatively simple vascular system: sporangia and paraphyses borne upon or sunk deeply into the leaf-surface: numerous annular cells, and a well-marked stomium. The gametophyte, where known, is of aberrant form, frequently bearing gemmae.

VITTARIA Smith

The name-genus of the Family comprises epiphytic Ferns with narrow grass-like leaves, their venation being reticulate without free vein-endings. The rhizome, which forks occasionally, bears clathrate scales: it is almost solenostelic, the wide leaf-gaps overlapping slightly. The transverse section shows a gutter-shaped stele, with internal and external phloem and endodermis. The leaf-trace is binary, one strand being derived from each margin of the gap. Near to the leaf-margin the spicular cells are found, which are characteristic of the family: they are elongated and indurated cells of the epidermis bearing spicules of silica. The sori of *Vittaria* are borne upon intra-marginal veins, varying in position in different species. In *V. lineata* (L.) Sm. they are deeply sunk in grooves of the lower surface, each with its relative vein below it (Fig. 740, *C*). The sporangia are few, with hair-like para-physes (*D*). Each sporangium has a vertical annulus with about 15 indurated cells, and a definitely four-celled stomium, with two-celled, thin-walled epi- and hypo-stomium (*E*). The spore-output is typically 64. The sporangial

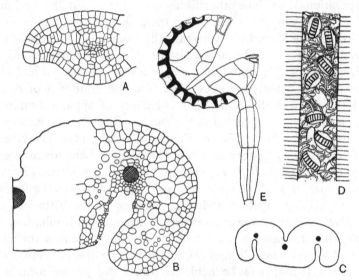

Fig. 740. *Vittaria lineata* (L.) Sm. *A*=section of young leaf showing marginal segmentation. *B*=an older leaf with the first sporangium in superficial groove, seated on a vein. *C*=section of a mature leaf, the sporangia omitted. *D*=portion of a soral groove in surface view, with sporangia and paraphyses. *E*=a sporangium. (*A, B* × 66: *D* × 75: *E* × 132.) *C*–*E* after Williams.

stalk is one-celled at the base, but more massive upwards. The development shows that the sorus is of superficial origin, and that there is no indusium. The leaf-margin develops with the usual segmentation (*A*), but the lateral flap of the blade becomes sharply curved downwards, so as to form a narrow groove; at its base the development of the sorus begins immediately above the underlying vein (*B*). Stomata are numerous on the epidermis lining the groove. Other species may be more massive in habit, and broader leaved than *V. lineata*: but the type is still essentially the same: except for the deep groove the soral condition is Gymnogrammoid. The general characters thus described for *Vittaria* are typical for the Family, though with variation of details which will be noted for the several genera.

MONOGRAMME Schkuhr

This genus includes depauperate examples of the same type of structure as *Vittaria*. Clathrate scales and spicular cells are abundant: but the thin rhizome, which forks occasionally, is protostelic, the simple leaf-traces coming off alternately from the margins of the slightly flattened stelar strap. The venation of the narrow, grass-like leaves is greatly simplified: it consists only of a mid-rib with an occasional lateral vein, right or left, that dies out often without fusion. The sporangia are of the *Vittaria*-type, with paraphyses: but the spore-counts give the lower typical numbers of 32 to 48, suggesting a derivative state.

ANTROPHYUM Kaulfuss

This genus also has simple leaves, but they are of firm and fleshy texture, with copious hexagonal areolae. Spicular cells and paraphyses are as before.

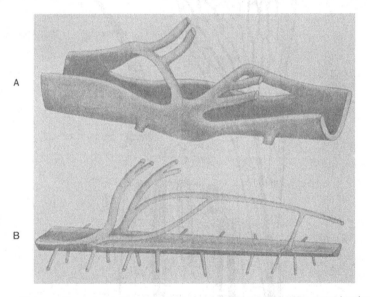

Fig. 741. Reconstructed vascular systems. *A*=of *Antrophyum lineatum*, showing the dorsiventral dictyostele with two closely inserted nodes. *B*=of *Anetium citri-folium*, with one node, and one internode. (After Williams.)

The anatomy shows a gutter-shaped dictyostele, with alternate leaf-gaps on its upper side, which overlap: sometimes there may be a closed solenostele (*A. plantagineum* (Cav.) Klf.). From the base of each leaf-gap arise the binary traces (Fig. 741, *A*). These widen out, and branch as they pass into the blade, with or without a distinct mid-rib being formed. The sori, which are borne above the veins, may be sunk in grooves, or superficial. They are more or less reticulate, as in *Hemionitis* (Fig. 742). The sporangia are as in *Vittaria*, but the spore-counts may vary from 48 to 64.

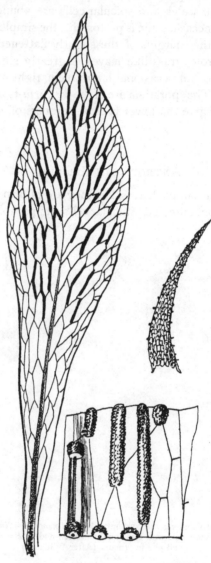

Fig. 742. *Antrophyum semicostatum* Bl. Leaf of natural size, showing reticulation, and sori: also a single scale detached. (After Blume, from Christ.)

HECISTOPTERIS J. Smith

H. pumila (Spreng.) J. Smith, the only species, is a minute depauperate Brazilian and West Indian epiphyte, which was originally included under *Gymnogramme*: but it was re-established as the sole representative of J. Smith's genus by Von Goebel (Fig. 743). The plant shows characters which help towards the interpretation of the whole family. The roots are long and bear frequent root-buds. The rhizome bears clathrate scales, and contains a simple protostele as in *Monogramme*, from which the leaf-traces arise as simple strands (see Williams, *l.c.* Fig. 24). The leaf is exceptional in

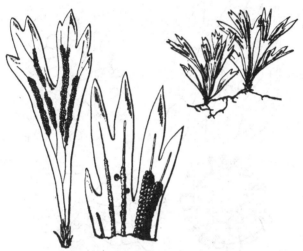

Fig. 743. *Hecistopteris pumila* (Spr.) J. Sm. To the right is a figure showing habit, natural size: to the left a leaf, and leaf-tip with sori, enlarged. (After Hooker, from Christ.)

being deeply cut from the margin, the dichotomous lobes (well provided with long spicular cells) being served by veins which are themselves dichotomous. No vein-fusions are found. Upon the veins the linear sori are borne superficially. The fronds reach a fertile condition very early: for instance, a sorus may be borne on a leaf which shows only its second dichotomy. The sporangia are of the usual type. The special interest of this Fern is that it presents a juvenile form of sporophyte, with precocious fertility (see comparison below).

ANETIUM (Kunze) Splitgerber

A. citrifolium (L.) Splitgb. is again a monotypic epiphyte of the western tropics. It has a slender rhizome, which branches dichotomously, bearing clathrate scales, and alternate leaves. The leaves may be 9 inches long and 2 inches broad, with reticulate venation, and free strands at the margin:

here numerous spicular cells are found. The sporangia are not restricted to the veins, but are borne also upon the areolae, thus giving an Acrostichoid character. On this account it was first named by Linnaeus *Acrostichum citrifolium.*

The vascular system is variable: it appears as a very attenuated dictyo-stele, with large leaf-gaps and perforations. Its most marked constituent is a strong meristele which follows the lower side of the rhizome. A binary leaf-trace supplies each leaf, the two strands arising near to the acroscopic end of the leaf-gap: they fork early (Fig. 741, *B*). The sporangia, which are

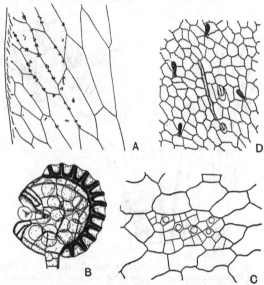

Fig. 744. *Anetium citrifolium* (Kunze) Splitgerber. *A* = portion of a frond, showing venation and sporangial distribution, with peripheral spicular cells. (× 4.) *B* = sporangium. (× 200.) *C* = group of small epidermal cells from which the sporangia arise. (× 200.) *D* = portion of the lower epidermis, showing stomata, a spicular cell, and glandular hairs. (After Williams.)

small but of the usual type of the family, are short-stalked, with an annulus of 12 to 14 indurated cells. They are borne in groups on special small-celled patches, which spring by division of 3 or 4 cells of the epidermis. These may lie either over the veins or on the surface of the intervening areolae (Fig. 744). The spore-counts point to a typical number of 32 or 48.

The prothalli of the three genera in which they are known are anomalous. Those of *Vittaria, Monogramme,* and *Hecistopteris* have been accurately described by Von Goebel (1888, 1896, 1924), and those of *V. lineata* by Britton and Taylor (1902). The vegetative region is irregularly lobed, and vegetative propagation by gemmae is similar to that already known in certain Hymenophyllaceae (Fig. 745). This may not improbably be a specialisation in both in relation to the epiphytic habit. The sexual organs, however, do not diverge from the standardised Leptosporangiate type.

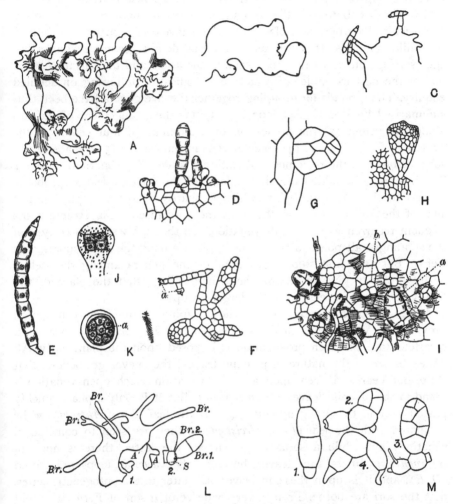

Fig. 745. Gametophyte of Vittarieae. *A–K*, the prothallus of *Vittaria lineata* (L.) Sm. (after Britton and Taylor). *A*, lower surface of prothallus. *B*, prothallus with eleven growing points. *C*, lobe of prothallus bearing gemmae. *D*, margin of prothallus with sterigmata. *E*, a gemma. *F*, a gemma with two withered antheridia (*a, a*) and a young prothallus from the margin of which three others have arisen. (× 25.) *G*, young prothallus arising from the margin of an old one. (× 80.) *H*, young prothallus with three growing points arising from a single cell of an old prothallus. (× 16.) *I*, underside of a lobe of a prothallus bearing archegonia. *J* and *K*, antheridia. (× 140.) *L* and *M*, the prothallus of *Hecistopteris pumila* (Spr.) J. Sm. (after Goebel). *L*, 1, prothallus with gemma-bearing lobes (*Br.*) and adventitious prothalli (*A, A*); 2, sterigma (*S*) and young gemmae (*Br.* 1 and *Br.* 2). *M*, 1, mature gemma; 2 and 3, gemmae which have formed young prothalli; 4, gemma, the end cells of which have grown out into short rhizoids. (After Williams, from authors quoted.)

COMPARISON

The Vittarieae as now constituted are believed to be a coherent and natural group of genera. But in their systematic history they have been arbitrarily distributed on the ground of external characters, while other Ferns which have now been placed elsewhere have been associated with them on similar grounds. It is only as the internal details become available that such sorting can be carried to a sound conclusion, and the Vittarieae offer one of the best examples of a natural classification following on detailed enquiry. The grounds for grouping together the five genera have been thus summarised by Von Goebel (1924, p. 93): (1) the arrangement of the sori along the veins: (2) the possession of a stomium of four cells: (3) the presence of paraphyses: (4) spicular cells in the leaves: (5) the absence of sclerenchyma in the rhizomes: (6) clathrate scales: (7) abnormal prothalli. To these may be added the simple form of the leaves, a feature probably related to the epiphytic habit; which is shared by other epiphytes. The sum of these characters, together with the specialised stelar structure, the frequent recurrence of reticulate venation, and the highly specialised type of the sporangia, all point to an advanced place among the Leptosporangiate Ferns. But as a first step, a grouping of the genera among themselves would probably give substantial help in deciding what the place of the family should be in relation to the larger groups.

A key to some general view of the grouping of the Family has been provided by Von Goebel in the suggestion that in *Hecistopteris* we see a dwarfed state, in which propagation is imposed upon the plant early. All its leaves are of the nature of juvenile leaves; they never get beyond that state, but bear sori even upon an open venation. Such open venation is found in the juvenile leaves of *Antrophyum*, but it is only a phase quickly passed over in the ontogeny (Fig. 746). If the leaf (*C*) of *Antrophyum* be compared with the leaf (*I*) of *Hecistopteris*, they are seen to coincide in venation: but while a sorus is present in the latter, there is none in *Antrophyum*. In its later leaves, however, the vascular loops are closed (*D*, *E*); and it is upon these, in leaves still later in the ontogenetic series, that the sori are borne. From a very wide comparison of Ferns it is held that phyletically an open venation preceded a closed one. If that criterion holds for the Vittarieae, then the sporophyll of *Hecistopteris* is nearer to an original type than those of either *Antrophyum* or *Vittaria*. Probably the distal cutting, present in *Hecistopteris* but absent in all the other genera, is also an ancestral character which they no longer retain.

If this reasoning be sound, then *Hecistopteris*, with its linear sori, will more nearly reflect a primitive state than the rest, notwithstanding the small size and simple structure of the plant as a whole. Nearest to it sorally would be

Vittaria and *Monogramme*, which also have linear sori. But *Antrophyum*, with its reticulate sori, would be more advanced, and *Anetium*, with its Acrostichoid sporophyll, would be the most advanced of the whole Family. It may probably be remarked that the vascular system accords ill with this conclusion: but the protostelic axis and undivided leaf-trace are juvenile features common to all sporelings. These the dwarfed *Hecistopteris* does not exchange for a more elaborate system such as appears in the adults of the larger genera: therefore a full anatomical comparison with them is impossible.

With *Hecistopteris* thus indicated as a relatively primitive type among this coherent group of genera, the question of the phyletic relations as a whole may be considered afresh. The instinct of Sprengel (1828) was

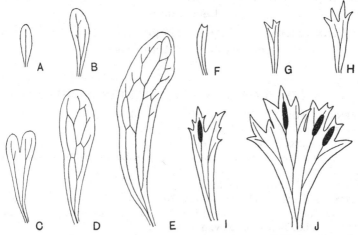

Fig. 746. *A–E*=juvenile leaves of *Antrophyum lineatum*. *F–J*=juvenile and fertile leaves of *Hecistopteris pumila*. (After Williams.)

probably quite correct in assigning to *Hecistopteris* a relationship with *Gymnogramme*, under his original name of *G. pumila*. But J. Smith (*Historia Filicum*, 1875, p. 178), finding that it did not appear to form a natural alliance with any section of that genus, gave to *Hecistopteris* generic rank. Now it may be held as a natural link connecting the Gymnogrammoid Ferns with the Vittarieae. The peculiarities which specially characterise these Ferns are probably consequences of habit: in fact the Vittarieae may be held to represent an epiphytic and rupicolous branch of specialisation from a Gymnogrammoid source. The most peculiar structural feature that is common to both Families is the spicular cells, which have been seen to recur in *Adiantum*, a Gymnogrammoid genus itself specialised along lines quite distinct. It would, however, be an error on the ground of this detail to press any close relation to the epiphytic types. But it may legitimately be pointed out that both have Gymnogrammoid sori, and both include types not only with open but also with coarsely reticulate venation.

The genera of Vittarioid Ferns may be placed roughly in sequence of specialisation as follows:

1. *Hecistopteris* J. Smith (1842) ...   .   1 species.
   (Open venation.)

2. *Monogramme* Schkuhr (1809) ...   .   15 species.
   (Occasional areolae.)

3. *Vittaria* Smith (1793) ...   .   46 species.
   (Closed venation.)

4. *Antrophyum* Kaulfuss (1824) ...   .   27 species.
   (Closed venation.)

5. *Anetium* (Kunze) Splitgerber (1840) ...   ...   1 species.
   (Closed venation and Acrostichoid sorus.)

BIBLIOGRAPHY FOR CHAPTER XLVII

747. SPRENGEL. Tent. Suppl. ad Syst. Veg. p. 31. 1828.
748. HOOKER. Second Cent. of Ferns, Plates 8, 70, 73, 79. 1861.
749. J. SMITH. Historia Filicum, p. 178. 1875.
750. VON GOEBEL. Zur Keimungsges. einiger Farne. Buit. Ann. VII, p. 74. 1888.
751. VON GOEBEL. *Hecistopteris*, Flora, LXXXII, p. 67. 1896.
752. DIELS. Natürl. Pflanzenfam. I, 4, p. 297. 1902.
753. BRITTON & TAYLOR. *Vittaria lineata*, Mem. Torrey Bot. Club, VIII. 1902.
754. GWYNNE-VAUGHAN. Solenostelic Ferns, II, Ann. of Bot. XVII, p. 718. 1903.
755. JEFFREY. Phil. Trans. B, CXCV, p. 119. 1902.
756. BENEDICT. Vittarieae, Bull. Torrey Club, XXXVIII, p. 153. 1911.
757. BENEDICT. Revision of the genus *Vittaria*, Bull. Torrey Club, XLI, p. 391. 1914.
758. VON GOEBEL. Vittariaceae und Pleurogrammaceae, Flora, Bd. 117, p. 91. 1924.
759. WILLIAMS. Trans. Roy. Soc. Edin. LV, p. 173. 1927.

CHAPTER XLVIII

GENERA INCERTAE SEDIS

A COMPARATIVE study of any large Class of Plants naturally akin, and represented by many families, genera, and species, will raise more questions of relationship by descent than can be finally resolved on the basis of present knowledge. Even a long and more or less consecutive geological history does not always suffice to decide problems of detailed affinity, though this provides the most powerful and direct check upon the results of comparison of forms still living. More than ever does difficulty arise when, as in the Class of the Filicales, there is evidence of widespread homoplastic modification, so that the final results of evolution from stocks originally distinct appear to resemble one another in certain leading features, though in others they proclaim a distinct origin. The balance of relatively late modification against archaic features may thus become highly debateable: a slightly greater weight accorded to the one and less to the other by different writers will probably lead to divergent conclusions. To the outside observer of such discussions the exact place assigned to one genus or another may appear to be tinctured by personal taste or idiosyncrasy as much as by scientific truth. The cynic may then say that problems so apparently insoluble are best left alone: but that would be a mere withdrawal from the natural field of science. Still certain problems may quite legitimately be left in abeyance, pending the discovery of new facts, or the introduction of new methods of attack.

These remarks apply to many genera of Ferns, which may be called "genera incertae sedis." Such genera have been accorded varying positions in the classificatory system by different writers, and the synonymy of many of them is a sufficient witness to the diversity of the views expressed. As to this, one general remark may be made. In the past the comparisons have for the most part been drawn between forms highly advanced, such as the Leptosporangiate Ferns. Little importance was attached by the early writers to the evolutionary aspect of questions of affinity. Before 1860 such an aspect hardly existed. Now, however, the outlook should constantly be phyletic: and the first question as to any genus of uncertain affinity will be its probable source by descent. How far are the characters used in comparison fundamental and archaic? how far have they arisen from relatively late and perhaps homoplastic adaptation? It is hardly necessary to explain that it is the pursuit of such questions which has led to the disruption of the old genera *Acrostichum* and *Polypodium*, as they were constituted in the *Synopsis Filicum*. It is now believed that the features upon which those genera were based are for the most part results of relatively recent homoplasy in stocks originally distinct. The new classificatory problem will be, how far can the method of examining facts historically rather than objectively be carried into effect? The nature of the difficulties which beset a study so conducted will be better understood by considering some examples in detail than by pointing out the many genera and species, the phyletic position of which is still held in the balance.

CYSTOPTERIS Bernhardi

The genus *Cystopteris* was established by Bernhardi in 1806, with the type *Polypodium fragile* Linnaeus, 1753. This Fern was subsequently known as *Cystopteris fragilis*, though Schkuhr (*Farnkräuter*, Taf. 54, p. 53) described it as *Aspidium fragile*. The genus *Cystopteris* now comprises 13 species: Sir W. Hooker notes it as a very natural genus, and may assuredly be considered a connecting link between the Davalliaceae and Aspidiaceae. He placed it between *Davallia* and *Lindsaya* (*Sp. Fil.* I, p. 196; *Syn. Fil.* p. 103). Presl ranks it in his Tribe III Aspleniaceae (*Tent.* p. 92): Mettenius in his Tribe IV Aspidiaceae (*Fil. Hort. Lips.* p. 96), close to *Onoclea* and *Woodsia*. Luerssen gives a whole page to its synonymy (*Rab. Krypt. Fl.* III, p. 451). Christ follows Mettenius (*Farnkräuter*, pp. 8, 280). Diels includes it in his Woodsieae-Woodsiinae (E. and P. I, 4, p. 159). But Christensen (Monograph of *Dryopteris*, I, p. 64), writing of his sub-genus I. *Eudryopteris*, remarks that "*Eudryopteris* in most characters agrees with *Cystopteris*.—To me it is probable that *Eudryopteris* and *Cystopteris* are closely allied to each other, and that it is unnatural to place them in two different tribes." In other words, he holds that the natural relation of *Cystopteris* is with Ferns where the sorus is typically superficial. The alternative is the relation pointed out by Sir W. Hooker, with *Davallia* and *Lindsaya*, where the sorus is typically marginal in origin. The question would then appear *primâ facie* to be one of developmental fact.

The fully developed sori of *Cystopteris fragilis* are clearly superficial, being inserted upon a vein which extends beyond the receptacle (Fig. 747). The indusium is basal, and forms an inflated investment. But in sections of the young leaf-margin traversing the sorus in a median plane, the actual margin and the receptacle appear in close relation to one another (Fig. 748, *A, B*): still these sections suggest the origin as superficial. But Professor Von Goebel describes, with drawings reproduced as Fig. 748, *C, D*, an earlier stage of the sorus in *Cystopteris montana*, as follows: "The placenta is initiated either from the leaf-margin, or in the nearest proximity to it, on the underside of the leaf, on which also the indusium grows out. But the development (or further development) of the leaf-margin is at first delayed, while the placenta swells strongly: later, however, it grows more actively and forces the sorus definitely to the lower surface" (*R*, Fig. 748, *C*, *D*). How nearly this corresponds to what is seen in similar sections of *Dryopteris filix-mas* appears on comparison of the drawings of *Cystopteris* with Fig. 748, *E*: in both,

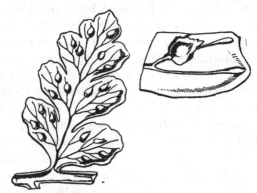

Fig. 747. *Cystopteris fragilis* (L.) Bernh., pinnule of a form from Tasmania, with its sorus: enlarged. (After Hooker, from Christ.)

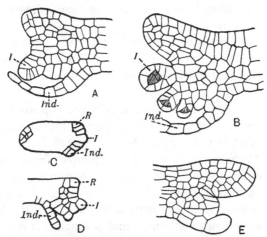

Fig. 748. Development of the sorus of *Cystopteris*. *A*, *B*= median sections of sori of *C. fragilis*, showing relation to the margin: *I*=first sporangium: *Ind.*=indusium. *C*, *D*= younger states of *C. montana*, after Von Goebel: *I*=first sporangium: *Ind.*=indusium: *R*=leaf-margin. *E*=a young sorus of *Dryopteris filix-mas* for comparison. (All highly magnified.)

however, the adult sorus is clearly superficial. The first sporangium (*I*) is followed by others in gradate sequence. The facts of development of the sorus in *Cystopteris* thus show a near similarity to those seen in *Lindsaya* (Chapter XXXVII, Fig. 603), which is held to be a Davallioid derivative: in fact a marginal type.

For any decision of relationship, not one range of facts only but the whole body of data must be taken into consideration. The opinion quoted above from Christensen as to the relations of *Cystopteris*, and in particular the identity of the dermal scales with those of *Eudryopteris*, which he notes, may be held as an embodiment of opinion based on external features; and it points definitely to a Dryopteroid affinity. How then does the anatomy accord with this? Schlumberger (*Flora*, 1911, Bd. 102, p. 410) has examined the vascular system of *Cystopteris*, but apparently in young plants, for he finds only a single strand of the leaf-trace, though in other respects the stelar structure is as in the Woodsieae. But Gwynne-Vaughan (*New Phyt.* IV, 1905, p. 215) found a dictyostelic structure with a binary leaf-trace, as in *Matteuccia*, or in *Dryopteris oreopteris*. In addition to this he describes for *Cystopteris* those peculiar involutions well known as existing in *Matteuccia*, which occur above each leaf-insertion, though here they are on a smaller scale. How far this peculiarity can be held as evidence of affinity is doubtful: none the less it is a common feature in plants alike on other grounds (Vol. I, p. 150). A further point to be noted is the absence of a perispore, though this is characteristically present in the Dryopteroid Ferns. But this peculiarity is shared by *Onoclea sensibilis*, though in *Matteuccia* a perispore is recorded as present (Hannig, *Flora*, 1911, Bd. 103, p. 340).

The prothalli of *Cystopteris* bear unicellular hairs, not pluricellular as in the Cyatheoid Ferns, and the lid-cell of the antheridia is undivided. They are thus of the standardised Leptosporangiate type, and do not help in the present argument.

Taking a general view, the balance of evidence appears to be in favour of *Cystopteris* ranking with the Cyatheoid rather than with the Davallioid derivatives: in fact that it sprang from a superficial ancestry, and with special relation to the Woodsioid and Dryopteroid Ferns. One feature that seems to have weighed more than it should in canvassing such a decision is the initial relation of the sorus to the leaf-margin. We are familiar with the transit of the sorus from the margin to the surface of the leaf, because in so many Ferns the originally marginal position was retained late in the evolutionary history, and evidence of the transit can be seen in several distinct phyla. But its frequent recurrence does not in any way preclude the converse trend. A type of sorus superficial through a long descent may be equally free to move from the surface towards, or may actually attain a marginal position. This may have happened in *Cystopteris*. With its real affinities still Dryopteroid, its sorus appears to have attained an approximate, or even an actual marginal position secondarily: in which case the natural affinity would still be with *Dryopteris*, and the Superficial Series.

ACROPHORUS Presl

Though this monotypic genus has already been discussed, and its place among the Dryopteroid Ferns recognised in Chapter XLI, p. 130, it is well to reconsider it here, as the question of its affinity has some similarity to that of *Cystopteris*, with which genus its close relation was recognised by Presl. Its single species *A. stipellatus* (Wall.) Moore had, however, been previously included by Blume in *Aspidium*. Hooker merged it in *Davallia*, as *D. (Leucostegia nodosa)* Hk. (*Syn. Fil.* p. 92). Against this Von Goebel has lately objected decisively (*Buit. Ann.* 1926, p. 99). Diels upheld the genus, placing it next to *Cystopteris* (*l.c.* p. 164), while Christ gives it near relation to *Struthiopteris* (*l.c.* p. 285). The question of affinity is here like that of *Cystopteris*, and its comparison helps to consolidate the conclusion as to the

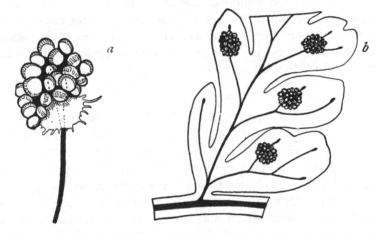

Fig. 749. *a*=sorus of *Acrophorus stipellatus* in surface view. (×35.) *b*=a pinnule of the same showing the veins continued beyond the sori. (×10.)

latter. Here the adult sorus is superficial, seated sometimes on an enlarged end of a vein (Fig. 749, *a*), but frequently its position is lateral (*b*): and the difference may be illustrated by the sori borne on a single leaf. The indusium is basal, as in *Dryopteris* or *Cystopteris*; but it is often minute or even vestigial. The stem and leaf-stalks are covered by chaffy Dryopteroid scales, as they are also in *Cystopteris*, and the vascular system of the upright stock is a dictyostele, but here with a highly divided leaf-trace. These features support the alliance of *Acrophorus* with *Cystopteris*, and they confirm for both the position assigned by Prantl in his Aspidiinae; that is, with the Dryopteroid Ferns.

MONACHOSORUM Kunze

Another genus long held as of doubtful affinity is seen in *Monachosorum* Kunze, which has been discussed in Chapter XXXVI, p. 13: here, however, in addition to other features, the existence of hairs and no scales appears decisive in assigning its place with the marginal series, as a derivative of the Dennstaedtiinae, probably along a line parallel to that of *Hypolepis*.

These three examples illustrate the critical questions that arise in placing isolated genera in relation to the greater series of Ferns. Those who attach less importance to the comparative history on which the distinction of the Marginales and Superficiales has been based, may see in them an argument against its validity. On the other hand, those who accept it will see in such problematical genera those very difficulties that were sure to arise as a consequence of progressive evolution, in Ferns of distinct affinity, under like conditions. It may be doubted, however, whether such relatively late developments in the phyletic history, as these genera present, can legitimately provide a basis for criticism of those more general groupings; for these have been founded on comparison of types of known antiquity, as shown by the early fossil records (see Chapter XL, pp. 113–118).

PROSAPTIA Presl

Another striking example of the difficulties that beset a natural classification of Ferns is seen in *Prosaptia*, a genus founded by Presl in 1836 to receive certain small species already described under *Davallia* by Swartz (*Tentamen*, p. 165, Pl. VI, Figs. 19, 25). The story how this genus, rightly distinguished by Presl as it now appears, was under the authority of Sir W. Hooker consigned for nearly a century to *Davallia*, has lately been told by Von Goebel in an essay which may be held as a model for the treatment of such critical questions (*Buit. Ann.* Vol. XXXVI, 1926; *Morph. u. biol. Studies*, pp. 148–158, Figs. 79–87). He tells how during that century opinion was divided between a Polypodioid and a Davallioid affinity, gradually hardening in the direction of the former. *Prosaptia* was compared more particularly in habit and in sorus with *Polypodium obliquatum* Blume. This Fern was included by Fée under his genus *Cryptosorus*, characterised by the young sorus being sunk in the parenchyma, and subsequently emerging on the surface by a narrow slit. *Prosaptia* has a similarly sunken sorus, but with the difference that here it lies close to the margin, and is surrounded by a funnel of protective tissue, in the formation of which the margin of the blade takes a part (Fig. 750). The general appearance of the sorus may well excuse a reference to *Davallia*, § *Scyphularia* Fée: but as Von Goebel shows conclusively, a wider comparison proves this to be fallacious.

The habit of *Prosaptia* closely resembles that of *Polypodium obliquatum*

(Fig. 588, *C–H*) : the peculiar twin bristles on the leaves, the clathrate scales of the rhizome, and the vascular anatomy also coincide. Even the superficial origin of the sorus, though it arises very close to the margin; the method of its overgrowth by tissue that is not of the nature of an indusium in the sense of *Davallia*; the details of the sporangia and spores, all point to *Polypodium obliquatum*. The difficulty lies in the near proximity of the sorus of *Prosaptia* in point of development to the margin, which results in 'a close imitation of *Davallia*. The whole story, which runs curiously parallel to that of *Cystopteris* in comparison with *Dryopteris*, shows that *Prosaptia* is not a Davallioid Fern. It is one of the superficial series in which the sorus, with special modifications, has approached secondarily to a marginal position. (For the systematic treatment of *Prosaptia* see Van Rosenburgh, *Malayan Ferns*, 1909, pp. 567, 613.)

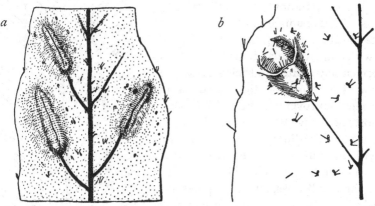

Fig. 750. *a = Polypodium obliquatum*, portion of a young fertile leaf, seen from below. The sori are hidden in long grooves, with only a narrow opening. *b = Prosaptia* sp. from Sumatra. Part of a leaf with young sorus, seen from below. It is clear that the lower part of the soral sheath belongs to the lower leaf-surface. Note the twin bristles on the leaf-surface of both. (After Von Goebel.)

TAENITIS Willdenow

The Taenitidinae, as grouped by Diels (E. & P. I, 4, p. 302), are palpably an unnatural association, from which *Heteropteris* Fée (= *Paltonium* Presl), *Hymenolepis* Kaulfuss, and *Christopteris* Copeland, may be detached on the ground of their venation, as well as on other features (see Chapter XLV). Of the remainder the leading type is *Taenitis*, a genus which affords one of the most open problems of comparison. This also has been recently treated by Von Goebel, but with indecisive results. It may still be held as a "genus incertae sedis."

The genus is characteristically Malayan. The synonymy of its best known and for long its only species, *Taenitis blechnoides* (Willd.) Sw., occupies nearly a page of Christensen's Index. The rhizome of this ground-growing Fern bears hairs, which are enlarged at the base into a cellular mass, sometimes flattened but not characteristically scaly. It contains a solenostele surrounding a sclerotic core: but transitions are found towards dictyostely. The erect leaves are a foot or more in length, and are simply pinnate, with leathery pinnae traversed by a mid-rib, and on either side a closed reticulum extending to the margin, but without any free vein-endings within the areolae. The petiole is traversed by two vascular straps, but the leaf-trace separates as a single sector from the stele: soon it divides into two by a perforation that dies out before the lowest pinnae are reached. The pinna-traces are of marginal origin.

The sporophylls resemble the sterile leaves: but each is traversed longitudinally by a coenosorus parallel to the margin, giving the appearance of having been imposed upon the reticulum without disarranging it, along a line about half-way between the mid-rib and the margin. There is no indusium (Fig. 751). The sorus consists of sporangia of mixed origin, associated with numerous glandular hairs, which are often closely related to their stalks. The glandular hairs precede the sporangia in development, and overarch them while young like a forest-canopy (Von Goebel). The sporangia are of an ordinary Leptosporangiate type, with annulus interrupted at the stalk, having 16–18 indurated cells. The stalk consists of three cell-rows, and the spores are tetrahedral, and appear to be without any saccate perispore: but they bear a peculiar solid equatorial ring (Hannig, *Flora*, Bd. 103, p. 341). The development of the coenosorus has not yet been traced from the beginning, but its origin can hardly be other than superficial. The sporangia arise from a shallow furrow, with no independent or regular underlying commissure: the coenosorus often follows short lengths of the veins, but it bridges over the gaps between them: thus it is not dependent upon the venation. In fact *Taenitis* presents a peculiar type of coenosorus: in particular it does not correspond to that of *Pteris* either in position or in vascular supply, but it seems nearer in its mode of origin to *Blechnum*. Not unfrequently it may be interrupted (Hook. & Grev., *Icones Fil.* Pl. LXIII; Von Goebel, *l.c.* Taf. X, Figs. 68, 72): such examples confirm the impression of the coenosorus as being imposed upon a pre-existent reticulum.

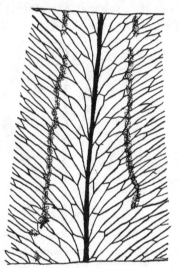

Fig. 751. *Taenitis blechnoides* (Willd.) Sw., venation of part of a fertile leaf-segment, showing an irregular commissure below the soral tract. (After Leonard.)

These characters indicate *Taenitis* as an isolated type. A careful consideration of them led Von Goebel to conclude that there are relations between it and the Pteroid Ferns, and that its naked coenosori were derived from an indusiate source: but he expressly excludes the suggestion that the genus is connected phylogenetically with the Pteridinae. In my own view a Blechnoid relation is almost equally probable: but it would be no surprise to find that a loose relation may ultimately appear with some of those Ferns which form that indefinite non-indusiate complex known as the Gymnogrammoid Ferns.

Deparia Hook. and Grev.

This genus was founded to receive a Fern collected by Macrae on the Island of Owhyhee, now known as *D. prolifera* Hk. Others have since been added: *D. concinna* Baker, from Peru, and *D. Moorei* Hk., from New Caledonia. The distinctive feature is that the sori, usually stalked, protrude from the margin of the frond, and are protected by a cup-shaped membranous indusium. *D. prolifera* is figured in Hook. and Grev., *Icones Fil.* Pl. 154: also in Hooker's *Exotic Ferns*, Pl. LXXXII: *D. concinna* in Hooker's *Sp. Fil.* I, Pl. 30, *B*, and *D. Moorei* in Hooker's *Exotic Ferns*, Pl. XXVIII. The vicissitudes of these species and of the genus itself need not be re-stated here: they are fully set out in Dr Thompson's Memoir (*Trans. Roy. Soc. Edin.* Vol. L, p. 837, 1915). It may suffice to say that the species appeared so diverse to Christ that he speaks of Hooker and Greville's "künstliche Genus *Deparia*." He placed *D. prolifera* with *Athyrium* Roth, while he included *D. Mathewsii* (= *D. concinna* Baker) with *Dennstaedtia* Bernh., and *D. Moorei* with *Aspidium* Swartz. He was followed by Diels, and so in Christensen's Index the genus *Deparia* is sunk.

There is no sufficient ground for doubting that *Deparia*, as founded by Hooker and Greville, is a good genus. The differences which appear to have weighed with Christ are such as are seen elsewhere within well-recognised genera, while the distinctive character of the sorus will be found sufficient for holding together the apparently diverse species. *D. prolifera* and *D. Moorei* present two types divergent chiefly in their vegetative characters, but they coincide in the structure of the sorus. *D. prolifera* has a short ascending stock, bearing closely disposed leaf-bases, both covered by scales. The petioles have a binary leaf-trace near to the base. The leaves are glabrous and pinnate, and the pinnae pinnatifid, with numerous projecting sori borne distally on the free veins. The indusium is shallow cup-shaped, and entire at the margin. These characters are such as to point clearly towards *Dennstaedtia*, with which this species shares the occasional presence of adventitious buds at the bases of certain pinnae. J. Smith (*Hist. Fil.* p. 265) compares the habit with *Athyrium*, but notes the similarity of the sorus to that of *Dennstaedtia* and *Microlepia*, though distinguished from them by its exserted position. This is, however, less marked in *D. concinna*, where the sori are partially sunk in the margin, while its habit resembles closely *Dennstaedtia adiantoides* (Willd.) Moore (*Syn. Fil.* p. 55). The onus of comparison will then fall especially upon the third species.

D. Moorei was first found in a dense wood by the side of the Copenhagen river, New Caledonia, and its habit should be viewed in relation to moist shade. The creeping rhizome bears crowded leaves with a slender rachis, bearing a broad deltoid lamina, thin but firm, pinnate below but pinnatifid above: the two lower pinnae are themselves again pinnate or pinnatifid. The whole affords a broad expanse with reticulate venation, as a rule without included vein-endings. Both scales and hairs are present. The sori are usually all marginal, seated on free vein-endings: but occasionally sori appear also upon the upper surface, at the ends of small included veins. In either case

Fig. 752. Two mature marginal sori of *Deparia Moorei* Hook. Enlarged.
(After J. McL. Thompson.)

they are slightly stalked, and the indusium is cup-shaped (Fig. 752). The sporangia are numerous, and long-stalked, with vertical interrupted annulus, and the sorus is of the "mixed" type.

The axis contains a perforated dictyostele with an ample pith. The leaf-gaps are very irregular in size and outline, as are also the numerous perforations. The result is a stelar skeleton of a type generally resembling that *Dryopteris*, but differing in detail (see Vol. I, Fig. 172). The leaf-trace consists of four or five strands, typically four: two larger adaxial strands arise some distance from the base of the gap, two smaller median strands from its base.

The receptacle of the sorus arises normally from the margin, while the upper and lower lips are of superficial origin, as in the Dicksonioid Ferns (Fig. 753, *A, B*). The receptacle shows slight signs of a gradate sequence of the sporangia, but almost at once it is followed by a mixed condition (*C*).

This is in essential accord with *Dennstaedtia*: but a difference appears in the fact that instead of the marginal sorus turning downwards with age, here it is directed definitely upwards. The origin of the superficial sori has been found by Thompson to be essentially the same, with a like relation to a vein-ending; a fact that bears upon the interpretation of their unusual position. If *Deparia* be adopted as a good genus, it is clearly related to *Dennstaedtia* through *D. prolifera*. The divergent peculiarities of the leaf of *D. Moorei* may be interpreted in relation to its shade-habit. A broader leaf-expanse and reticulate venation are signs of specialised advance; they are

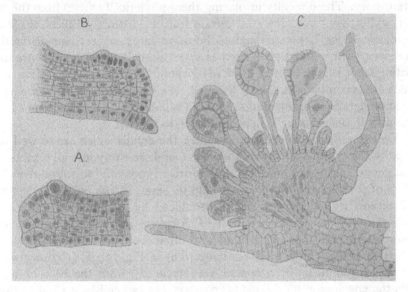

Fig. 753. Young sori of *Deparia Moorei* Hook., showing their marginal origin. In *A* the wedge-shaped cell marks the centre of the receptacle, with the upper and lower indusial lips arising superficially. *B* shows the same more advanced. *C* is a sorus almost mature, of mixed type, and definitely turned upwards. The figures have all the natural orientation. Highly magnified. (After J. McL. Thompson.)

seen also in the shade-loving *Hypoderris* among the Woodsioid Ferns, in *Onoclea* as against *Matteuccia*, and in *Christensenia* among the Marattiaceae. Similarly *D. Moorei* would appear to be a condensed shade-form of *Deparia*. With that condensation it may be held that the occasionally included veins, stopping short of the margin which they all reach in the originally open venation, offer a physiological opportunity for the formation of sori here not marginal, but appearing in an anomalous position on the upper surface. An analogy, though not an exact one, of transfer of sori to the upper surface is seen in *Polystichum anomalum* (Vol. I, p. 225). In neither case should any far-reaching phyletic or systematic weight be attached to the anomaly. *Deparia*, including *D. Moorei*, should therefore resume its natural place as a substantive genus derivative from the Dennstaedtioid Ferns.

SALVINIA Adan. and AZOLLA Lam.

One of the outstanding problems in the Natural Classification of the Filicales is found in the Hydropterids. A reasonable probability of a relation of the Marsiliaceae to the Schizaeaceae has been stated in Vol. II, Chapter XXIII. But the Salviniaceae, with their two genera *Salvinia* and *Azolla*, still remain as "*incertae sedis.*" Few organisms have ever been subjected to so exact an analysis as these, both as regards structure and development. There can be no doubt of their Fern-nature, notwithstanding their pronounced heterospory. The difficulty in placing them phyletically arises from the fact that, while they have preserved certain features distinctly Filical, these are standardized for Leptosporangiate Ferns at large rather than distinctive. Moreover, their vegetative system is so reduced in accordance with their floating habit, that there is little left in it beyond the initial segmentation upon which exact comparison can be founded. We have learned that comparison based on details of segmentation by itself is liable to pierce the hand of him who uses it, unless there is other evidence to support it.

There is no need to recapitulate here the details which are so well rendered in the original memoirs, and have found their way into all textbooks[1]. The general facts are indeed fully known. Synonymy, so suggestive elsewhere of possible affinity, does not help us here. *Azolla*, with its 5 species, still retains its old generic name introduced by Lamarck in 1783; and *Salvinia*, with its 13 species, that of Adanson in 1763. It is the soral characters that offer the best ground for comparison. In *Azolla filiculoides* Von Goebel finds the female sorus constructed as in Fig. 754 (i) (*Organographie*, II, 2, 1918, p. 1132), with a conical receptacle (*P*) from the base of which rises the enveloping indusium (*Id*) composed of two cell-layers, with a distal pore. The receptacle, with its vascular supply, bears distally a single megasporangium (*Ma*), while below, in gradate sequence, are microsporangia which are arrested at an early stage, but show the usual segmentation. On the other hand the male sori may bear an abortive megasporangium distally. There appears to be some inconstancy in the occurrence of these in individual sori, but Von Goebel finds both states in *A. filiculoides* (*l.c.* p. 1132, footnote). He interprets the facts as showing a reduction in number of the megasporangia as compared with that of the microsporangia, and a separation of the originally bisexual sori into those that are male or female. The separation appears to be complete in *Salvinia*: this genus is peculiar among Ferns in the branched pedicels of its microsporangia (compare Luerssen, *l.c.* Fig. 186).

[1] For references to the literature see Luerssen, *Rab. Krypt. Fl.* III, 1889, p. 595. Diels, E. and P. I, 4, 1902, p. 381. Campbell, *Mosses and Ferns*, 3rd edn. 1918. Strasburger, *Textbook*, Engl. Edn. 1921, p. 381.

In their distal marginal position upon the leaf-segments the sori of the Salviniaceae correspond to those of the Hymenophyllaceae, or to *Thyrsopteris*. This position was observed by Pringsheim for *Salvinia*, but it has been explained in greater detail by Von Goebel (*l.c.* pp. 1060, 1193). Here the dorsal leaves act as floats and serve nutrition: the sori are borne upon the ventral, submerged leaves, which arise late on the young plant, and are richly branched. They may be designated the sporophylls, and the distinct male and female sori

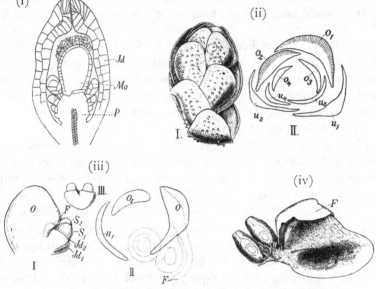

Fig. 754. *Azolla filiculoides* Lam. (i) Longitudinal section through a megasorus. *Id*=indusium: *Ma*=megasporangium: *P*=placenta: enlarged. Above the megasporangium *Anabaena*-threads are seen: below it are rudiments of abortive microsporangia. (ii) I=habit of a shoot seen from above, enlarged. II=transverse section through a bud: three pairs of leaves are marked *o*, *u*, etc., being the upper and lower lobes of the respective leaves. The position of the palisade parenchyma is shown by shading on the upper leaves. (iii) I=sporophyll dissected free, seen superficially: *O*=upper lobe: *F*=rudiment of its flange: S_1, S_2=rudiments of megasporangia: Id_1, Id_2 their indusia. II=transverse section through two leaves, the left sterile, O_1 its upper, *O* its lower lobe. Right a fertile leaf focused in upper and lower planes, the latter dotted. *O*=upper lobe, *F*=its flange covering the two megasori. III=the lower lobe dissected free: it is wholly used up in the formation of two megasori: the indusia appear as circular walls. (iv) Sporophyll of *Azolla filiculoides*, flattened out. To the left are two megasori, to the right the upper lobe: *F*=the flange-like outgrowth, below the mucilage-cavities are visible. (After Von Goebel.)

are borne in bunches of 3 to 8, each terminal upon an abbreviated leaf-segment (compare Luerssen, *l.c.* Fig. 185). Their exact position in *Azolla*, which is different in detail from that in *Salvinia*, has lately been re-examined by Von Goebel (*l.c.* pp. 1062, 1132). The leaves are arranged in two rows, and each leaf is two-lobed. The upper lobe is nutritive, and harbours the well-known *Anabaena*-colony: the lower consists normally for the most part of a single cell-layer (Fig. 754, (iii) II). It is this lobe which bears the pairs of sori, which are covered by a cowl-like sheath. The lobe undergoes an early branching, each branch ending in a sorus. Thus each sorus represents an abbreviated leaf-segment resulting from a branching of the second order. Each is protected by its own indusium, but both are covered also by the cowl, which springs from the upper lobe. The terminal position is in fact the same as in *Salvinia* (Fig. 754, iii, iv).

The sporangia of the Salviniaceae have no annulus: the number of spore-mother-cells as shown by the microsporangia is 16, but in the megasporangia it is only 8, while only a single megaspore comes to maturity (see Vol. I, p. 267).

From such facts it is difficult to attach the Salviniaceae definitely to any Family of living Ferns. It seems inevitable that they sprang from a type that was homosporous, but the absence of an annulus removes one of the most reliable features for comparison, while the segmentation of the sporangium and the exact number of the spore-mother-cells give general rather than precise indications of relationship. The basipetal succession of the sporangia upon the elongated receptacle, and the basal indusium, together with the distal position of the sori upon the leaf-segments, constitute the best grounds for comparison. These all point towards the gradate Marginales in general, and if to any one Family of them, probably to the Hymenophyllaceae. Provisionally this relation may be regarded as a possibility: but in the absence of forms connecting these highly specialised aquatics with other Filical types the question of a more exact relation must be left open.

Finally, Seward remarks that "there is no evidence contributed by fossil records which indicates a high antiquity for the Hydropterideae. It is unsafe to base any conclusion on the absence of undoubted Palaeozoic representatives of this group; but the almost complete absence of records in pre-Tertiary strata is a fact which may be allowed some weight in regard to the possible evolution of the heterosporous Filicales at a comparatively late period in the earth's history" (*Fossil Plants*, Vol. II, p. 477).

The examples of Ferns of uncertain affinity, brought under discussion in this Chapter, may serve to indicate some of the difficulties that beset any attempt at a complete grouping of this complicated Class. The list of "*genera incertae sedis*" might be greatly extended: the result would still be the same, viz., while determining with reasonable probability the relationship of some, to leave others definitely unattached. But the difficulties thus apparent in Ferns are probably less grave than those encountered in a similar attempt for other Classes of Plants. These habitually offer less distinctive features for comparison than do the Ferns, and they are still more deficient in palaeontological history. The chief source of uncertainty in the construction of a phyletic classification lies in that standardisation of structure of the more modern types which naturally follows on homoplastic adaptation. It is this which has deprived the archegonium and antheridium, as well as the sporangium and even the sorus itself, of their comparative value in the advanced Leptosporangiate Ferns, though these provide more ample material for comparison in the ancestral forms. In a measure it is the same with the vegetative system of the sporophyte, and in still greater degree of the gametophyte.

But if the deficiency of differentiated detail is felt in the comparative study of Ferns, what shall we say of the Mosses, or of the Fungi, with their multitudinous genera and species, so highly standardised in so many respects, and with a marked deficiency in their palaeontological record? Or finally of the Angiospermic Plants, in which the ovule and pollen-sac, or the pollen-grain and embryo-sac, are severally so highly standardised, while their vegetative system is so open to homoplastic adaptation? The student of phyletic classification who devotes himself to the Filicales may, even at the close of a Chapter on "*genera incertae sedis,*" consider himself fortunate by comparison. He may be pardoned for believing that his chosen field is perhaps the most favourable of all Classes of Plants for the pursuit of his distant "El Dorado"—a Natural Classification that shall include all known types.

CHAPTER XLIX

SUMMARY OF RESULTS

A BRIEF Organographic History will now be given of the probable steps that have been taken in the evolution of the Filicales. That history does not purport to be an essay in Taxonomy: but if it has been correctly gathered from the available facts, both of the present and the past, it should itself provide the main outlines of a Systematic Grouping of the Class. The story will necessarily be discontinuous, and there can be no certain beginning: except such as may be reconstructed from the general characteristics of those Filical types which we regard as primitive, and their comparison with those of organisms held as lower in the scale of vegetation. The aim will be to trace the successive steps of their advance and modification, keeping constantly in view the biological aspect of those changes which appear.

ALTERNATION

The most general characteristic of all is the alternating life-cycle, for not only is this present in all Ferns, but also in all Archegoniate plants, while it extends in some form or another to all plants that reproduce sexually. This is not the place to discuss its nature or its origin. Here we accept it as an objective fact, which lies at the back of all higher organography, not only of Ferns, but also of land-living plants generally, and of most Algae and Fungi. But whereas in the Thallophytes the events of the cycle appear less stable, and are less definitely related to distinct types of somatic development, in the Archegoniate plants generally, and in the Ferns in particular, the cycle is defined organographically. Though exceptions are numerous, the normal alternation in Ferns is between the simpler sexual gametophyte and the more complex non-sexual sporophyte. The high degree of stability in the succession, the organisation, and the proportion of those phases is a fundamental feature in the organography of Ferns, as it is of Vascular Plants at large. But in the Algae, and especially in the less highly organised types of them, that stability is less marked: the facts indicate that in this there is a difference between sub-aerial and aquatic vegetation—but it is a difference of degree rather than of kind.

The reason for this greater stability of the alternating generations in land-living plants may properly be sought in the difference of environment. It is the leading function of the gametophyte to bear gametangia, with syngamy as the final result: each fusion initiating a new sporophyte. Spore-production is the leading function of the sporophyte, each spore initiating a new gametophyte. If these functions be carried out in the uniform medium of water, as in the Algae, the two somata may be and often are alike: in them syngamy is commonly effected through the medium of water, and it can happen at any time, while the carpospores that are its ultimate result float naturally away to new stations under conditions that are uniform, discharged by generations that may be uniform also. Thus in the Algae the problems of propagation and dissemination are easily solved. In all primitive land-plants we find syngamy through the medium of water rigidly maintained, and no normal cycle is complete without it: but for them the time and place of this regularly recurrent syngamy are restricted, since external liquid water is not constantly available. On the other hand, the carpospores, when detached, are spread through the medium of air: they must as a rule be scattered as dry dust if new stations are to be taken up. Thus in the Archegoniatae, which may be described as the amphibians of the Vegetable Kingdom, the one critical event of the life-cycle depends on water, the other on air, for its effective discharge.

The regularly stabilized alternation, and the marked difference in size, form, and structure of gametophyte and sporophyte are best interpreted in relation to these conflicting requirements. In the Filicales the gametophyte is relatively small, and usually evanescent: as a rule it bears gametangia early, and after producing one embryo it dies: but that embryo is stationary at the spot where it was produced. The sporophyte develops as a perennial plant, with a much more definite hold upon the soil, and an elaborate aerial shoot: this is able to produce successive crops of spores: each spore may be air-borne to a distance, and in favourable circumstances will germinate in its new station. The divergent features of the two somata and their constant alternation are clearly adapted to this separation of the functions of syngamy and dissemination, conducted as they are in different media. Their difference may be held as a natural consequence of the amphibial life, which has guided a less standardised alternation into a special and constant channel, as now seen in the Archegoniatae generally and in the Ferns in particular.

The interest in the two alternating generations or somata centres in the propagative organs which they bear. Von Goebel has suggested the propriety of adopting a uniform terminology for them: on the one hand *gametangia* are borne upon the gametophyte, and are liable on sexual differentiation to appear respectively as *mega-gametangia* (archegonia), and *micro-gametangia* (antheridia): on the other hand *sporangia* are borne upon the sporophyte, and though all are primitively alike, they are liable on sexual differentiation to appear as *mega-sporangia* and *micro-sporangia* (Vol. I, p. 290). The origin of the two types of sporangia from a common source is a familiar topic of the textbooks: the common origin of the gametangia is less obvious, being suggested by detailed comparison rather than by demonstration

for Archegoniate Plants. The argument based on comparison with the
Phaeophyceae was first stated in 1903 by B. M. Davis (*Ann. of Bot.* XVII,
p. 477; see also Schenck, *Engler's Jahrb.* 1908, Bd. XLII, p. 1), and illustrated by diagrams here quoted as Fig. 755. His suggestion that the antheridium and the archegonium represent derivatives from a common source
received support from Holferty's demonstration of bi-sexual archegonia in
certain Mosses (Vol. I, Fig. 280): and more recently by the discovery of indeterminate sex-organs in *Lycopodium lucidulum* by Spessart (*Bot. Gaz.* 1923,
Vol. LXXIV, Pl. XVI, Figs. 60, 61). Exceptional states observed by Miss Lyon

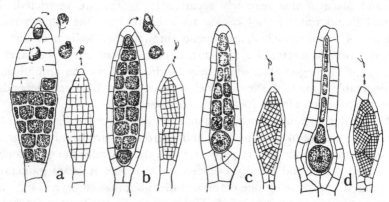

Fig. 755. Diagrams illustrating the possible evolution of the archegonium and antheridium
from the plurilocular sporangium. *a* = plurilocular sporangia with large and small
gametes discharged from the apex, after the habit found in certain Phaeophyceae (e.g.
Chilionema Nathaliae, Ectocarpus virescens, etc.). *b* = plurilocular gametangia of a
hypothetical algal type, which has adopted a terrestrial habit. The outer layer of gamete-
mother-cells has become sterilised as a protective capsule enclosing the fertile tissue.
The gametes are differentiated in sex, but both are still motile. *c* = plurilocular gametangia of somewhat higher hypothetical forms, at the level of heterogamy. Sterilisation
has proceeded so far in the female gametangium that only a few gametes are matured
at the base of the organ, and these are eggs. *d* = simple types of archegonium and
antheridium: the female gametes are reduced to one, while the number of male gametes
is greatly increased, and these cells are smaller and more highly specialised than in the
earlier conditions. (After B. M. Davis.)

and others also accord with the view that the sex-organs of the Archegoniatae are differentiated gametangia (*Bot. Gaz.* 1904, Vol. XXXVII, p. 280).
There is then reason for recognising in the propagative organs of both
generations a parallel progression in sexual differentiation; but apparently
that in the gametangia preceded by long ages that in the sporangia. This
parallel gives added interest to the comparison between the antheridia and
sporangia of Ferns in respect of bulk, which may in the future take an
important place in their phyletic comparison (Vol. I, p. 290).

It may still remain an open question whether the nearest points of phyletic
contact of the Algae with the Archegoniatae are to be found in the Red,
the Brown, or the Green Algae: analogies may be traced with each of these.
But whatever real relationship there may have been, indications of it should

be sought among lowly and generalised forms rather than among the more specialised Algal types. Similarities of vegetative form between the larger types and the most primitive Archegoniatae suggest homoplasy rather than homogeny. The comparison of the gametangia stands on a different plane, and the differences can be correlated with the environment. Sub-aerial life entails more exact protection of the gametes and of the zygote than is needed in aquatic life: the reaction of the organism is seen in the cellular protective walls of the gametangia, absent in most Algae, but constantly present in the Archegoniatae, the origin of which may well have been by sterilisation of superficial cells. Whether or not this was so, the gametangia of archegoniate plants are so protected now, and in particular the archegonium, which is the distinctive female organ of early land-living plants. Its protective function does not end with syngamy: the young embryo cannot, as in so many Algae, be cast adrift to take care of itself. It is retained on the parent, encapsulated in the archegonium and nursed there—a matter of the greatest importance for successful sub-aerial life. An internal embryology, however, restricts the development of the embryo: in particular a filamentous form, such as is common in freely germinating Algae, would be inconvenient. There are evidences of this primitive construction in the young sporophyte of Ferns, and particularly the suspensor in certain primitive types may be held as such: but they have been partially or completely eliminated, and the embryo of most Ferns is from the first a massive structure.

The production of carpospores in the Bryophyta is carried out without delay in the sporogonium, the necessary nutrition being drawn mostly from the persistent gametophyte: and the capsule is distal. But where the gametophyte is small and evanescent, as it is in the Ferns, the exigencies of nutrition fall early upon the embryo itself, with the consequence of an interpolated vegetative system, and spore-production is delayed. Still in certain relatively simple vascular plants the distal position of the sporangium, which was probably its original place, is maintained (Psilophytales, *Sporogonites*, *Stauropteris*). In many primitive Ferns a similar distal or marginal position of the sporangia is a feature not readily yielded up to the advantage of better protection: it persists in the marginal sporangia or sori so prevalent among the earliest of them.

By the effective nutritional scheme of the sporophyte the Ferns have often developed to large size, and have provided for a large output of spores, which with few exceptions are all alike, homosporous. The success of their propagation depends upon multiplying chances rather than upon that refinement of detail which comes with heterospory. Those Ferns which adopted the latter device never established themselves as real competitors with the Seed-bearing Plants. In point of fact the Filicales present to us a singularly successful Class, the most successful in fact of all those with homosporous propagation. Probably their conservatism in point of their method of fertilisation restricted them, as confirmed amphibians, to those habitats where syngamy through external water could be efficiently carried out. A special interest which they present is then in seeing how, subject to this restriction, they have been able in a long evolutionary history to make the best of an obsolete method.

In drawing to a close this preliminary sketch of the relation of the two generations to the environment, it is best to admit at once that the weight of comparison has necessarily fallen upon the sporophyte. This is partly due to its higher complexity providing a more ample field of fact, showing greater phyletic inertia, and with progressive sequences standardised relatively late: but it also arises from the fact that the simpler gametophyte is highly plastic in its vegetative structure, while the sexual organs, and particularly the archegonium, were standardised relatively early. Partly, however, or even mainly it is due to the deficiency of specific fact: for not only is the knowledge of the gametophyte of the fossils practically a blank, but the same may be said for many of the advanced Leptosporangiates, owing to the difficulty or neglect of cultivation; but more particularly to the lack of accurately determined material collected in the field. This weakness may be overcome by time and care, but none the less it exists now.

On the other hand, if the summary of phyletic results that follows is to be concise, much of the varied detail on which conclusions have been based must needs be omitted, and attention focused on the most consistent comparative features. These are now, as they have always been, those of the sori and sporangia. That such evidence as they afford does not stand alone may readily be seen in any specific case by reference to the text of Vols. II and III: and this will frequently be aided by citations of Chapter or Page.

SUMMARY OF PHYLETIC GROUPING.

There are two ways in which we may attain some conception of the primal representatives of the Filicales: either by direct observation of early fossils, or by the comparative study of those living Ferns which we have reason to hold as themselves primitive. The best results may, however, be expected to follow from a combination of both. This was attempted in Vol. I, Chapter XVII, where, on the basis of a wide comparison, an archetype of the sporophyte for the Class of Ferns was sketched as a word-picture. It would consist of a simple upright shoot of radial symmetry, possibly rootless, dichotomising if it branched at all, and with the distinction of axis and leaf ill-defined. The leaf, where recognisable as such, long-stalked with distal dichotomy, tending in advanced forms towards sympodial dichotomy. All the limbs of the dichotomy would be narrow, and separate from one another. The whole plant would be relatively robust as regards cellular construction, and traversed by conducting strands with a solid xylem-core. The surface might be glabrous, or invested with simple hairs. The solitary sporangia would be relatively large and distal in position, with thick walls and a simple method of dehiscence: and each would contain numerous homosporous spores. The archetype thus specified presents a real similarity to plants which have actually lived and are now well known, viz. the Psilophytales of the Devonian Period. From some such type of vegetation we may conceive the true Ferns, as distinct from the Pteridosperms, to have sprung. There need be no attempt to link any type of the Filicales directly to any type of the Psilophytales as offspring or progenitor. What is important is to bear in mind while studying the earliest Filical types that one of the oldest terrestrial Floras included vascular plants such as those which were found in the Rhynie Chert.

The Coenopteridaceae may be regarded as the characteristic Ferns of early Palaeozoic times. Of these *Stauropteris* appears as the most archaic type (Vol. II, p. 28). Though plentiful in the Coal Measures, it is suggestive that no axis has yet been associated with the upward-growing, slightly bifacial rachis, with its alternating pairs of appendages. These themselves branch again, so as to form a feathery plexus of delicate terete branchlets. Some of them bear terminal sporangia, which differ from typical fern-sporangia in being radially constructed, and opening by a distal pore. The anatomical structure suggests on the one hand the Psilophytales, on the other a Zygopterid character. If there actually was no axis, nor yet roots, the similarity to the Psilophytales would be impressive. But alternatively the question might then be asked whether *Stauropteris* could be adopted as a true Fern. (For a full discussion see Scott, *Fossil Plants*, 3rd edn., Part I, pp. 329, 413.)

The relation of *Stauropteris* to the Zygopterideae is generally admitted: in the latter the Filical characters are clearer, for there is a definite relation of axis and leaf, while the plant is rooted in the soil: but here again the leaves are strangely complicated, often bearing four rows of appendages (Vol. II, Chapter XVIII). The sori command special interest, for they appear as distal tassels of massive, dorsiventral, annulated sporangia, with lateral dehiscence. In *Etapteris* these are all separate, each on its own vascular stalk: but in *Corynepteris* they are closely grouped into radial sori marginally seated upon the narrow leaf-segments. In these three Zygopterids we may see three probable steps in the origin of the radiate uni-seriate sorus, which plays so distinctive a part in later types of Ferns, viz. the solitary distal sporangium with terminal pore; the pedicellate tassel, and the compact sessile sorus; both of these last have lateral dehiscence.

The Botryopterideae were relatively small plants with upright or creeping shoot, rooted in the soil. Here the structure of the elongated rhizome may sometimes be so similar to that of the petiole, as to suggest their representing unequally developed branches of a dichotomy (*B. cylindrica*), thus throwing a side-light on a possible mode of origin of the leaf. The vascular system is very simple in accordance with their small size, and the sporangia bear a general resemblance to those of *Etapteris* (Scott, *l.c.* p. 344).

The Coenopteridaceae thus show among themselves a considerable range of size, form and structure, and collectively afford suggestions of value in relation to various features seen in more typical, but yet primitive Ferns. Their early occurrence, coupled with the small size and simple structure of some of them: their prevalent dichotomous branching: the indeterminate distinction of axis and leaf: the simple though large hairs of *Botryopteris*, enlarged basally to stiff bristles in *Zygopteris*: their primitive sporangia, and numerous homosporous spores collectively indicate generalised types. Their complex leaf-structure based on bifurcation of narrow segments, and with circinate venation: the vascular structure, simple in small types but increasing in complexity with size: the grouping of the sporangia from the solitary distal capsule to the radial sorus; these are all pointers towards more definitely Filical types. The Coenopteridaceae may fairly be held as Ferns, and as representing more than any other known organisms an approximate source from which the whole Class of the Filicales has originated. On the other hand, that problematical plant *Stauropteris* gives the most valid basis for comparison with the Devonian Psilophytales: together they suggest that an upright, profusely branched indeterminate frond may have provided the raw material for the elaboration of the Fern-leaf, and that a solitary distal sporangium of radial structure, almost Bryophytic in its character, heralded the aggregated sori with dorsiventral sporangia seen in most of the true Ferns.

The Ophioglossaceae (Vol. II, Chap. XIX) appear as a blind evolutionary series, in the sense that they do not link phyletically with any living Family of Ferns. But they present certain features of advance parallel to those seen elsewhere. In contrast to the Coenopteridaceae, their gametophytes are well known. Fungal infection affects both generations, particularly their massive underground mycorhizic prothalli, and this has probably promoted their survival. Both generations are constructed on a massive scale, essentially Eusporangiate, as are all of the Coenopterids. But the single initial cell of the stem, and the presence of a suspensor in *Helminthostachys* and in a section of the genus *Botrychium*, may perhaps be held as traces of far-off filamentous origin. These two genera, from their lobed lamina, open venation, coherent stele and leaf-trace, and the presence of a suspensor, appear to be relatively primitive: while *Ophioglossum*, with its blade usually entire, reticulate venation, dictyostelic axis, its sometimes divided leaf-trace, and absence of a suspensor, may be held as advanced. A peculiar line of reduc-

tion of the leaf is seen in *O. intermedium*, which attains its limit in the mycorhizic *O. simplex*, where the sterile blade is absent (Vol. II, Fig. 345). The
fertile region in them all is held as of pinna-nature: the sexual organs, and
sometimes also the sporangia, are deeply sunk: the sporangia are largest in
the advanced genus *Ophioglossum*, a fact that may be related to the mycorhizic state. The Family is held to consist of imperfectly modernised relics
of an extinct Palaeozoic Flora, such as is represented by the fossil Coenopteridaceae.

All the sporangia of the Ophioglossaceae are distal or marginal, thus
retaining what may be regarded as a primitive position. The parts which
bear them in the Ophioglossaceae are narrow, and often branched. It is,
however, a wide-spread fact of experience among Ferns that where the
fertile blade is expanded the sori are liable to assume a position not at the
margin, but on the surface of the blade. Indeed it may be stated generally
that where this is so the position has been acquired secondarily. An illustration of this change may be seen in the living representatives of that ancient
Family, the Osmundaceae (Vol. II, Chapter XXI). *Osmunda* normally bears
its sporangia in marginal tassels on the narrow sporophyll, showing in this
the primitive state of the Coenopterids. *Todea* bears them superficially upon
the expanded pinnules, a state which may be held as derivative. But ab-
normally *Osmunda* may itself illustrate the transition, in leaves that are
described as metamorphosed (Vol. II, Fig. 420). This transition may have
happened early or late in descent in Ferns at large, for instance in the
Marattiaceae and Gleicheniaceae the sori are superficial; and Ferns of both
of these types, with sori already superficial, existed in Palaeozoic times.
Others have retained the marginal position to the present day, for instance
the Schizaeaceae and Hymenophyllaceae: while others again may be caught
in the act of transition, for instance the Pteroid Ferns. A broad distinction
thus exists between those which adopted the change early (*Superficiales*), and
those which retained the marginal position, or only departed from it relatively
late (*Marginales*). In certain great phyla this distinction dates back to the
Palaeozoic Period.

The Marattiaceae (Vol. II, Chapter XX) are illuminating in this relation.
The sorus in them all is radiate-uniseriate, derived from a type like that of
the Coenopterid *Corynepteris*, which is itself a compact tassel, seated at or
near to the margin of a narrow leaf-segment (Vol. II, Fig. 334). A broadening
webbed leaf, with such sori intra-marginal and slightly elongated, would
give the condition seen in *Angiopteris* if the sporangia were still separate;
or of *Marattia* if they were fused to form synangia (Vol. II, Fig. 392, *A, C*).
Extension of the former along the veins would result in the state of
Archangiopteris, of the latter that seen in *Danaea*; while further widening of
the blade now reticulate, and with the segregated sori scattered over the

surface, would result in *Christensenia*, a Fern showing adaptation to forest shade (Fig. 392, *B*, *D*, *E*). These Marattiaceae, however interesting thus for comparison, appear to be a blind branch phyletically, without further derivatives. Nevertheless the comparison of their living sporophylls illuminates the progression from marginal to superficial sori.

While the Osmundaceae, which date back to the Permian Period, also appear to bridge the distinction between *Marginales* and *Superficiales*, there are two other great series of known antiquity; the one strictly Marginal, the Schizaeaceae; the other strictly Superficial, the Gleicheniaceae. The three types thus named probably represent, as nearly as any others that are known, the progenitors of most of the modern Ferns. *It is believed that they have remained distinct from Palaeozoic or early Mesozoic time, and represent phyletic lines that were pursued apart throughout the intervening ages* (see Vol. II, p. 231, and also its Preface).

The Schizaeaceae (Chapter XXII) are signalised by a very variable vegetative system: on the one hand we see the primitive form and anatomy of *Schizaea* and *Lygodium*, on the other the more advanced features of *Anemia* and *Mohria*. But they all agree in bearing solitary sporangia of strictly marginal origin. These may be shunted during individual development to a position more or less distinctly superficial, upon the lower surface: but still they all spring from marginal cells; while a false margin, or in *Lygodium* certain superficial growths, supply a protection, giving the character of a primitive indusium.

The Gleicheniaceae on the other hand have a very uniform vegetative system (Chapter XXIV). They are primitive in their anatomy, and have their sporangia grouped in radiate sori, which are markedly superficial, and without any specialised protection whatever. The general type of sorus is that of *Corynepteris* on the one hand, or of *Angiopteris* on the other, though in the latter it is slightly elongated. The superficial position suggests that there had been an early slide to the surface of the widening blade: that position is here as strictly maintained as the marginal position is in the Schizaeaceae. A minor Family, the Matoniaceae (Chapter XXV), of early Mesozoic origin, shares the soral characters of the Gleicheniaceae; but with advances in anatomy and venation, and it differs in the presence of a unique distal indusium and low spore-output.

All the Ferns mentioned so far have one feature in common. It is that there is no succession of the sporangia of the single sorus in time of origin: either these are produced solitary (monangial), or where a plurality of sporangia exists constituting a sorus, they are simultaneous in origin. It is this which defines the Simplices, and it may be regarded as characteristic of the Ferns of the Palaeozoic Age. Their collective characters have been discussed in Chapter XXVI, together with the question of the phyletic distinctness of the Marginales and Superficiales. Both on grounds of comparison and of Palaeontological evidence, the Simplices may be held as comprising all those most primitive types of which we have any knowledge; and they include all known Eusporangiate Ferns.

As already stated, the Marattiaceae and Ophioglossaceae cannot readily be linked with any known derivative forms. But the remaining Families of the Schizaeaceae, Osmundaceae, and Gleicheniaceae, which are themselves less definitely eusporangiate than the rest, severally show features in common with other Families of Mesozoic or of more recent time. Later types may be linked with one or another of these three ancient Families, of which they appear to be derivative descendants. At least three main stirps may thus be distinguished among modern Ferns, and from very early times each of these is believed to have pursued a phyletic course apart.

MARGINAL DERIVATIVES

The Hymenophyllaceae (Vol. II, Chapter XXVII, p. 234) are probably an ancient stock of specialised hygrophytes. Their form, simple anatomy and venation, and their large spore-output suggest this, while the strictly marginal position of the sori and the filamentous gametophyte point towards the Schizaeaceae. They were probably of early origin, but it is difficult to link them definitely downwards with any one Family of the Simplices: upwards they appear to end as a blind line, unless it is right to regard the Salviniaceae as specialised aquatic and heterosporous derivatives from them (Chapter XLVIII, p. 260)—a loose attachment like that of the Marsileaceae to the Schizaeoids (Chapter XXIII).

The most marked distinction of the Hymenophyllaceae from the Simplices is the strictly marginal, gradate sorus (Vol. I, p. 212). In them we see in its most perfect form the receptacle with basal intercalary growth, bearing a long continued basipetal sequence of sporangia. The type from which this may have originated would probably be found among the protostelic Simplices, with which relation the sporangial structure and high spore-output would accord, occurring as these features do in a simplified hygrophytic stock. In close relation with them are the Loxsomaceae, also blind upwards, but with possible forerunners in Jurassic time (Chapter XXVIII).

From the point of view of the phyletic origin of typically modern Ferns, the Dicksoniaceae, which have also gradate and marginal sori, command a wider interest (Chapters XXIX, XXX). They comprise both dendroid and creeping forms, and like almost all the early marginal types they have only hairs as dermal appendages. In accordance with their larger size they are solenostelic or dictyostelic, sometimes also with medullary vascular tracts as well. The gradate sori are protected by a basal indusium, more or less distinctly two-lipped, and the strictly marginal receptacle bears sporangia, the first of which originate from marginal cells: they are followed by others in gradate sequence. They are short-stalked and massive in *Thyrsopteris*, a type which is believed to have had a Jurassic origin: but in *Dicksonia* and *Dennstaedtia* they are longer stalked, and more delicate. The spore-output is typically 64, and the spores are without perispore.

Two important modifications of the sorus make their appearance in this Family. One is a definite ontogenetic bias towards the lower surface, so that the mature sorus faces downwards: the other is the departure from the basipetal sequence, by interpolation of younger sporangia without strict order between those earlier formed. Both of these, together with the interruption of the annulus at the insertion of the stalk, are seen in *Dennstaedtia*, pointing towards a superficial type, so common in advanced Leptosporangiate Ferns. The most interesting genus in this connection is *Hypolepis* (Chapter XXXVI). Here with consistently Dennstaedtioid habit, anatomy, and dermal hairs, not only is the sorus deflected to the lower surface and of a fully mixed type, but the lower indusium is partially or completely aborted: in *H. repens* the upper indusium is even merged in the flattened blade itself; the lower vestigial or absent; the receptacle is flattened, and apparently superficial, with "mixed" sporangia. In fact *Hypolepis* merges into *Polypodium*, giving a phyletic transition from Ferns with a marginal, gradate, two-lipped sorus to a type with sorus superficial and mixed, and apparently without any indusium at all. Thus the Dicksoniaceae have given rise to Ferns which have been ranked under that advanced Leptosporangiate type that has been designated *Polypodium*.

Two derivative series may be traced from a Dicksonioid source: viz. the *Davallioid Ferns*, in which the sori for the most part retain their identity and marginal position: and the *Pteroid Ferns*, which show soral fusion so as to form linear coenosori, while the receptacle tends strongly to become superficial. The first of these centre round *Davallia* (Chapter XXXVII). They mostly retain a Dennstaedtioid habit, but with dermal scales, and often a highly segregated vascular system. The sorus is mixed, often with signs of a gradate state at first, and a flattened receptacle. But in some the receptacle may slide to the lower surface, and not unfrequently the individuality of the sorus may also be lost by fusion into coenosori, as in *Nephrolepis* and *Lindsaya*. These Ferns appear to have led to no further development.

The Pteroid Ferns illustrate a progressive sweep of greater extent and importance for comparison with other phyla. It may be held to start from *Pteridium* and *Paesia*, which have marginal, bi-indusiate fusion-sori, and it leads by gradual steps to the fully Acrostichoid state seen in *Acrostichum aureum*. This may not be a simple phyletic line, but the comparisons are broad and cogent (Chapter XXXVIII). The sequence involves parallel progression from solenostely to advanced disintegration: from hairs to scales: from open venation to reticulate: from marginal sori to superficial: from bi-indusiate to uni-indusiate, and from gradate sori to mixed: from a convex receptacle to one flattened and extended over the whole of the lower leaf-surface. Comparison of habit in the earlier terms of the series leaves no room for doubt that the Pteroids are of Dennstaedtioid affinity, and the series leads finally to that cosmopolitan Fern, which, with two others, remain the sole claimants to the old generic name of *Acrostichum* (Vol. III, p. 58).

Two features present in the primitive state are retained by the Marginales with peculiar tenacity, but finally relinquished: viz. the marginal position of the sorus, and dermal hairs. It is a fact of diagnostic interest that the dendroid Dicksoniaceae retain both, while the dendroid Cyatheaceae have relinquished both. This is not the sole ground for separating the former phyletically from the latter: but either of these points would suffice at a glance to distinguish a Dicksonioid Fern from any true Cyatheoid, the latter having superficial sori, and plentiful chaffy scales.

SUPERFICIAL DERIVATIVES

There is ample evidence of the existence of Ferns in Palaeozoic time with sori seated on the lower surface of the flattened blade. With the Coenopteridaceae before us to point to the distal or marginal position as primal, and the fact that more than one series of living Ferns may be seen in course of transfer of the sorus from the margin to the surface, it is legitimate to conclude that the superficial types of Palaeozoic time had also acquired the superficial position secondarily, though at a very early period. This conclusion is not, however, essential to our argument; the undoubted fact of their early existence is sufficient. *Oligocarpia* (Vol. II, p. 211) gives an example, from the Coal Measures, of a Fern with a superficial sorus of Gleicheniaceous type. The oldest authentic records of Gleicheniaceae are however, from the Keuper of Switzerland: but they are plentiful throughout the Mesozoic Period, those early records running parallel with but separate from those of the marginal Schizaeaceae, as they do to-day.

The Gleicheniaceae (Vol. II, Chapter XXIV) are a Family of uniform habit, with bifurcating usually protostelic rhizome, and leaves often possessing unlimited apical growth and of Pecopterid type, with open venation: their radial unprotected sori consist of few large sporangia, with high spore-output, and longitudinal dehiscence. All these characters, seen particularly in *Dicranopteris*, mark them as relatively primitive (Chapter XXIV). *Eu-Gleichenia*, *Stromatopteris* and *Platyzoma* are specialised xerophytes. A peculiar phyletic interest attaches to the sub-genus *Eu-Dicranopteris*, which bears hairs only, while scales are present in other Gleichenias. Solenostely is characteristic of *D. (Gl.) pectinata* (Willd.) Pr., while this species and *D. (Gl.) linearis* (Burm.) Clarke have in each sorus 6 to 12 smaller sporangia, with a lower spore-output, crowded in a close hemisphere. These features indicate advance; but the sorus of these species has reached the point of mechanical dead-lock, for the sporangia often lack space to open and shed their spores. None of the known Gleicheniaceae have surmounted this difficulty (Vol. II, p. 204, Figs. 486–488). Attention is thus directed to the Cyatheaceae, which also have superficial sori, often unprotected (*Alsophila*), and with habit, anatomy, and venation not markedly different in the simpler types from the more advanced Gleicheniaceae. They have adopted a gradate sorus and lateral dehiscence, which solves the dead-lock mechanically. The next question will be whether all Cyatheaceae share this. Exceptions are found in Presl's genera *Lophosoria* and *Metaxya*, latterly merged in *Alsophila*; each is represented by a single species. These genera are now re-established as substantive genera, and are included in the new Family, Proto-Cyatheaceae (Chapter XXXII). They appear to take their places at the base of certain distinct phyletic lines, which have met the mechanical difficulty in different ways: *Lophosoria* leading to the Cyatheoids, and *Metaxya* to a less extensive series which may be styled the Metaxyoid Ferns (Chapters XXXIII and XLVI).

Lophosoria is really one of the Simplices, for all the sporangia of its sorus arise simultaneously on a receptacle corresponding in position to that of *Gleichenia* (Fig. 548). The upright axis is covered with hairs and slightly dictyostelic, but solenostely is prominent in its horizontal runners. It is in fact a compact upright shoot closely related anatomically to *D. (Gl.) pectinata*, with which also the fertile pinnae agree (Figs. 547–550). The sorus, however, compares rather with that of *D. (Gl.) linearis* (Figs. 487, 550): but essential differences lie in the lateral dehiscence of the sporangia, and in the smaller spore-output. The large size of the sporangia, however, and the ill-differentiated stomium, suggest that *Lophosoria* represents an amendment on the sporangia of *Gleichenia*: they open outwards, and the spores are readily shed. It may be held as one of the advanced Simplices, with near relation to *Gleichenia*, but with features of *Alsophila* seen in the upright habit and lateral dehiscence. Hitherto it has been included in *Alsophila*. Clearly it is a synthetic type, leading to the Cyatheaceae rather than actually one of them.

The Cyatheoid Ferns, while maintaining the habit of *Lophosoria*, and often developing it to large size with advances in vascular structure, introduced two further changes, viz. the broad chaffy scales in place of hairs, and a gradate sorus and smaller sporangia: in some of them there may also be a basal, partially or completely cup-shaped indusium. All of these are physiologically probable amendments (Chapter XXXIII). The prototype of the scale, with its often massive base, is seen in *D. (Gl.) pectinata* (Fig. 475), and in *Lophosoria* (Fig. 547). The gradate sorus of *Alsophila* follows naturally on intercalary elongation of the receptacle (Fig. 564): an argument in favour of the origin of the basal indusium as a new formation, originating from dermal appendages, has been advanced in Vol. II, p. 307, and it is more fully stated in Chapter XL, on the Woodsieae. Its extreme variability in *Alsophila* and *Cyathea*, noted by Christ, supports this view (*Farnkräuter*, p. 323). The smaller size, lower spore-output, and greater precision of structure of the sporangia in the Cyatheoids are in accord with the introduction of the gradate sequence, and find their homoplastic parallel in the Hymenophyllaceae.

The Woodsieae (Chapter XL), which may for the most part be regarded as arctic and mountain congeners of the Cyatheoids, form an important bridge to the very large series of the Dryopteroid Ferns. They introduce two fresh features among the Superficiales, viz. a mixed condition of the sorus, and its lop-sidedness or zygomorphy. The radial sorus of *Woodsia* is gradate, with a basal indusium, the various forms of which readily accord with an upgrade origin of a coherent cup, derived from hairs such as are seen in a like position in Gleichenioid Ferns, in *Lophosoria*, and in *Alsophila* (see p. 114). The gradate state characteristic of *Woodsia* is hardly more than indicated in *Diacalpe* and *Peranema*, for profuse interpolation of sporangia soon intervenes upon the widened receptacle, giving the state typically present in *Dryopteris*. Finally *Hypoderris* is a specialised shade-form, with widened reticulate blade: but the mixed sorus with basal indusium is still radial as in *Woodsia*, from which it is probably a derivative.

In *Peranema* the receptacle, at first radial, as it is in all earlier Superficiales, soon becomes lop-sided and stalked (p. 113). Its indusium, which covers the receptacle completely, is unequally developed, being absent on the side next the leaf-margin: thus a narrow slit is left facing outwards from the midrib, while the receptacle itself is tilted over in the same direction, and covered by the stronger growing side of the indusium. This zygomorphy of the sorus corresponds to that general in Dryopteroid Ferns, while the habit, anatomy, and chaffy scales of *Peranema* all point to its synthetic position between *Woodsia* and *Dryopteris*. Probably *Cystopteris* and *Acrophorus* find their natural affinity here (Chapter XLVIII).

The view at which we arrive by following the phyletic line thus traced is that the reniform Dryopteroid sorus is a zygomorphic derivative of a superficial radial sorus, protected by an indusium originally basal, and derived by modification of dermal appendages. A comparatively slight modification of it, by extension of the receptacle right and left so as to encircle the stalk of the indusium, and fusion of its edges, gives the "*indusium superum*" as seen in *Polystichum* (Von Goebel, *l.c.* p. 1150). In other Ferns of this affinity various degrees of abortion of the indusium are seen leading to its complete absence: as in *Meniscium* or *Phegopteris*. The latter, long included as a section of *Polypodium*, consists of Ferns that are ex-indusiate Dryopteroids. Lastly, the genera *Polybotrya*, *Stenosemia*, together with the pinnate species of *Leptochilus*, are Dryopteroids in which the individuality of the ex-indusiate sori has been lost by extension of fertility over an extended leaf-area, giving the Acrostichoid character (pp. 131–135).

It thus appears that, following the Superficial Series upwards, the sorus was first simple, radial, and exposed: it then became gradate, and acquired a basal indusium: then followed the mixed condition, zygomorphic in relation to the leaf-margin, with first a reniform, then an orbicular indusium: next followed the ex-indusiate or Polypodioid state: and finally the Acrostichoid. The parallel between this and what has been seen in the Marginal Series is obvious though not exact. It is closest between *Dryopteris* on the one hand, and such marginal derivatives as *Nephrolepis* and some species of *Lindsaya*, though detailed comparison in other features would negative a real affinity. On the other hand, there can be no doubt that the Polypodioid state of *Hypolepis* and the Acrostichoid state of *A. aureum* were attained by lines phyletically distinct from those ending in *Phegopteris* or *Polybotrya*. These are examples of that homoplastic likeness that frequently recurs in distinct phyla of Ferns.

A collateral step in the progressive series of changes that started from the non-indusial radial sorus is seen in the Asplenieae, which may be held as derivative from the Nephrodioid type of sorus (Chapter XLII). In *Didymochlaena* the sorus appears as a horse-shoe, elongated so as to follow the course of the underlying vein. The primary receptacle corresponds to the point of strongest curvature, here directed towards the leaf-margin. If the fertility at that region were lost, but retained on one side or the other, or on both of the elongated shanks of the curve, the result would be the state seen in *Asplenium* or *Diplazium*. Intermediate conditions are frequent in *Athyrium*, but they are particularly well seen in *Diplazium lanceum* (Fig. 672). Finally, the indusium itself may be abortive, giving the Pseudo-Gymnogrammoid state of *Ceterach* or of *Pleurosorus*. Possibly the Acrostichoid state of *Rhipidopteris* may have had its origin from such Ferns as these, by spread of fertility over the reduced surface of the fertile blade.

Starting afresh from the Onocleoid Ferns, *Matteuccia* and *Onoclea*, which are best separated as another distinct line of derivatives from the Cyatheoid type, a phylum may be traced parallel to that of the Nephrodioids, but still distinct (Chapter XLIII). A recently described species, *Matteuccia intermedia* C. Chr., serves as the starting point. It is a coarse-growing Fern with massive dictyostelic stock, binary leaf-trace, very scaly surfaces, leaves dimorphic, open venation, and superficial non-indusiate sori ranged in regular rows, one on either side of the mid-rib. The sori are gradate, and the sporangia relatively large. These characters point to a relation with *Alsophila* and *Lophosoria* rather than with *Woodsia*. It stands alone among the Onocleoids in being non-indusiate: protection is afforded by the strongly reflexed leaf-margin. It is but a step from this to linkage of the lines of sori by a vascular commissure, and the Blechnoid type results, as in *B. tabulare* or *capense*, the habit and anatomy of which are closely similar to those of *Matteuccia*. All of these Ferns are of coarse and xerophytic habit. They suggest the origin of the Blechnoid Series, which forms a prominent feature in the Southern Hemisphere, and particularly in Polynesia (Chapter XLIV).

The large genus *Blechnum*, with its linear coenosori, introduced several new and peculiar features, leading finally to states parallel with, but phyletically distinct from those of the Nephrodioid Ferns. The most marked external feature is the origin of the longitudinal "flanges," right and left, along the lines of greatest curvature of the reflexed pinnae. These assume, by gradual steps illustrated by specific comparison, a vascular system of their own, and a definite photosynthetic structure: in fact they become substitutionary laminar flaps, but in point of descent they are new structures (p. 165). The sori originally gradate become mixed: the receptacle is flattened, and tends in some derivatives to spread widely over the lower surface: in this a climax is reached in *Stenochlaena* and *Brainea*: in the latter the falsely indusial leaf-margin is abortive, and a fully Acrostichoid condition is reached. All this happened without any true indusium being present at all: all the protective flaps are specialised leaf-margins.

On the other hand, the coenosori may be interrupted, breaking up into short lengths not necessarily corresponding to the original sori. Each is covered by a short length of the "false" indusium. This is seen normally in *Woodwardia* and *Doodia*: sometimes on a widening pinna more than one row of them may be seen on either side of the mid-rib. A still more remarkable result appears in *Blechnum punctulatum* Sw. var. *Krebsii* Kze. Here with a widening fertile pinna, the coenosorus shows varying states of sinuous curvature outwards towards the margin, and of interruptions at the points of strongest curvature, leaving portions of the coenosorus facing one another in the characteristic manner of *Phyllitis* (*Scolopendrium*). The anatomy as well as the gradual intermediate steps show that this is the true interpretation of *Phyllitis*. Consequently this genus is not immediately related to *Asplenium*, which has been traced from a Nephrodioid source (pp. 192–198).

In the preceding paragraphs the Cyatheoid, Nephrodioid, and Blechnoid Ferns have all been traced from a superficial origin such as the Gleicheniaceae afford, with *Lophosoria* as an illustrative connecting link. But *Metaxya*, the other genus of the Proto-Cyatheaceae, was left aside. The mechanical dead-lock of the advanced Gleicheniaceous sorus was resolved in *Metaxya* by spreading the flattened receptacle out along the surface of the vein, but without any gradate sequence. Here again the sporangia, though numerous, arise simultaneously: thus technically it is still one of the Simplices (Chapter XLVI). *Metaxya* is a creeping Fern with solenostelic structure and undivided leaf-trace, and it bears hairs, but no scales. The sporangia have a four-rowed stalk, a slightly oblique annulus, and lateral dehiscence. This monotypic genus does not accord with other Cyatheaceae, though it has habitually been included in the Family. It stands by itself. But in habit and soral characters it may probably be linked with two old genera, *Syngramme* and *Elaphoglossum*, which share its venation and certain other features; they differ, however, in having vein-fusions though without included free endings. The sporangial stalks are three-rowed, and in *Elaphoglossum* interpolated sporangia lead to a "mixed" state. In *Syngramme* the sori are elongated, but in *Elaphoglossum* they are fully Acrostichoid, being spread over the whole leaf-surface. These comparisons are suggestive rather than demonstrational, and the Metaxioid derivatives will require accurate re-investigation from this point of view.

It is well known that Ferns of Dipteroid type were prevalent in the Mesozoic Period: but it has hitherto been insufficiently realised that they have left behind quite a considerable number of striking descendants living at the present day, such as *Platycerium, Neocheiropteris,* and *Cheiropleuria.*

The relations of *Dipteris* to *Matonia* are certainly real: both may be held as related by descent to the Gleicheniaceous stock. The Dipteroid Ferns (Chapter XLV) are throughout non-indusiate, and most of the living species of *Dipteris*, as also the related Mesozoic fossils, rank as Simplices. But many of the later derivatives possess a mixed sorus. Another and more effective method of increase in their spore-production was secured by extension of the individual receptacle, which is often accompanied by a special sub-soral vascular system running in a plane parallel to that of the blade (diplodesmic). This is in fact a special feature in Ferns of Dipteroid origin. They show a considerable range of progressive characters: originating from a type with narrow bifurcate leaves, having one row of hemispherical sori on either side of the mid-rib, after the Gleicheniaceous type. From this they have passed over by webbing and expansion to broad irregularly lobed blades, with the characteristic *Anaxeti* venation: also from protostely, through solenostely, to the highly perforated polycyclic state seen in *Platycerium*: also from dermal hairs to scales. But it is the diplodesmic soral expansion that gives their most marked character: this is accompanied by a direct passage from the simple to the mixed state, and from an oblique to a vertical annulus. In *Platycerium* the sori that constitute the large fertile patches retain something of their original identity, though greatly extended in length and branched. In *Neocheiropteris* and *Pleopeltis* the sorus is also extended, but it retains the circular or oval form. In *Cheiropleuria* (with its strangely mixed characters), in *Christopteris* and in *Hymenolepis* the fertile area extends over the whole surface of the blade, and the identity of the sorus is lost, as in any fully Acrostichoid type (pp. 213–224).

Comparison as detailed in Chapter XLV leaves no room for doubt that the Dipteroid derivatives comprise a considerable body of Ferns, whether fossil or living: and that the series has pursued a course of its own, independent of other phyla, from the Triassic Period to the present day. It originated from a source related to other early superficial Simplices, and ends in types which have been referred to the old comprehensive genera, *Acrostichum* and *Polypodium*. Such a progression is thus homoplastic with others that passed through a similar though phyletically distinct history.

There remain a considerable number of genera of Ferns commonly of small size, with their unprotected sori more or less distinctly superficial, and often elongated, following the course of the veins. These genera have been associated habitually with *Gymnogramme*, and in point of fact many of them have been from time to time included in that old and comprehensive genus (Chapter XXXIV). They were grouped by Diels with the Pterideae, as Pterideae-Gymnogramminae: but this presumes for them a secondary origin by abortion from an indusioid type, of which there is no evidence: they all appear to have had unprotected sori from the first. In the Osmundaceae, and certain Schizaeaceae, together with *Plagiogyria* we see relatively primitive Ferns which had non-indusiate sori, and it is in relation with these that their natural place may be sought.

The Gymnogrammoid Ferns fall into four natural groups: (i) those which have primitive characters such as *Llavea* and *Cryptogramme*: (ii) those which centre round *Gymnogramme* and *Hemionitis*: (iii) the well-marked genus *Adiantum*: and (iv) a group of specialised xerophytic types associated with *Cheilanthes* and *Notholaena*. There is no sufficient ground for assuming that these four groups were of common origin, for the features on which they are associated together are negative rather than positive. *Llavea* and *Cryptogramme* have usually been held as related to *Plagiogyria*: and this, with a further reference to *Todea*, may be held as a probable source for them. Perhaps also for *Ceratopteris* and *Jamesonia*, though with less certainty of reference as between an Osmundaceous and a Schizaeoid origin (Chapter XXXIX, p. 72).

The second group presents a more difficult problem: they no doubt form a coherent body of genera, and are probably related to the first group: but at present their definite reference must be left uncertain. The same may be said of the genus *Adiantum* which forms the third group (Chapter XXXIX, p. 78).

The fourth, or Cheilanthoid group, with its sparse sporangia borne close to the protective margin, and the habit so often suggestive of *Mohria*, has long been compared with the Schizaeaceae, a comparison which detailed study of both generations has strengthened. Though this falls short of actual demonstration of affinity, there is reasonable probability that *Notholaena* and *Cheilanthes* are Schizaeoid derivatives from the type of *Mohria*, their sporangia having passed from the margin to the surface of the blade, in a manner analogous to the like change in the Pteroids. With these genera *Pellaea* and *Doryopteris* may be linked, while *Trachypteris* and *Saffordia* present a type like the latter, but with an Acrostichoid spread of the sorus over the surface of the blade. Superficially this is like what has been seen in the Pteroid Ferns; but the two series, the one indusiate and the other non-indusiate, appear to have progressed along homoplastic lines, separate but parallel, or even convergent. The Gymnogrammoid Ferns as a whole probably represent a plexus of non-indusiate phyletic lines, all traceable back to Ferns with marginal sporangia larger than their own, such as are seen in the ancient Families of the Osmundaceae and Schizaeaceae (Chapter XXXIX, pp. 92–97).

As epiphytic derivatives from the central group of the Gymnogrammoids, the five genera now included as the Vittarioid Ferns take their natural place. They appear to be forms arrested or specialised in relation to their habitat. In *Anetium* a step towards an Acrostichoid state is again seen (Chapter XLVII).

It must not be assumed that by essaying some clearer phyletic grouping of Ferns the writer has undertaken to place all genera in some probable relation by descent. Anyone who makes the attempt will very soon find problems rising before him that are insoluble for want of precise data: or questions in the resolution of which a personal estimate of the importance of relevant facts necessarily takes the place of proof. Illustrations of this are given in Chapter XLVIII, which indicate the sort of difficulties that are apt to arise. They also suggest the wide intervals that lie between possibility, reasonable probability, and demonstration. That Chapter was intentionally introduced to show that the present work is a mere *Tentamen*, not in any sense a finished task. Some of the examples chosen illustrate how a better knowledge of detail may lead to an assured conclusion (*Prosaptia, Deparia*), or to a definite bias in a certain direction (*Cystopteris*): others leave alternative views still in suspense (*Taenitis*): others again show how slight are the present grounds for any definite opinion (*Azolla, Salvinia*).

The analysis contained in Vol. I, in respect of the twelve criteria of comparison used in the phyletic treatment of Ferns, has led to the recognition of states respectively primitive and relatively advanced for each of them. For instance, the progressions from equal dichotomy to dichopodial, and finally to monopodial branching: from protostely to vascular disintegration: from dermal hairs to scales: from open to reticulate venation: from more complex to simpler cellular construction: from a marginal to a superficial position of the sporangia: from simple to gradate and mixed sori: from complex and sunk to simpler and stalked sporangia: from a large spore-output to a smaller: from larger to smaller antheridia: from embryos with a suspensor to those with none—all of these as well as others are essentially independent steps, though they commonly march together. So usual is this that when a discrepancy occurs the attention of the observer is at once drawn to it. On the other hand, the discrepancies show that some degree of independence of the several criteria actually exists. For instance, in *Ophioglossum* and in *Christensenia* stelar disintegration is associated with a massive Eusporangiate structure; in *Cheiropleuria* a protostele, an undivided leaf-trace, and dermal hairs are associated with an Acrostichoid sporophyll. Such examples show that parallel progressions in respect of the several criteria, though usual, are not obligatory. Nevertheless the conclusion that follows is that, subject to exceptions, the organisation of a Fern has evolved as a whole, its general progress being expressed in a plurality of characters which appear distinct from one another, and may be subjected to separate comparative treatment. There is in fact a general drift of organisation of the individual, affecting as a rule all its parts. The organism behaves in its evolution as an integer, or whole.

Further, the progressions in respect of the several criteria are not restricted to any single phylum: they are exhibited with a high degree of uniformity in a plurality of phyla, which palaeontology shows to have been broadly distinct from one another in descent from very early periods. The chief lines of descent of Ferns are already suggested by Palaeozoic Fossils, but they became for the most part clearly defined in the early Mesozoic Rocks. From that time onwards, since the several phyla have maintained their characteristic features, their evolution must have been independent and homoplastic, however similar the steps of advance in the several phyla may appear to be. Moreover, the progress in respect of these has not been merely parallel, but at times actually convergent. It is this fact which has led so frequently to those systematic difficulties which have found expression in a wide synonymy. Its results centre round the old genera *Polypodium*, *Acrostichum*, and *Gymnogramme*. In their old extended sense none of these represented of necessity any real kinship: the genera comprised Ferns that show conditions or states of the sori now known to have been acquired by types of quite distinct affinity. In fact these are not genera at all in the phyletic sense. For instance, it is now recognised that the Acrostichoid state has been arrived at along fully half a dozen different phyletic lines. After the representatives of these have been severally allocated to their natural places, the old genus in the strict systematic sense retains only three species, which are themselves advanced Pteroid derivatives.

The fact is that advances shown in characters such as those above enumerated have not been restricted to the individual, the species, the genus, or even to the family. They have been liable to affect a plurality of Families, or even the whole Class. The most striking instance of all is found in those steadily progressive reductions of mass, of complexity in cell-structure, and of spore-output which mark the passage from the Eusporangiate to the advanced Leptosporangiate state. These were not executed in one phylum only, but independently in many. There has been, in fact, a great Class-Progression, carried out by homoplastic steps, originated and executed independently in the several phyla of Ferns. Collectively those independent steps constitute a general progressive drift which presents a very complicated problem to the evolutionist. They raise the question of causality in a peculiarly complex form. In particular we may ask, Were the causal stimuli that brought it all about external or internal: or is it not possible that both may have interacted to produce the results which we see? Such questions arising out of the phyletic study of Ferns cannot be held as applying to them alone. They suggest that similar progressive phenomena involving wide homoplasy will emerge in other Classes of organisms when subjected to similar phyletic study.

CHAPTER L

EVOLUTIONARY BEARINGS OF THE RESULTS

A COMPARATIVE study of the Class of the Filicales extending over nearly half a century affords some justification for stating how certain of its results appear to the author to be related to evolutionary theory. It has been said that the essence of Evolution is unbroken sequence, and in this respect the Filicales are not found wanting. From the earliest fossil records of the Coenopteridaceae to the present day we may hold that Ferns have been Ferns: and that those we see now living have been derived in continuous sequence from pre-existent Ferns. The theme of this work has been evolutionary progress within the Class thus early defined, rather than its ultimate origin.

The lines of enquiry into the progressive evolution of the Class have been chiefly morphological, pursued by study of the external form and internal structure of the adult: also by following the ontogeny: while palaeontological data, the evidence best fitted to indicate actual successions, has been accepted as a valid check upon the results acquired from comparison. On the other hand, though physiological and genetic experiment may be introduced to illuminate the problem, this can only be carried out upon organisms now living; hence it cannot safely be held as reconstructing evolutionary history: it provides a basis for estimating the probability of earlier events rather than a demonstration of them. Even Mendelian analysis conducted experimentally has not given us the origin of species, as Bateson himself has admitted. Intercrossing acts as a distributing agency of characters, but it does not create them. All of these methods of enquiry should be coordinated, and the student of Descent should be cognisant of them all: but Comparative Morphology, checked by the positive facts of Palaeontology, must still be the chief foundation on which to base phyletic conclusions. As Darwin said, it is "the most interesting Department of Natural History, and it may be said to be its very soul." Applying the results of Comparative Morphology, aided as above, to the phyletic study of so circumscribed a Class as the Filicales, we may expect not only to approach a Natural Classification of them, but also to obtain some definite opinions which will illuminate the problem of Evolution at large.

The facts detailed in these three Volumes, summarised in Chapter XLIX, have led to the general conclusion stated in its last paragraph. The Filicales have been shown to comprise a skein of phyletic lines: the early origin of three of these is proved by the facts of Palaeontology. There is reason to believe that those main phyla, already characterised as they were in Palaeozoic or in early Mesozoic time, have maintained their separate identity to the present day, through persistent genera or their direct derivatives. Wide homoplastic advance has affected these several phyla independently of one another. Its results have been found to constitute a great Class-Progression or Phyletic Drift, expressed in many apparently independent structural changes. It leads from the Eusporangiate state characteristic of Palaeozoic time, by a plurality of distinct lines of detailed modification, to the advanced Leptosporangiate state characteristic of Present-Day Ferns.

How from the evolutionary point of view are we to regard such a Class-Progression, carried out in a plurality of distinct lines, and involving many heritable changes which can only have been achieved independently of one another in the several phyla which show them? We may take refuge in the beneficent word "adaptation," and may exercise ingenuity in accounting for what we see by fitting effect with supposed cause. But it would be better to attempt some more searching analysis of the influences effective in producing the evolutionary results which we see. The most important of these operative in the evolution of living things may be grouped under four heads:

(i)   A general initiative present in all organic life to develop.

(ii) Stimuli and limiting factors that shape and control the results.

(iii) Syngamy and Mendelian Segregation that distribute those results.

(iv) Natural Selection which determines survival or failure.

Of these (iii) and (iv) need not be discussed here, for they do not produce new features: they only distribute, select, or annihilate them. The immediate interest will lie in the origin and modification, rather than in the manipulation or destination of characters. Here, therefore, we need only consider (i) and (ii).

The consequences of the all-pervading initiative of living things to develop are seen primarily in an increase of size of the individual: secondly, in an increasing complexity of its form and structure: thirdly, in its variation of detail as shown by comparison of individuals and races. It is upon these variations that the theory of evolution is based, and the question at once arises of their origin and transmission. On the latter point variations have been ranged in two categories: (*a*) *fluctuating variations* which appear to leave no permanent impress so as to affect the reproductive cells; consequently they are held as not being hereditary: (*b*) *mutations* which are heritable, the qualities which they express having been in some way stamped upon the gametes, so that they are transmitted to the offspring. The line of distinction between these categories has often been sharply drawn: but in Plants there is reason to doubt whether the difference between them is anything more than an expression of limited experience: in fact, whether variations in the first instance apparently of the fluctuating type may not turn out to be in the long run heritable. A Class of Plants with a very long and consecutive history, such as the Ferns, gives an exceptional opportunity for testing this point.

The question of the origin of heritable characters is still quite an open one for Plants. The fact that in Animals the germ-cells are segregated early from the somatic cells does not affect the question, for in Plants such early segregation does not occur. In them, that is in Plants, the tissues, still undifferentiated as vegetative or propagative, are for long exposed to whatever the conditions of the individual life may have been before its gametes are specialised. This suggests that Plants would be particularly favourable subjects for critical observation. The prevalence in them of parallel development, or even of convergence in respect of characters that appear to be adaptive, suggests that the changes upon which they have been built were not produced at random. In particular, Ferns with their striking instances of homoplasy offer favourable opportunity for forming an opinion whether or not heritable changes of an adaptive character may have been promoted or actually determined in their direction or quality in some way by the conditions, external or internal. Moreover, in Ferns, with their long fossil history, these need not have acted within the restricted time-limits of present experiment. The wide latitude of geological time has been available for evolution to proceed. Though we may grant that the direct and immediate effect of external or internal conditions in producing adjustment of structure is not immediately and visibly heritable in the individual, nevertheless it is possible that the effect of such influences continued through long ages may become apparent in the race. It will be well, however, to realise that such a relation of inherited characters to the directive influence of conditions exercised over long periods of time is not recognised in certain quarters, in particular by the followers of Weismann.

On the other hand, the opinion of de Vries, writing more specially in relation to plants, may be translated thus (*Mutationstheorie*, Vol. I, p. 144): "And so I come to the conclusion that the Mutation Theory demands a mutability of organisms in all directions. Neither Palaeontological nor Systematic facts are irreconcilable with this view: and the construction of ordinary or collective species from groups of elementary species, whose characters diverge from one another in every direction, indicates clearly an earlier mutability in all directions." The results of intensive study of Ferns detailed in these Volumes appear to make the demand for mutability *in all directions* needless for them. Both externally and internally the changes in certain well-marked instances show a definite trend, readily related to the conditions, rather than a "mutability in all directions."

No change in the Ferns points more clearly to a decision as to the origin of an inherited character than the shifting of the sorus from the margin of the sporophyll, which comparison shows to have been its original position, to a place on the lower surface which may be held as derivative, and physiologically advantageous. There is reason to believe that this change has happened independently in many distinct phyla, sometimes early in the evolutionary history (Marattiaceae, Gleicheniaceae): sometimes it appears to be now in transition (Pteroid and Schizaeoid Ferns). That the change is associated with a widening of the leaf-blade is suggested by a comparison of the closely related genera, *Osmunda* and *Todea* (Vol. I, Fig. 210). The Schizaeaceae are also very suggestive, for in them, though the sporangia all originate from marginal cells, they acquire in the course of the individual development a superficial position (Vol. I, Figs. 213, 214). But the most illuminating results come from the Dicksonioid-Pteroid series, in which the comparison yields many transitional steps. Starting with the ancient Jurassic type of *Thyrsopteris*, the gradate sori are marginal upon the narrow leaf-segments, and remain so till maturity (Vol. II, Figs. 527, 529). In *Dicksonia* and *Cibotium* the receptacle and the first sporangia are seen to be still marginal (Figs. 534, 535), but the bi-labiate and gradate sorus is deflected downwards in the course of individual development on the widened blade (Fig. 530). In *Hypolepis* the lower indusium is absent or only vestigial, and the receptacle, with its mixed sporangia, is not marginal even in origin, but superficial; and it is spread over the lower surface of the leaf, thus giving a structure indistinguishable from the superficial sorus of *Polypodium* (Chapters XXIX, XXX, XXXVI, Figs. 584–587). These closely related Dicksonioid-Dennstaedtioid Ferns thus illustrate steps in a "phyletic slide" of the originally marginal sorus to the lower surface of the blade. The protection thus derived gives the biological justification for the change, which has become heritable. It is significant that along with the change of position go also the progressive change to a mixed sorus, and abortion of the lower indusium.

A second series which shows a like progression is found in the Pteroid Ferns (Chapter XXXVIII). These are again Dicksonioid derivatives, but they constitute a separate phylum distinguished from the first by prevalent soral fusions to form linear coenosori. The series starts from *Pteridium* and *Paesia*: both of these genera have a marginal receptacle and a two-lipped sorus, as in the Dicksonioids. As in them also the sorus is deflected downwards in the course of the individual development, while the lower indusium is inconstant (Fig. 218). A very beautiful intermediate state is seen in *Histiopteris incisa* (Fig. 219): here the orginal relation of the receptacle to the margin is variable, but with a strong initial bias towards the lower surface; and the lower indusium is abortive. In *Pteris* itself the lower indusium is again absent, but here the flattened receptacle is clearly intra-marginal, the first sporangia springing from the lower leaf-surface (Fig. 220). Thus the "phyletic slide" is again complete; and it is carried out in a sequence systematically independent of that of the Dennstaedtioid Series. These two series are held as demonstrating real progressions, each within a distinct natural circle of affinity. Traditional systematic grouping together of the Ferns quoted in either sequence, as also their rich synonymy, show the near kinship within either of these parallel sequences. Further, the species quoted exhibit parallel progressive changes in their vegetative characters also, while in either case there is a passage of their sori from the gradate to the mixed state. These facts indicate that the sequences are natural and progressive, not only in one but in many distinct characters. The identity of the sorus is assumed throughout. The change in its position is held to be adaptive, and that change has become hereditary. It involves readily intelligible steps in homoplastic amendment. It is believed to illustrate the inheritance of progressive structural characters biologically advantageous, in plants akin, developing without any restriction of time-limit, and carried out in a plurality of series phyletically distinct. Only a few outstanding examples have been quoted here: but there is evidence of a progressive slide of the sorus from the margin to the surface of the blade having taken place in yet other distinct phyla. *The general conclusion which follows from this comparative discussion is that a change, first seen as an ontogenetic adjustment in more primitive Ferns, is liable to become an accentuated and inherited feature in more advanced members of the same natural affinity. In other words, that ontogenetic adjustments—so-called fluctuating variations—continued without any time-limit, may become hereditary.*

The incidence of a imiting factor may also be effective in establishing heritable characters such as have been widely used in classification, as the next instance will show. The initiative to develop, resulting in increasing size of the individual, brings it up against such limiting factors. In Ferns this becomes obvious in relation to the increasing size of the conducting tracts (Chapter X). Since under suitable surface control these tracts supply material to the tissues they traverse, and since all nutritive supply from them is a unction of surface, the question of proportion of surface to bulk in the conducting tracts is a critical one. As the size increases, supposing the originally simple form of the conical stele to be maintained, the bulk of it will vary as the cube of the linear dimensions, while the surface varies only as the square.

Consequently a critical point will always be coming nearer, and may ultimately be reached, where the surface area will be functionally insufficient. All Ferns begin their individual development with a small, solid, vascular tract: it enlarges conically upwards as the plant produces successively larger leaves. In the absence of any secondary thickening, each plant is thus constantly approaching that limit of size when further development would be functionally impracticable. The difficulty can be overcome by change in form of the stelar column. Any change of form from the cylinder or cone would give an increased proportion of surface to bulk. Such changes are actually illustrated by comparison of sections of the stele in certain allied fossil stems of different sizes, all drawn to the same scale (Fig. 175). The smallest steles are approximately cylindrical: the larger steles become fluted, or in transverse section stellate; the involutions being deepest where the size is greatest. The change in form certainly does result in an increase in the proportion of surface to bulk, beyond what the proportion would have been if the enlargement went along with a simple conical form. Results similar in principle though different in detail may be seen in all the larger modern Ferns (Vol. I, Chapter X). Successive sections from the same individual Fern-stem, as it enlarges upwards, serve to illustrate the changes which affect the proportion of surface to bulk of the vascular tracts (Fig. 178). Similar reactions appear in many other plants and plant-parts, provided that secondary thickening does not step in and vitiate the problem. In fact the principle is of wide application. In the long run it is held to account causally for that curious disintegration of the vascular system which is so marked a feature in the larger, and particularly the more advanced, Ferns.

The form of the vascular system is so far distinctive and constant in character that it is made use of ever more and more in the systematic comparison of Ferns (Chapters VII–X). Its features have become hereditary, though subject to variability in detail. A protostelic Fern, such as *Gleichenia* or *Cheiropleuria*, remains protostelic: a solenostelic species or genus, such as *Loxsoma*, is constantly solenostelic: a polycyclic type, such as *Matonia*, regularly becomes polycyclic as it matures to the adult state: every sporeling of a typically dictyostelic Fern assumes that character as it develops. The relation between the form of the vascular tracts and the limiting factor of size may in each be clearly traced, though the necessary adjustment may differ in detail from one phylum or genus to another: but in point of fact those adjustments do become within limits fixed, and are transmitted as such in descent. *Here again a modification, primarily of the nature of an ontogenetic adjustment, has become an accentuated and inherited feature in the more advanced Ferns, and has even acquired high diagnostic value in the hands of the systematist.*

How then are we to regard such hereditary features as those of the soral slide, and the vascular fluting, decentralisation, and disintegration, from the point of view of Descent? All of them suggest that features of adjustment imposed upon a succession of individual lives by conditions, sometimes external sometimes internal, have become hereditary. Naturally the reply may be made that probably mutations favourable to the perpetuation of the imposed character may have made that character permanent. If we grant that, do we not thereby simply admit that the distinction between fluctuating variations and mutations is not absolute? In other words, that fluctuating variations repeatedly imposed adaptively upon successive generations, without any near limit of time, are liable to become mutations? It is difficult to see any other rational explanation of the wide-reaching facts of homoplastic adaptation, shown in exceptional profusion in the ancient Class of the Filicales, but evident also in Plants at large.

This conclusion is, however, directly opposed to the opinion of those who hold that characters impressed upon the individual life are not heritable. A very fair and philosophical balance of their position as against the Mnemic Theory of Semon, with a definite conclusion in favour of the latter, is to be found in the Address of Sir Francis Darwin, as President of the British Association in 1908. The Mnemic Theory proceeds on the conception that, as a consequence of stimulus imposed by conditions of life, a record or engram is impressed on the organism. Sometimes the engram may be recognised only functionally: but Sir Francis remarks in further explanation: "As I have attempted to show, morphological changes are reactions to stimulation of the same kind as these temporary changes. It is indeed from the morphological reactions of living things that the most striking cases of habit are to be found." The examples given above from the Filicales appear to be illustrations of this.

Semon assumed that when a new character appears in the body of an organism a new engram is added to the nuclei of the part affected; and that further disturbance tends to spread to all the nuclei of the body, including those of the germ-cells; and so to produce in them the same change. But this can only be made efficient by prolonged action. He laid great stress upon the slowness of the process of building up efficient engrams in the germ-cells. Direct experiments bearing on the inheritance of acquired or impressed characters have so far given indecisive or negative results. But all of these experiments range within the narrow limits of laboratory time. If, however, reference be made to a sequence of events conducted with the latitude of geological time, and the effect appears to be positive, it would seem right to give such positive conclusions precedence over the negative evidence of experiments limited within a brief period. This is the interest which is presented by the phyletic facts relating to Ferns, when traced from early geological times to the present.

There is thus reason to believe in some secular establishment in the Ferns of new inherited features that are adaptive, from adjustments originally ontogenetic. The effect of this should be to relax the statement of a general and rigid negative. Any student of Nature who is duly impressed by the infinite possibilities of life will look with distrust on a general *dictum* that imposes a limit upon them. The utmost that can legitimately be said in opposition is, that the positive evidence of inheritance of adaptive adjustments of the individual has hitherto been insufficient to afford general proof: and that is very different from a negation. On the other hand, to most botanists a sharp antithesis between somatic cells and germ-cells, in respect of their receptivity for the impress of characters, seems inherently improbable for Plants. Comparative study of such primitive Plants as the Algae suggests that the zoospores and gametes were originally alike, the latter being probably specialised forms of zoospores: that is, of somatic cells. Even in the Higher Plants the distinction between somatic and germ-cells cannot be drawn by the most exact microscopic analysis till a late stage of development has been reached; while it is a common experience for active somatic cells to be aroused to regeneration, and so ultimately they may produce germ-cells. The commonness of fluctuating variations, and the extreme sensitiveness seen in the adjustment of developing plants to their environment, prove how readily the somatic cell is influenced. But the difficulty in obtaining experimental evidence of the inheritance of such adjustments shows how resistant the germ-cells are to fresh impressions. In this, however, there is no justification for holding the difference as absolute. The facts derived from the comparative study of Ferns certainly indicate that it is not. They favour the acceptance of some form of Mnemic Theory as a working hypothesis for Plants, until a final decision shall have been obtained.

It should, however, be realised that this discussion touches only one facet of the infinitely varied problem of Evolution, though that facet is of prime importance. The complexity of the whole problem appears to show that no single solution of it is probable: but rather that the mechanism of Evolution has been, and is, as complex as the consequences are diverse. Nevertheless if we recognise the secular origin of hereditary changes carried out with wide homoplasy, and consider its cumulative effect in form and structure, this would afford some elucidation of parallel development in respect of individual criteria, in phyla distinct from one another since early Palaeozoic time. Such detailed steps would bear their own value in contributing to more general conclusions.

There lies beyond this still the obscure question of causality. Does the initiative to develop an adaptive feature spring from within the organism, or from without: or is it some interaction of both which produces the inherited result?[1] The existence of so wide a common trend of change as is seen in the parallel progressions of Ferns, in respect of the several criteria of comparison not only individually but also collectively, is a fact that strongly suggests some defining tendency within the organism, promoting its development not only in individual features, but also as a whole. Modification or even stimulus due to the incidence of external conditions, however effective, does not suffice to give a full explanation of so broad a sweep of cognate change as that from the Eusporangiate to the Leptosporangiate type. The facts appear to suggest for the Filicales not only detailed modification influenced from without, but also some more general bias or tendency of initiative within the organism itself: not fortuitous, nor yet a vague mutability in all directions, as De Vries would suggest, though this need not necessarily be ruled out: but specifically directive, upon the results of which limiting factors, whether environmental or internal, have acted in defining and shaping the heritable details. Such a conception, prompted as it is by intensive study of a wide area of fact, may go far towards explaining that majestic and highly polyphyletic progression which leads from the Palaeozoic Coenopteridaceae to the Leptosporangiate Ferns of the Present Day. Such a general tendency of initiative applicable for the great Class of the Filicales would suggest a still wider and more varied application in the Evolution of Organic Life at large.

If such a tendency be held as included in "the nature of the organism," and the results are hereditary as the evidence from the Ferns indicates that they are, then this position appears to be in accord with the expressions of Darwin, in Chapter V of the *Origin of Species*, where he says: "In all cases there are two factors, the nature of the organism, which is much the most important of the two, and the nature of the conditions. The direct action of changed conditions leads to definite or indefinite results. In the latter case the organisation seems to become plastic and we have much fluctuating variability. In the former case the nature of the organism is such that it yields readily, when subjected to certain conditions, and all or nearly all the individuals become modified in the same way."

[1] The argument stated in this Chapter accords with the views so well expressed by Vines, in his Article on "Morphology of Plants" in the *Encyclopaedia Britannica*, 11th edn. 1910–11: "In endeavouring to trace the causation of adaptation, it is obvious that it must be due quite as much to properties inherent in the plant as to the action of external conditions; the plant must possess adaptive capacity. In other words, the plant must be irritable to the stimulus exerted from without, and be capable of responding to it by changes of form and structure. Thus there is no essential difference between the 'direct' and the 'indirect' action of external conditions, the difference is one of degree only. In the one case the stimulus induces indefinite variation, in the other definite; but no hard-and-fast line can be drawn between them." Inasmuch as specific examples are now given, showing the establishment of definite adjustments, in the first instance ontogenetic, as inherited characters, the view of Vines is thereby advanced from the phase of theoretical discussion towards that of demonstration.

BIBLIOGRAPHY FOR CHAPTER L

There is no need to quote in detail here the extensive literature on Evolution. The following selection of titles of recent works will suffice, if the references which they severally contain be followed up.

760. DE VRIES. Mutationstheorie. 1901.
761. Sir FRANCIS DARWIN. Presidential Address, British Association. Dublin. 1908.
762. VINES. Encyclopaedia Britannica, 11th Edn., Article on "Morphology of Plants."
763. SEWARD. Hooker Lecture, Linnaean Journal, XLVI. 1922.
764. SCOTT. Extinct Plants and Problems of Evolution, Macmillan. 1924.
765. BATESON. Birkbeck Address, "Progress of Biology," Nature, May 3 and 10. 1924.
766. YALE UNIVERSITY PRESS. Organic Adaptation to Environment. 1924.
767. JENNINGS. Prometheus, Kegan Paul, Trench, Trubner & Co.
768. Evolution. A Collective Work, Blackie. 1925.
769. SMUTS. Holism and Evolution, Macmillan. 1926.
770. GRAHAM KERR. Evolution, Macmillan. 1926.
771. PATTEN. The Memory Factor in Biology, Baillière, Tindall and Cox. 1926.

INDEX

Homoplasy, 284, 292
Humata, 20, 37
  H. heterophylla, 20
Hydropterids, 260
Hymenolepis, 192, 213, 232, 255, 280
  H. spicata, details of, 223 (Fig. 731); habit of, 222; *venatio anexeti* of, 222
Hymenophyllaceae, 273
Hypoderris, 99, 119, 277; a synthetic type, 113; habit of, 101 (Fig. 646); habit and structure of, 104; soral condition of, 104; sorus of, 101 (Fig. 646); young sporangia of, 104 (Fig. 649)
Hypolepis, 5 *sqq.* (ch. XXXVI), 15, 81, 92, 274, 278, 288; habit of, 7; sorus of, 7
  H. Bernh., synonymy of, 5 (Fig. 582)
  H. nigrescens, mature sorus of, 9 (Fig. 585)
  H. repens, pinnule of, 9 (Fig. 586); vascular system of, 8 (Fig. 583); young sorus of, 10 (Fig. 587)

"Incertae sedis," genera, 3
Incidence of size-factor, 31
Index Filicum, 4
Indusioid margin of *Pellaea intramarginalis*, 93 (Fig. 638)
Indusium, abortion of, 7, 12, 129, 288; abortion of lower, 47; basal, 112; elimination of, 51; elimination of lower, 12, 23; orbicular, of *Polystichum*, 128; various types of, 117; of *Arthropteris*, abortion of, 36; of *Aspidium (Polystichum) lobatum*, 129 (Fig. 660); of *Athyrium alpestre*, 145; of *Ceterach*, 147; of Cyatheaceae, 115; of *Matonia*, 114; of *Phyllitis*, 196; of *Pleurosorus*, 147; of *Polystichum*, 114; of Shield Fern, 124; of *Woodsia*, 102, 103 (Figs. 647, 648)
Indusium inferum, morphology, 114; of Cyatheoids, 118; of Superficiales, 114, 116; origin of, 117, 118
Indusium superum of *Aspidium*, 127; of *Polystichum*, 127
Inherited feature, ontogenetic adjustment becomes an, 290; secular establishment of, 292
Initiative to develop, 286
Internal embryology, 267

Jamesonia, 68, 73, 93, 97, 281; an Acrostichoid, 70; habit of, 69 (Fig. 628); hairs of, 70; spore-counts of, 70
  J. verticalis, fertile pinna of, 70 (Fig. 629)
Juvenile leaf of *Cheiropleuria*, 202 (Fig. 711); of *Platycerium Veitchii*, 202 (Fig. 711)
Juvenile leaves of *Antrophyum*, 247 (Fig. 746); of *Hecistopteris*, 247 (Fig. 746)

Laboratory time, limits of, 291
Laccopteris, 228

Land Plants, alternation stabilized in, 265
Lastraea, 132
Leaf of *Antrophyum semicostatum*, 242 (Fig. 742); of *Diacalpe*, 106 (Fig. 650); of *Peranema*, 106 (Fig. 650); of *Syngramme borneensis*, 235 (Fig. 735); Sphenopterid, 141
Leaf-blade of *Dictyoxiphium*, 34
Leaf-margin, relation of the sorus to, 22
Leaf-segments of *Histiopteris incisa*, 49 (Fig. 614)
Leaf-trace of *Histiopteris*, 48; of *Platycerium*, 209
Leaves, of *Lindsaya*, 32; of *Platycerium*, 208; of *Stenosemia aurita*, 135 (Fig. 667)
Leptochilus, 131, 137, 277; venation of, 132 (Fig. 663)
  L. cuspidatus, 132; Acrostichoid state of, 132, 133 (Fig. 664); young sporophyll of, 134 (Fig. 665)
  L. tricuspis, 218
  L. varians, 218
Leptolepia, 13, 15
  L. Novae Zelandiae, 13, 16
Leptosporangiate Ferns, 1
Leptosporangiate state, 284
Limiting factors, 286
Lindsaya, 31, 38, 274, 278; leaves of, 32; rhizome of, 31 (Fig. 600 *bis*); sorus of, 32; sorus, development of, 33; sporangia of, 34 (Fig. 604); -structure of stele, 24
  L. lancea, marginal region of a pinna of, 32 (Fig. 602); pinna of, 32 (Fig. 601)
  L. linearis, stele of, 31 (Fig. 600); young sorus of, 33 (Fig. 603)
Lindsayopsis, 29
Linkage of sori, 24
Litobrochia, 48, 56
Llavea, 6, 72, 81, 93, 97, 281; sporangia of, 65; vascular system of, 65
  L. cordifolia, habit of, 66 (Fig. 624)
Lomaria, 163, 165 (Fig. 688), 192; soral linkage of, 192; type of pinna of, 167
"Lonchitidinae" of Prantl, 41
Lonchitis, 61; an intermediate type, 46
  L. aurita, pinna of, 48 (Fig. 612)
Lophosoria, 111, 115, 116, 276
Lower indusium, abortion of, 47, 54; elimination of, 12, 23; of *Anopteris hexagona*, 55
Loxsomaceae, 273
Luerssenia, 121, 127, 137
Lygodium, 80, 96, 272

Maidenhair Ferns, 78, 80, 94
Male Shield Fern, 121; anatomy of, 122; habit of, 122
Marattia, 271
Marattiaceae, 271

CAMBRIDGE: PRINTED BY W. LEWIS, M.A., AT THE UNIVERSITY PRESS

Printed in the United States
By Bookmasters